Topics in Applied Physics   Volume 37

Topics in Applied Physics / Founded by Helmut K. V. Lotsch

Thermally Stimulated Relaxation in Solids

Edited by P. Bräunlich

With Contributions by
P. Bräunlich   L. A. DeWerd   J.-P. Fillard   J. Gasiot
H. Glaefeke   P. Kelly   D. V. Lang   J. Vanderschueren

With 142 Figures

Springer-Verlag Berlin Heidelberg New York 1979

Professor Dr. *Peter Bräunlich*

Department of Physics, Washington State University,
Pullman, WA 99164, USA

ISBN 3-540-09595-0 Springer-Verlag Berlin Heidelberg New York
ISBN 0-387-09595-0 Springer-Verlag New York Heidelberg Berlin

Library of Congress Cataloging in Publication Data. Main entry under title: Thermally stimulated relaxation in solids. (Topics in applied physics; v. 37). Bibliography: p. Includes index. 1. Solid-Thermal properties. 2. Thermally stimulated currents. 3. Exoelectron emission. I. Bräunlich, P., 1937–. QC176.8.T4T45   530.4′1 79-18071.

Monophoto typesetting, offset printing and bookbinding: Brühlsche Universitätsdruckerei, Gießen
2153/3130-543210

Preface

Thermally stimulated relaxation (TSR) by charge transport is only a small segment of the rich spectrum of solid-state processes which require an activation energy to proceed. This book deals specifically with electronic and ionic transport phenomena observed during nonisothermal temperature scans: thermoluminescence, thermally stimulated conductivity, thermally stimulated particle emission, and thermally stimulated depolarization. In selecting these topics we were guided by the need for critical appraisal of their utility as methods to obtain information on defects and impurities in nonmetals.

For historical reasons a number of loosely related phenomena are collectively called exoemission. Although only thermally stimulated exoemission of particles is of interest in the context of this volume, we choose to present it within the framework of all exoemission research.

The subjects are discussed on the basis of a mainly phenomenological or, at best, semiempirical theory of relaxation kinetics. Important applications in the dosimetry of ionizing radiation and in geological and archaeological research are described.

An enormous wealth of material on these topics has accumulated in the literature over the last three decades. A publication rate of over 500 per annum during the time this volume was prepared bears witness to a presently still very active field of research. Nevertheless, it appears to have reached a critical stage. Difficulties in unambiguous interpretation of experimental data have been recognized in recent years and a growing awareness has evolved of inherent limitations associated with most of the techniques for quantitative trap level spectroscopy.

This book was written to focus the attention of the specialists as well as the newcomers to the field onto these problems and to point out the need for novel experimental and theoretical approaches which fully utilize the potential of nonisothermal relaxation phenomena as investigative tools in solid-state physics.

Important first steps have already been taken. The development of junction capacitance spectroscopy for semiconductors is one example. Even though its most elegant experimental techniques do not require heating the sample according to a controlled heating program but rather are performed under isothermal conditions, in many cases rapid measurements can be made during a temperature scan and thus they resemble, at least formally, traditional thermally stimulated relaxation experiments.

Rather than writing a report on recent advances or a thorough review of experimental findings, we felt the purpose of this volume would best be served by choosing a largely tutorial style. Late developments, that for this reason could not be cited in the text (published mainly in 1977–1978), are listed in a bibliography.

Pullman, WA, August 1979   *P. Bräunlich*

Contents

5. Exoemission
By H. Glaefeke (With 24 Figures) 225

Contributors

Bräunlich, Peter
   Department of Physics, Washington State University,
   Pullman, WA 99164, USA

DeWerd, Larry A.
   Departments of Radiology and Human Oncology,
   Medical Physics Division, 3321 Sterling Hall, University of Wisconsin,
   Madison, WI 53706, USA

Fillard, Jean-Pierre
   Université des Sciences et Techniques du Languedoc, Place E. Bataillon,
   F-34060 Montpellier Cedex, France

Gasiot, Jean
   Université des Sciences et Techniques du Languedoc, Place E. Bataillon,
   F-34060 Montpellier Cedex, France

Glaefeke, Harro
   Wilhelm-Pieck-Universität Rostock, Sektion Physik, Universitätsplatz 3,
   DDR-25 Rostock, German Democratic Republic

Kelly, Paul
   Division of Physics, National Research Council Ottawa,
   Ontario K1A OS1, Canada

Lang, David Vern
   Bell Laboratories, Murray Hill, NJ 07974, USA

Vanderschueren, Jacques
   Lab. de Chimie Physique, Université de Liège, Sart-Tilman,
   B-4000 Liège, Belgium

1. Introduction and Basic Principles

By P. Bräunlich

With 15 Figures

Two basic conditions have to be fulfilled for the occurrence of thermally stimulated relaxation (TSR) processes:

a) The system must be removed from statistical thermodynamic equilibrium and exist in a state which requires the reactants to surmount a free-energy barrier in order to move toward the reestablishment of equilibrium.

b) The system must be in contact with a temperature reservoir that provides the thermal energy necessary to activate the relaxation process.

Of interest for the study of thermally activated relaxation from a nonequilibrium steady-state situation back toward thermal equilibrium are therefore the ways to remove the system from equilibrium and the phenomena that can be measured or monitored during the relaxation process and thus utilized to characterize the occurring reactions. These processes encompass an cnormous wealth of biological, chemical, and physical phenomena. Herein only those are considered which involve the redistribution over available energy states of electronic or ionic charge carriers in semiconducting and insulating solids during the relaxation process. Specifically selected are thermally stimulated charge transport and luminescence phenomena and thermally stimulated emission of charged particles from solid surfaces.

Theoretical descriptions of these phenomena are almost entirely based on the so-called absolute rate theory borrowed from chemical reaction kinetics [1.1]. Phenomenologically introduced reaction rates are determined experimentally on the basis of reasonable kinetic models.

1.1 Perturbation of the Thermodynamic Equilibrium

It is instructive to recall some of the fundamental relations of chemical reaction kinetics before we venture into the discussion of formally related phenomena in solid-state physics.

Consider the following reaction between ideal gases A, B, C, and D of concentrations n_A, n_B, n_C, and n_D:

$$A + B \leftrightarrow C + D. \tag{1.1}$$

In thermodynamic equilibrium the law of mass action states

$$K \cdot n_A n_B = n_C n_D;$$ (1.2)

K is the equilibrium constant given by

$$K = \exp\left(\frac{\Delta S}{k}\right) \exp\left(-\frac{\Delta H}{kT}\right).$$ (1.3)

ΔS is the change in entropy and ΔH the change in enthalpy. The entropy of the system is a measure of the total number of quantum states available to it. Thus, K is determined by the thermodynamic properties of the system which include the available energy levels and the temperature. Since $\Delta H - \Delta(TS) = \Delta G$, the change in standard Gibbs free energy, the equilibrium constant can be evaluated from a knowledge of the change in free energy and vice versa. This is one of the most powerful relations in the practical application of chemical thermodynamics.

The conditions to be fulfilled for the reestablishment of thermodynamic equilibrium of a system after perturbation is that the "path" between states is "sufficiently open". That is, once the system's equilibrium is perturbed, the return to equilibrium must be possible with a finite probability at $T \neq 0$. The processes occurring during the establishment of equilibrium in a perturbed system are the subject of interest in reaction chemistry which is concerned with the rate of chemical change and with the measurement of reaction rates to elucidate the mechanisms involved in the system's approach to thermodynamic equilibrium. These kinetic processes are not uniquely restricted to chemical reactions but also apply in quite precise analogy to such phenomena as thermal ionization of gases or defect states in solids, polarization or magnetization of materials, annealing of defects, diffusion and flash desorption, to mention just a few [1.9], and, therefore, it appears justified to apply the wealth of physical insight available from chemical kinetics to the subjects selected for the present volume.

Most of the answers to the question as to how much one can learn about a reaction from kinetic considerations alone have been given before by chemical kineticists. The following statement by *Noyes* [1.2] is as valid for electron kinetic reactions in solids as it is for chemical reactions:

"The *mechanism* of a chemical reaction is the detailed combination of reversible and irreversible elementary processes whose net consequence is the reaction of interest. Kinetic data furnish the most direct clues for discovering the mechanism in any specific example. However, in principle an infinite number of microscopic mechanisms can be proposed to explain any set of macroscopic observations. A molecular mechanism can never be proved in a rigorous sense, but kinetic measurements can often disprove a mechanism and show it is not consistent with observations. Confidence increases with the volume of data consistent with a simple interpretation. Mechanisms of a small but significant number of reactions now seem to be established beyond question."

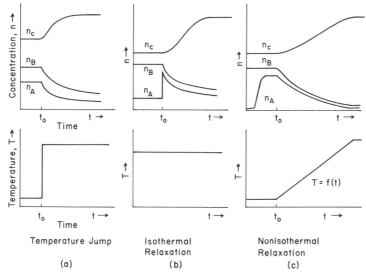

Fig. 1.1a–c. Relaxation of a system of reactants after perturbation of the thermodynamic equilibrium by a sudden increase in temperature (**a**), a sudden increase in the concentration of one reactant at a constant temperature (**b**), and an increase of the concentration of one reactant at a low temperature with subsequent linear temperature rise (**c**)

The equilibrium of a system may be perturbed by changing the concentration of the reactants, the temperature, pressure, electric or magnetic field (including electromagnetic fields), etc. (Fig. 1.1). The establishment of a new equilibrium condition during and after the perturbation may be monitored by the measurement of the concentration of the involved species and the results can be utilized for the study of the involved chemical or physical reactions. Since the temperature has, in the majority of cases, a most pronounced effect on the reaction rates, it has to be carefully controlled. Two basic types of relaxation techniques are used:

a) Isothermal relaxation: the perturbation is implemented at a constant temperature which is, whenever possible, selected to assure experimentally convenient relaxation times.

b) Nonisothermal relaxation: the system is perturbed at a sufficiently low temperature to reduce the probability to establish a new statistical equilibrium. Subsequently, the temperature is increased according to a well-controlled heating program $T(t)$, thus increasing the reactions rates and the relaxation of the system can be monitored as a function of temperature and time.

Technique a) is most successfully used not only in the study of chemical reactions [1.3, 4], but also in electronic reaction kinetics in solids [1.5–7], most notably in the recently developed technique of deep-level transient spectroscopy [1.8]. Nonisothermal relaxation is employed in the studies of thermally stimulated luminescence, conductivity, exoemission, polarization, and depolarization.

1.2 Formal Description of Kinetic Observations

The rate at which a perturbed system reacts during the relaxation process has frequently been described in a rather formal way, justified solely on the basis of an "acceptable" fit of experimental observations, by an equation of the form

$$\frac{d[n_D]}{dt} = \alpha[n_A]^m[n_B]^n[n_C]^l, \tag{1.4}$$

where n_D is the concentration of the product resulting from the reaction between the reactants n_A, n_B and n_C, and α is the so-called specific reaction rate constant or rate coefficient. A reaction of the form (1.4) is visualized as involving elementary processes, taking place in single steps, in which one, two or possibly more species of the reactants take part. These steps are called unimolecular, bimolecular, etc., and the reaction rates accordingly are said to be of first, second, third order and so on. The reaction of (1.4) is of m-th order in n_A, n-th order in n_B, and l-th order in n_C.

Sometimes it is possible to describe the change of the concentration of a reactant n_D by a simple equation of the form $d[n_D]/dt = \alpha[n_D]^l$, where $\alpha = \nu \exp(-E/kT)$, E is an activation energy, and ν is the so-called frequency factor. This approach is justified whenever empirical data can be fitted by an Arrhenius equation.

Clearly, such a formal "order-kinetics" of the reaction or relaxation process is only a phenomenological description of measured rates that can be related to actually involved physical processes only in very simple cases (see Sect. 2.6.4) [1.9,10]. The order of a process is an empirical quantity, the experimental determination of which does, in general, not permit an unequivocal interpretation of the details of the reaction. In addition, no reaction will obey (1.4) close to steady-state equilibrium when $d[n_D]/dt$ approaches zero despite the fact that neither any of the concentrations nor α is zero [1.2].

1.3 The Principle of Detailed Balance

Equation (1.4) applies only to systems that are far removed from thermal equilibrium so that the reaction essentially occurs only in one direction. It is assumed that only the product n_D is formed from the reaction of the reactants n_A, n_B and n_C. Eventually the reverse reaction, namely the decomposition of n_D into reactants, becomes more and more probable until the forward and reverse reaction are in a state of detailed balance.

This principle of detailed balance is a result of microscopic reversibility of chemical, electron kinetic and other reactions. It is prerequisite for the establishment of thermal equilibrium which requires that forward and reverse reaction rates are identical for every individual reaction path possible.

In essence, a reaction made possible at a temperature T by increasing the concentration of one or more reactants and thus removing the system from equilibrium, will continue until new concentrations of the reaction product and the reactants are established. If the reaction proceeds isothermally, the equilibrium constant is unchanged. In nonisothermal reactions it varies with T and an equilibrium is finally established in which all concentrations as well as K assume new values. The principle of detailed balance if of fundamental importance to establish helpful relations between reaction constants and equilibrium constants [1.11], since both at the initial thermal equilibrium and at the new equilibrium after the relaxation of the perturbation the net forward and reverse reaction rates are zero.

1.4 Theoretical Approaches to Reaction Kinetics

Kinetic theories of chemical and other (e.g., electronic recombination) reactions attempt to explain the rates at which the processes proceed and to describe their behavior as a function of such variables as temperature, time, and, of course, concentration of the reactants and products.

Two approaches toward theoretical interpretations of such reactions are taken:

I) It is attempted to describe the reaction in terms of elementary processes, the theory of which results in a rate equation with known cross sections for the elementary processes. The collision theories of simple bi- and mono-molecular gas reactions are examples [1.12, 13].

II) A statistical thermodynamic theory of absolute reaction rates is applied which assumes that the reaction is a continuous process and that the concentrations of the reactants as well as the products can be calculated by statistical methods. Characteristic for this theoretical approach is that one configuration, the activated complex or transition state, is crucial. Once it has been attained by passage through an energy barrier (activation energy), the probability for completion of the reaction is assumed to be large and, at the same time, the equilibrium (or better steady-state) distribution of the reactants capable of reaching this state is not disturbed. This then is precisely why in a nonisothermal experiment the heating rate must be sufficiently slow because otherwise statistical steady-state arguments cannot be used to calculate the ratio of the concentration of reactants in the transition state and any other of their possible energy states [1.14, 15].

Naturally, a complete theory of the reaction which provides a detailed description of the rate constant and the cross sections is preferable. However, it is available only for a few cases [1.12, 13]. The more prevalent situation is characterized by a number of possible microscopic mechanisms, such that all that is available for the interpretation is a set of different plausible phenomeno-logical rate equations with unknown rate constants. Through statistical arguments certain relations between rate constants and concentrations can be assumed at best. Usually, one or several transitional states are possible with one or several different unknown activation energies. Through experiments some mechanisms and rate equations can be disproved. But, as pointed out by *Noyes*, a rigorous proof for the validity of a certain mechanism can, in general, not be found through this procedure [1.2]. An acceptable fit of experimental data with solutions of the rate equations, obtained with experimentally determined values of activation energies and some of the reaction rate constants, yields the conclusion that the absolute rate theory is consistent with experiment. But this should not be mistaken as proof.

The theory of thermally stimulated electronic and ionic relaxation phenomena discussed in this book is in a state of sophistication comparable to chemical reaction kinetics. The detailed physical mechanism of only a few of these thermally stimulated reactions is known [1.8, 16]. The bulk of attempts of theoretical explanation is based on phenomenological rate theories.

In the following section we outline the precise analogies between chemical reaction kinetics and thermally stimulated relaxation phenomena in solids. The type of phenomena we consider encompasses the thermally stimulated redistribution of electrons and holes over available energy states in insulators and semiconductors after perturbation of the thermal equilibrium.

1.5 Equilibrium Electron Statistics for Insulators and Semiconductors

To outline the analogy of chemical reaction kinetics and electronic recombination kinetics in nonmetallic solids, we concentrate first on the derivation of the "law of mass action" for the equilibrium concentration of carriers. Thereafter the population of impurity levels in thermal equilibrium is described and subsequently nonequilibrium steady-state situations which can be attained during and after externally inflicted perturbations of the equilibrium are discussed.

1.5.1 Law of Mass Action in Intrinsic Materials

Consider an intrisic‚ insulator or semiconductor in thermal equilibrium. Thermal excitation of valence electrons, n_v, creates holes of density p and free electrons n_c which can recombine. This "chemical" reaction can be described by

$$n_c + p \leftrightarrow n_v + p_c, \tag{1.5}$$

where we have denoted p_c as the density of holes in the conduction band (equivalent to the density of unoccupied electron states). The determination of the equilibrium concentration of the reactants in (1.5) as a function of temperature is easily accomplished by applying Fermi–Dirac statistics to the appropriate set of electron levels of the semiconductor or insulator

$$n_c = \frac{1}{V} \int_{E_c}^{\infty} dE \, g_c(E) [e^{(E-E_F)/kT} + 1]^{-1},$$

$$p = \frac{1}{V} \int_{-\infty}^{E_v} dE \, g_v(E) [e^{(E_F-E)/kT} + 1]^{-1}. \tag{1.6}$$

Here $g_{c,v}(E)$ is the energy density of levels in the conduction and valence band, respectively, and E_F the chemical potential, usually called the "Fermi level", even though, in contrast to metals, no real electron level may exist at that energy in an insulator or semiconductor. V is the volume of the material and E_c and E_v are the band edge energies. Equation (1.6) is valid not only for pure (intrinsic) materials, but also in the presence of defects and impurities which affect the values of n_c and p solely through an appropriately changed value of E_F. Since E_F almost always lies in the forbidden gap, one can assime $E_c - E_F \gg kT$ and $E_F - E_v \gg kT$ even for energy gaps as small as a few tenths of an eV and temperatures higher than room temperature. Under these conditions $n_v \approx N_v$ and $p_c \approx N_c$, and (1.6) reduces to

$$n_c = N_c e^{-(E_c - E_F)/kT}$$

$$p = N_v e^{-(E_F - E_v)/kT} \tag{1.7}$$

where

$$N_c = \frac{1}{V} \int_{E_c}^{\infty} dE \, g_c(E) e^{-(E-E_c)/kT},$$

$$N_v = \frac{1}{V} \int_{-\infty}^{E_v} dE \, g_v(E) e^{-(E_v-E)/kT}, \tag{1.8}$$

are the density of electron/hole states in the conduction/valence band.

The product of the electron and hole densities in (1.7) yields the law of mass action

$$n_c p = N_c N_v e^{-E_g/kT}. \tag{1.9}$$

The dependence on the Fermi level has vanished. The analogy with (1.2) is complete if one replaces the free energy ΔG with the width of the forbidden gap $E_g = E_c - E_v$. Since in (1.9) only energies within a few kT of the band edges contribute to the integral (because of the exponential factor), the parabolic band approximation is justified [1.16]:

$$g(E)_{c,v} = \frac{V}{2\pi^2} \left(\frac{2m^*_{c,v}}{\hbar^2} \right)^{3/2} |E - E_{c,v}|^{1/2}. \tag{1.10}$$

Thus, the volume densities of electron and hole levels become

$$N_c(T) = \frac{1}{4}\left(\frac{2m_c^* kT}{\pi\hbar^2}\right)^{3/2}$$

$$N_v(t) = \frac{1}{4}\left(\frac{2m_v^* kT}{\pi\hbar^2}\right)^{3/2}$$

(1.11)

where $m_{c,v}^*$ are the effective masses of electrons and holes, and $E_{c,v}$ the energies at the edges of the conduction band and valence band, respectively.

For intrinsic semiconductors $n_c = p$, and (1.9) can be written as $n_c = p = (N_c N_v)^{1/2} \exp(-E_g/2kT)$. Therefore, knowledge of the band gap is sufficient to calculate the concentration of electrons and holes in an intrinsic semiconductor at any temperature T. With the aid of (1.7) the electrochemical potential E_F is found to be $E_F = E_v + \frac{1}{2}E_g + \frac{1}{2}kT\ln(N_v/N_c)$. Choosing the origin of the energy axis at $E_v = 0$, (1.9) finally becomes for $E_g \gg kT$

$$n_c = p$$
$$\approx (N_c N_v)^{1/2} \exp(-E_F/kT).$$

(1.12)

Extrinsic semiconductors contain defects and impurities which prominently determine the concentration of free electrons and holes. The knowledge of the concentration of these defects and their energy states within the forbidden gap suffices in principle to calculate the equilibrium concentration of free carriers at any temperature. In practice this knowledge is of course not available a priori and must in general be obtained experimentally. The study of thermally stimulated relaxation phenomena is one of the experimental techniques which, if carried out with caution and in conjunction with other independent techniques, can provide much of the desired data.

1.5.2 Law of Mass Action in Extrinsic Materials

The law of mass action may also be formulated for extrinsic semiconductors. In thermal equilibrium a Fermi level can again be postulated so that

$$n_c = N_c\{1 + \exp[(E_c - E_F)/kT]\}^{-1}$$

and

(1.13)

$$p = N_v\{1 + \exp[(E_F - E_v)/kT]\}^{-1}.$$

Its position within the forbidden gap is now given by the concentration of the impurities and defect which can act as donors or acceptors of electrons and

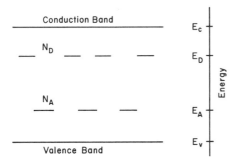

Fig. 1.2. Energy levels of an extrinsic semi-conductor containing donors of concentration N_D and acceptor of concentration N_A

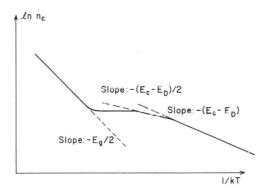

Fig. 1.3. Free electron concentration n_c of a partially compensated n-type semi-conductor $(N_D > N_A)$ as a function of temperature

holes (or electron and hole traps) [1.16, 18]. We will discuss these terms in more detail below. Equations (1.13) are undetermined as long as E_F is not known. They form an example of the common practice in semiconductor statistics to simply replace unknown parameters by others more convenient to handle. This will become evident in the statistical treatment of nonequilibrium steady-state situations in extrinsic semiconductors and insulators (Sect. 1.6).

In certain cases it is possible to eliminate E_F and to express n_c (or p) as a function of $E_c - E_D$ (or $E_A - E_v$), where E_D is the energy level of the donor and E_A that of the acceptor (Fig. 1.2). Let us consider a partially compensated extrinsic semiconductor containing a density of donors, N_D, and of acceptors, N_A, and assume $N_D > N_A$. The density, n, of electrons remaining in the donors is

$$n = N_D \{1 + g \exp[(E_D - E_F)/kT]\}^{-1}, \qquad (1.14)$$

where g is the degeneracy factor [1.16, 17]. Since the donor supplies electrons to the acceptors, only $N_D - N_A$ electrons are available for distribution between the donor level and the conduction band, and consequently $n_c + n = N_D - N_A$. The density of ionized donors, $N_D - n$, becomes with (1.14), $N_D - n = N_D \{1 + g^{-1} \exp[-(E_D - E_F)/kT]\}^{-1}$, which yields the following quadratic equation in n_c:

$$n_c(N_D - n)/n = n_c(n_c + N_A)/(N_D - n_c - N_A)$$
$$= gN_c \exp(-E/kT), \tag{1.15}$$

where $E \equiv E_c - E_D$.

For $E_c - E_F \gg kT$, this equation is readily solved to obtain an expression for the free carrier density in terms of N_D, N_A and the temperature. The temperature behavior of n_c is shown in Fig. 1.3. At temperatures low enough so that $n_c \ll N_A$, n_c becomes proportional to $N_c \exp(-E/kT)$ and, at higher temperatures, merges into a region for which its variation with T is given by $\exp(-E/2kT)$. Finally at high temperatures, the density of intrinsic carriers increases well above the available extrinsic carrier density and (1.9) holds again just as in an intrinsic semiconductor.

In the preceding we have used a notation for donors and acceptors common in semiconductor physics. In the physics of thermally stimulated relaxation in insulators, N_c is used instead of N_D, emphasizing the nature of a donor impurity as an electron trap, and A instead of N_A, stressing that the acceptor acts as a recombination center (Sect. 1.6.3).

Of interest to the study of thermally stimulated relaxation is of course a system that is not in thermal equilibrium but has been brought, as a result of some external perturbation, into a state that may be discribed as a non-equilibrium steady state. If the perturbation, e.g., by absorption of photons of energy $\omega\hbar \geq E_g$, is carried out at sufficiently low temperature, not all carriers, raised to higher levels from their equilibrium distribution over available states, will immediately relax back. Due to the presence of donors and acceptors, or, in another language, electron and hole traps and recombination centers, some of the charge carriers get attached to these defect or impurity levels such that their density does not correspond to statistical equilibrium at that temperature. Ionization of traps or, in other words, thermal release of trapped electrons and holes is required to bring about the redistribution of these carriers back to the equilibrium distribution. An activation energy $E = E_c - E_D$ or $E_A^* = E_A - E_v$ has to be provided to ionize electron traps and hole traps, respectively.

1.6 Nonequilibrium Steady-State Electron Statistics for Insulators and Semiconductors

In this section steady-state distribution functions for electrons and holes which arise as a result of an external perturbation of thermal equilibrium are discussed. They provide the initial condition for the thermally stimulated relaxation phenomena of electronic carriers described in Chaps. 2, 4 and, to some extent, Chap. 5.

The mathematical treatment of this situation is taken from the classic paper by *Shockley* and *Read* [1.11] for the case of a single trap level and from its extension to arbitrary distributions of traps by *Simmons* and *Taylor* [1.18].

Shockley–Read statistics is the basic statistical approach for such materials as Ge and Si because semiconductor technology has advanced to such a degree of sophistication that very pure intrinsic materials can be produced to start with. They can then be back-doped with the desired impurity so that the recombination and thermal generation processes are indeed controlled by one or very few different single trapping states. The resulting relative simplicity in the kinetic theory is one of the reasons why the technique of thermally stimulated relaxation has been so immensely successful in these materials.

The problems are far more complicated in crystalline insulators where trapping levels are generally distributed throughout the forbidden gap. This is even more the case in polycrystalline dielectrics and in amorphous materials [1.19–21]. Precisely for this reason our knowledge of the recombination and thermal excitation kinetics in insulators is less advanced. Even though thermally stimulated relaxation processes are easily measured or demonstrated in some form or other in virtually all insulating materials, their interpretation is in many cases not even attempted. Practically all statistic-kinetic approaches to interpret experimental data are based on oversimplified single trap models [1.22].

In the following discussion we assume with *Simmons* and *Taylor* [1.18] that an arbitrary distribution $N(E)$ of electron levels exists between the top of the valence band at E_v and the bottom of the conduction band at E_c. The state of occupancy determines whether a level acts as a hole trap or electron trap. When the level is unoccupied, it is ready to receive a hole and is, therefore, a hole trap. A definition of a level acting as a recombination center is given in Sect. 1.6.3. The charge state of a trap is not considered in this treatment. Instead, emphasis is placed on the cross section for electron or hole capture, which implicitly encompasses the charge of the state. For example, a positively charged unoccupied monovalent state is likely to have a large cross section for electron capture.

The purpose of this section is to describe the distribution of electrons and holes over available states during the perturbation and during the subsequent relaxation process. The sequence of the typical TSR experiment starts with a system in termal equilibrium at temperature T_1, proceeds with the establishment of a nonequilibrium steady state during the perturbation, e.g., generation of electron–hole pairs via photon absorption at the same temperature T_1, and the gradual reestablishment of thermal equilibrium at T_1 or, after completion of a heating process, at a higher temperature T_2.

The key to the occurrence of thermally stimulated phenomena during this redistribution process, is, of course, the fact that all real semiconductors and insulators possess quantum states in the forbidden gap and that capture and thermal emission transitions of holes and electrons take place between the bands and these states.

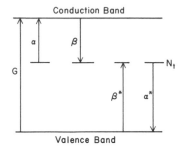

Conduction Band

Valence Band

Fig. 1.4. Schematic energy diagram of an insulating or semiconducting material containing defect states (traps) of density N_t. During perturbation of the thermodynamic equilibrium (e.g., by absorption of light with $\omega\hbar \geqq E_g$) free electrons and holes are generated with a rate G. Possible transitions between the transport bands and the traps are thermal emission of electron (α) or holes (α*) and capture (trapping) of free electrons (β) and free holes (β*); α, α*, β, and β* are the coefficients for thermal emission and capture, respectively

1.6.1 Shockley–Read Statistics

Consider first a single nondegenerate trap level of density N_t at energy E_t (Fig. 1.4) as a special case of an arbitrary trap distribution. According to *Shockley* and *Read* [1.11], the capture rate of electrons n_c in the conduction band by the trap is

$$r_\beta = \langle v_n \rangle S_n n_c N_t [1 - f(E)]$$
$$= \beta n_c N_t [1 - f(E)] , \qquad (1.16)$$

where $\langle v_n \rangle$ is the average thermal velocity of the electrons, S_n the capture cross section, $f(E)$ the probability that the trap level is occupied by an electron, and β the capture coefficient. In thermal equilibrium

$$f(E) = \{1 + \exp[(E_t - E_F)/kT]\}^{-1} . \qquad (1.17)$$

In the nonequilibrium steady-state situation, $f(E)$ is given by a similar expression in which the equilibrium Fermi level E_F is replaced by a quasi-Fermi level. Knowledge of $f(E)$ at any time during the thermally stimulated process together with knowledge of the respective capture and thermal emission rates completely determines the statistics of the system under investigation.

The rate of thermal release of trapped electrons is

$$r_\alpha = \alpha N_t f(E) , \qquad (1.18)$$

where α is the coefficient for thermal emission of electrons from the traps into the conduction band. Similarly, the capture rate for holes from the valence band is

$$r_\beta^* = \langle v_p \rangle S_p p N_t f(E)$$
$$= \beta^* p N_t f(E) , \qquad (1.19)$$

where p is the hole density in the valence band, $\langle v_p \rangle$ their average thermal velocity, S_p the capture cross section for holes, and β^* the capture coefficient.

The rate of hole emission into the valence band is given by

$$r_\alpha^* = \alpha^* N_t [1 - f(E)] \tag{1.20}$$

where α^* is the coefficient for thermal release of holes from the trap into the valence band. In thermal equilibrium the net rate of capture and emission of electrons and holes is zero (principle of detailed balance). It is therefore $r_\beta = r_\alpha$ and $r_\beta^* = r_\alpha^*$, which yields with (1.13)

$$\alpha = \langle v_n \rangle S_n N_c \exp[-(E_c - E_t)/kT],$$
$$\alpha^* = \langle v_p \rangle S_p N_v \exp[-(E_t - E_v)/kT]. \tag{1.21}$$

Let us now discuss the probability of occupation for a trap in the case of a nonequilibrium steady-state situation, arising, for example, during exposure of the sample to light quanta of sufficient energy to create electron–hole pairs. When the solid is uniformly illuminated, a steady-state is established such that creation of electrons and holes is balanced by capture and thermal release at the respective trapping levels. For an arbitrary distribution $N(E)$ of nondegenerate trapping states per unit energy and volume this may be expressed as [1.23]

$$\frac{dn_c}{dt} = G - \int_{E_v}^{E_c} c_n N(E) [1 - f(E)] dE + \int_{E_v}^{E_c} \alpha N(E) f(E) dE$$

$$\frac{dp}{dt} = G - \int_{E_v}^{E_c} c_p N(E) f(E) dE + \int_{E_v}^{E_c} \alpha^* N(E) [1 - f(E)] dE \tag{1.22}$$

where $c_n \equiv \langle v_n \rangle S_n n_c$, $c_p \equiv \langle v_p \rangle S_p p$. G is the net generation rate of electron–hole pairs. In the steady-state equilibrium $dn_c = dn_p = 0$ and

$$\int_{E_v}^{E_c} N(E) [-c_n(1-f) + \alpha f + c_p f - \alpha^*(1-f)] dE = 0.$$

This is valid for any distribution of trapping states $N(E)$. Therefore the term in the bracket is zero and

$$f(E) = \frac{c_n + \alpha^*}{\alpha + \alpha^* + c_n + c_p}. \tag{1.23}$$

The probability of occupation in thermal equilibrium is replaced by this new distribution function or occupational function that characterizes the nonequilibrium steady state *during* the ongoing stimulation (perturbation). We will show below that (1.23) can be cast in a form similar to the Fermi–Dirac distribution function where the equilibrium Fermi level is replaced by suitable quasi-Fermi levels. Note that $f(E)$ is independent of the form of stimulation, provided, or course, that the stimulation process does not alter $N(E)$, i.e., no new defect levels are created in the stimulation process.

1.6.2 Quasi-Fermi Levels

The use of quasi-Fermi levels for the description of occupation functions is quite common in solid-state statistics [1.5, 6, 24, 25]. It is based on the assumption that in the nonequilibrium steady-state situation the relative populations of electrons over certain available levels are indeed, in close approximation, determined by a thermal equilibrium between these levels. For example, the populations of electrons in the conduction band and in electron traps can be characterized by an occupational function $f = \{1 + \exp[-(E - E_f)/kT]\}^{-1}$ and a quasi-Fermi level E_f such that

$$n_c = N_c \exp[-(E_c - E_f)/kT]. \tag{1.24}$$

Similarly, the quasi-Fermi level for holes in the valence band is E_p and

$$p = N_v \exp[-(E_p - E_v)/kT]. \tag{1.25}$$

Both the statistical thermodynamic theory of chemical reaction kinetics [1.12, 13] and the theory of thermally stimulated relaxation kinetics in semiconductors and insulators successfully utilize the concept of quasi-thermal equilibrium despite the fact that, on the average, the reaction (or relaxation) proceeds in only one direction. The requirement for this is of course that the reaction (e.g., the recombination of electrons, thermally released into the conduction band) does not significantly disturb the quasi-equilibrium between free and trapped electrons maintained by recapture of thermally released electrons.

The introduction of quasi-Fermi levels through these occupational or distribution functions does, in principle, not add anything to our knowledge of electron or hole densities as a function of energies and temperature in the various levels, since these new Fermi levels can only be determined through the measurement of the occupation densities. However, they provide a convenient description of the establishment of the nonequilibrium steady state during ongoing perturbation and the reestablishment of thermal equilibrium during the thermally stimulated relaxation experiment.

Simmons and *Taylor* [1.18] introduced two additional fictitious levels, E_f^n and E_f^p, the quasi-Fermi level for trapped electrons and holes, respectively, which permit one to express the occupational function f of traps in a Fermi–Dirac form. This can be done, however, only for a given species of traps such that each species is associated with its own quasi-Fermi levels E_f^n and E_f^p.

Naturally, if a large number of different traps is involved then characterizing the statistical properties of these traps via individual quasi-Fermi levels is impractical and it becomes more convenient to use (1.23) which provides a single occupational function for the system.

In an arbitrary distribution $N(E)$ of traps each set of traps with unique capture cross sections S_n for electrons and S_p for holes will be characterized by its own distribution function and set of values for E_f^n and E_f^p. Due to the particular form of (1.23) it is, however, possible to define a species of traps by that set of traps that is characterized by a particular value of $\langle v_n \rangle S_n / \langle v_p \rangle S_p$. By inspection of (1.23) it can be seen that a species of traps is defined by a unique function $f(E)$. Both S_n and S_p may depend on E as well as on T [1.8].

For the development of the quasi-Fermi level concept for traps, it is possible to neglect α^* in (1.23) at all energies above the equilibrium Fermi level E_F and, conversely, α at $E < E_F$:

$$f(E) = c_n/(\alpha + c_n + c_p); \qquad \alpha^* \ll c_n, c_p, \alpha; \qquad (1.26)$$

$$f(E) = (c_n + a^*)/(\alpha^* + c_n + c_p); \qquad \alpha \ll c_n, c_p, \alpha^*. \qquad (1.27)$$

These occupational functions may be rewritten in a modified Fermi form by defining a quasi-Fermi level E_f^n via

$$c_n N_c \exp[-(E_c - E_f^n)/kT] = n_c(c_n | c_p), \qquad (1.28a)$$

and a quasi-Fermi level E_f^p via

$$c_p N_v \exp[-(E_f^p - E_v)/kT] = p(c_n + c_p), \qquad (1.28b)$$

to yield with (1.26)

$$f(E) = \frac{c_n}{c_n + c_p} \{1 + \exp[(E_t - E_f^n)/kT]\}^{-1} \qquad (1.29)$$

and

$$1 - f(E) = \frac{c_p}{c_n + c_p} \{1 + \exp[(E_f^p - E_t)/kT]\}^{-1}. \qquad (1.30)$$

In this form, the occupation functions are now recognized as those for electrons and holes, respectively, and E_f^n and E_f^p as quasi-Fermi levels for trapped electrons and holes. If we consider, for example, a distribution of traps $N(E)$ belonging to a single species, it is easily seen that the electron states are filled with electrons according to a Boltzmann distribution above E_f^n, and occupied to a constant level given by $f(E) = c_n/(c_n + p)$ below E_f^n. Similarly, the

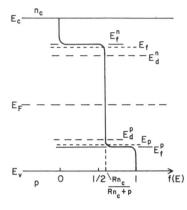

Fig. 1.5. A typical occupational function $f(E)$ for an arbitrary distribution of traps [1.18]. E_f and E_p, defined by (1.24) and (1.25), are the quasi-Fermi level for free electrons and holes, respectively. E_f^n is the quasi-Fermi level for trapped electrons and E_f^p that for trapped holes. defined by (1.29) and (1.30). Shown as well are the electron and hole demarcation levels E_d^n and E_d^p (see Sect. 1.6.3); $R \equiv \langle v_n \rangle S_n / \langle v_p \rangle S_p$

states are empty (occupied with holes) according to a Boltzmann distribution below E_f^p and filled with holes to a constant level given by $f(E) = c_p/(c_n + c_p)$ above E_f^p (Fig. 1.5). Under nonequilibrium steady-state conditions $E_f^n > E_f$ and $E_f^p < E_p$. In thermal equilibrium, all quasi-Fermi levels coincide with the equilibrium Fermi level E_F.

1.6.3 Classification of Trapping States Based on Electron Statistics

Defect states in the forbidden gap may act as electron traps or hole traps depending on their states of occupancy. The introduction of quasi-Fermi levels for trapped electrons and holes allows the classification of trapping states as shallow or deep traps. The involvement of electron states in the reaction process decreases with increasing energy from the quasi-Fermi level for trapped electrons up to the edge of the conduction band, or with decreasing energy from the quasi-Fermi level for trapped holes down to the top of the valence band. Free carriers falling into one of these trapping states will be reemitted with high probability back into the band from which they came from. These states are empty and are termed shallow traps. Obviously the temperature at which the nonequilibrium steady state is established together with the degree of occupancy of a trapping state will determine whether or not it acts as a shallow trap. For that reason we may refer to this definition of a shallow trapping state as a "statistical" one in contrast to the distinction between shallow and deep trapping states provided by physical arguments: Shallow traps are characterized by a very small ionization energy which is of the order of the phonon energies (e.g., substitutional impurities from groups III or V have ionization energies around 0.01 eV in Ge and around 0.04 to 0.02 eV in Si) [1.16]. Deep traps are those whose ionization energy is many times that of a phonon and, consequently, radiationless capture of a free carrier may involve multi-phonon transitions [1.8]. Naturally, at temperatures above several K, shallow trapping states are ionized (empty), in which case both definitions of shallow traps are

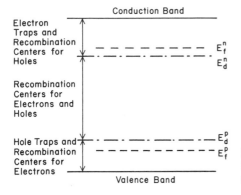

Electron Traps and Recombination Centers for Holes

Recombination Centers for Electrons and Holes

Hole Traps and Recombination Centers for Electrons

Conduction Band

E_f^n

E_d^n

E_d^p

E_f^p

Valence Band

Fig. 1.6. Demarcation levels E_d^n and E_d^p of an insulator or semiconductor with an arbitrary distribution of traps [1.5]

identical. On the other hand, at elevated temperatures deep trapping states may turn into "statistically" shallow traps.

For a discussion of all TSR processes that involve the redistribution of electrons and holes over the electron states within the forbidden gap, it is useful to provide a criterion that permits the classification of such states as traps or recombination centers. We have previously used the generic term "trap" for all states in the gap. However, a clear distinction is possible as to whether, at a given temperature and for a given occupational function, the probability for a carrier to be released from the defect state into one of the two allowed bands is larger than the probability for capture of a free carrier. This is done by introducing so-called demarcation levels, E_d^n, for trapped electrons and, E_d^p, for trapped holes. Defining, according to *Rose* [1.5], E_d^n as the energy level in a trap distribution $N(E)$ as the level at which the electron has equal probabilities of being thermally released into the conduction band or of recombining with a free hole, one obtains from (1.21) for $r_\alpha/r_\beta^* = 1$ and $r_\alpha^*/r_\beta = 1$

$$E_d^n = E_v + (E_c - E_p) + kT \ln \left[\frac{\langle v_p \rangle S_p N_v}{\langle v_n \rangle S_n N_c} \right] \text{ and similarly} \tag{1.31}$$

$$E_d^p = E_c - (E_f - E_v) + kT \ln \left[\frac{\langle v_n \rangle S_n N_c}{\langle v_p \rangle S_p N_v} \right]. \tag{1.32}$$

It is now obvious that all states between E_d^n and E_v are recombination centers for electrons, and all states between E_d^p and E_c are recombination centers for holes. It can be shown that, in first approximation, E_d^n is somewhat smaller than the quasi-Fermi level for free electrons and E_d^p somewhat larger than the quasi-Fermi level for free holes [1.18]. This situation is illustrated in Fig. 1.6.

1.6.4 Filling Diagrams

The purpose of discussing occupational functions and of defining a variety of special levels such as quasi-Fermi and demarcation levels is to provide a simple

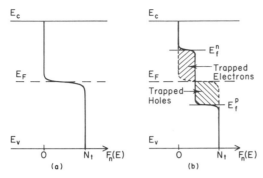

Fig. 1.7a, b. Filling diagram $F_n(E) = f(E)N(E)$ of a solid before (**a**) and after (**b**) perturbation of the statistical thermodynamic equilibrium. The cross-hatched sections are equal in area, indicating that the density of trapped electrons above the equilibrium Fermi level E_F is equal to the density of trapped holes below it [1.18]. It is assumed that a uniform density N_t of traps (defect states) is present in the solid

language suitable to efficiently characterize the nonequilibrium steady-state situation of a semiconductor or insulator during perturbation of thermal equilibrium. Filling diagrams, which are simply a plot of $f(E)$ as a function of energy in the energy interval of the band gap, are a convenient means of illustrating not only the initial condition for a thermal relaxation process but may actually be used to display the changes of the occupational function during the relaxation process itself [1.26].

Consider $f(E)$ for electrons in the case of some arbitrary trap distribution $N(E)$. The thermal equilibrium is characterized by the Fermi level E_F (Fig. 1.7) and the nonequilibrium steady-state condition during or immediately after uniform illumination of the solid by one quasi-Fermi level each for trapped electrons and holes. Under normal illumination levels (for which the density of free carriers is much smaller than that of trapped carriers) the cross-hatched sections below and above E_F are equal in area indicating that as many electrons are trapped above E_F as have been removed from levels below E_F. This is a direct consequence of the charge neutrality condition.

It is instructive to use filling diagrams for the illustration of the steady-state condition during or immediately after illumination at different generation rates G or temperatures. At a fixed temperature, an increase in the light intensity increases the free carrier densities and thus causes the quasi-Fermi levels as well as the demarcation levels to move closer toward the respective energy bands. The associated changes in the occupational function $f(E)$ for electrons are schematically shown in Fig. 1.8. At a fixed rate G, a decrease in temperature will move these levels in the same direction (Fig. 1.9).

Filling diagrams may also be used to display the actual filling of available levels. The filling of states with electrons is $F_n(E) \equiv N(E) f(E)$ and, correspondingly, that with holes is $F_p(E) \equiv N(E) [1 - f(E)]$. Some examples are shown in Fig. 1.10. Figure 1.10a is the case of four discrete trap levels belonging to the same species [1.18]. The density of states $N(E)$ is shown together with the occupational function $f(E)$ and the electron and hole fillings, $F_n(E)$ and $F_p(E)$.

Figure 1.10b illustrates the case of a uniform trap distribution and Fig. 1.10c that of an exponential distribution as a special case of a nonuniform trap distribution, corresponding to the "density of states tail" typical for amorphous semiconductors [1.27].

1.6.5 Nonequilibrium Steady-State Relaxation

In this section we discuss the relaxation of the system after the external perturbation G is removed at time $t = 0$. On first sight the kinetics may appear to be adequately described by (1.22) with $G = 0$ and the usual interpretation of the coefficient c_n, c_p, α and α^* obtained from Shockley–Read statistics [1.11]. However, one has to remember that the system is never completely isolated but rather it is in equilibrium with a temperature reservoir which is visualized, for example, as the radiation field composed of black body energy of density $\varrho(E)$ at temperature T. Thus, switching off the light source does not remove the black body radiation field and, consequently, the relaxation process proceeds in the presence of it. In (1.22) no explicit mention was made of this black body energy density, because it simply contributes to G. Nevertheless, the band-to-trapping state transitions take place in the radiation field $\varrho(E)$. With this simple concept, *Marlor* [1.28] has shown that the inherent limitations of the traditional phenomenological approach are removed and such parameters as capture cross sections and attempt-to-escape frequencies v can be interpreted in terms of the Einstein coefficients.

Before proceeding any further, we have to complete this picture of the system consisting of free and trapped electrons in the crystal matrix by adding to the interaction of photons and electrons that of phonons and electrons. In a phenomenological way this may be achieved by representing the energy density of the phonon field as $\varrho_p(E)$. Of course, the spectral distribution of this density depends on the individual crystal and is a function of temperature as well. The detailed theory of the interaction of the electrons in defect states with the phonons is presently still an area of intense research activity [1.8]. The situation is complicated by the fact that, in contrast to the interaction with photons, it is in general not sufficient to only consider single-phonon interactions. Except for shallow trapping states (which require the acquisition or disposal of only one phonon in order for the electron to be released from it or trapped by it) most states of interest here are deeper and either a cascade of single-phonon transitions [1.29, 30], or actual multiphonon transitions have to be taken into account. Some of the aspects of this problem are discussed in Chap. 2.

The electron–photon interaction together with spontaneous transitions provides, of course, for radiative transitions. Nonradiative transitions involve the absorption or emission of phonons.

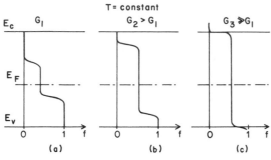

Fig. 1.8a–c. Steady-state occupational function $f(E)$ during perturbation with different but constant generation rates G [1.18]

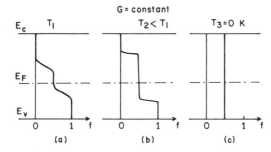

Fig. 1.9a–c. Occupational function $f(E)$ for a constant perturbation with a generation rate G at different temperatures. At $T = 0\,K$ no relaxation is possible and all defect levels are occupied with equal probability after a steady-state equilibrium is established [1.18]

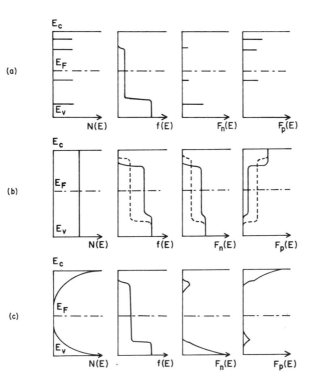

Fig. 1.10a–c. Caption see opposite page

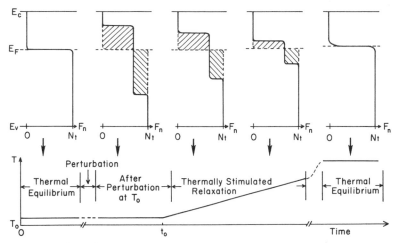

Fig. 1.11. Schematic representation of the filling levels $F_n(E)$ for a uniform trap distribution N_t during a typical thermally stimulated relaxation experiment. The trap system starts from thermal equilibrium at T_0 and is brought to a nonequilibrium situation at the same temperature by temporarily switching on a perturbation G. T_0 is chosen low enough so that the relaxation process is prevented to proceed. During a temperature scan $T(t)$, $F_n(E)$ gradually changes toward a new equilibrium. Note that the Fermi level at the initial equilibrium at T_0 is lower compared to the equilibrium reached at the elevated temperature

After removal of the source of external perturbation the system is left in a nonequilibrium state the relaxation of which may be considered to occur in two phases: a) the excess free carrier density decays via recombination with those recombination centers that have the largest cross section and b) further relaxation toward thermal equilibrium proceeds via thermal release of carriers from the traps into the bands and subsequent recombination with recombination centers. The rate limiting step of this second phase is the thermal release of trapped carriers. Its probability can be enhanced by increasing the temperature. Precisely this is done in TSR experiments. In the process, the quasi-Fermi levels and demarcation levels move toward the equilibrium Fermi level (Fig. 1.11). The relaxation kinetics is customarily described by assuming simply:

  I) band–band transitions are relatively unlikely to occur as compared to free carrier recombination with recombination centers [1.31];

◄ **Fig. 1.10a–c.** Occupancy-filling energy diagrams for several different trap systems: **(a)** Four discrete trap levels belonging to the same species, that R is the same for each level and, thus, only one function $f(E)$ describes the occupancy. The filling levels are $F_n(E)$ and $F_p(E)$; **(b)** Uniform trap distribution of two different species $R_1 < R_2$. In this case two functions $f(E)$ are required (dotted and solid lines) as well as two different filling levels of the two trap species with electrons and holes; **(c)** Exponential distribution of traps belonging to one species and corresponding occupational and filling functions

II) transitions between trapping states for electrons above E_d^n and for holes below E_d^p are nonradiative transitions and involve the emission or absorption of phonons only;

III) transitions between recombination centers and the two bands may be radiative transitions involving the emission or absorption of photons;

IV) the system, that is the nonequilibrium distribution of electrons and holes over available energy states, is in thermal contact with black body radiation of density $\varrho(E)$ and phonons of energy density $\varrho_p(E)$; these densities are the equilibrium densities at the temperature T of the solid.

Equations (1.22) can now be written as

$$
\frac{dn_c}{dt} = - \int_{E_v}^{E_d^n} [a(E) + b(E)\varrho(E)] \, Vn_c \, N(E) \, [1 - f(E)] \, dE
$$

$$
+ \int_{E_v}^{E_d^n} b(E)\varrho(E) V N_c \, N(E) \, f(E) \, dE
$$

$$
+ \int_{E_d^n}^{E_c} \alpha(E) N(E) f(E) \, dE - \int_{E_d^n}^{E_c} \beta(E) n_c N(E) \, [1 - f(E)] \, dE . \tag{1.33}
$$

A similar equation is obtained for dp/dt.

Here $a(E) = a(\omega)$ and $b(E) = b(\omega)$ are the Einstein coefficients for spontaneous and induced transition probabilities, V the volume of the solid. State degeneracies have again been neglected. We have refrained from rewriting $\alpha(E)$ and $\beta(E)$ in terms of phonon transition probabilities but instead left them in the form $\alpha = \langle v_n \rangle S_n N_c \exp[-(E_c - E_t)/kT]$ and $\beta = \langle v_n \rangle S_n$, noting that the capture cross section S_n is now that of a single or multiphonon transition. We further note that the filling of states is implicitly time dependent during ongoing relaxation. This fact reduces the utility of (1.33) to a formal and, in general, intractable rate equation which requires drastic simplification if one attempts to describe isothermal or thermally stimulated relaxation in an insulator.

The success of thermally stimulated relaxation techniques to obtain information on trapping states in the gap depends critically on whether or not the experiment can be performed under conditions that justify (1.33) to be reduced to simple expressions for the kinetic process. Historically, the kinetic theory of such thermally stimulated relaxation phenomena in bulk insulators and semiconductors as thermoluminescence, thermally stimulated current, polarization, and depolarization, have been interpreted by simple kinetic equations which were arrived at for reasons of mathematical simplicity only and which had, in most cases, no justified physical basis [1.32–34]. The hope was to extract information on traps (activation energies, thermal release probabilities, capture cross sections, etc.) by fitting experimental curves to these oversimplified kinetic descriptions [1.22, 35]. Naturally, the success of this approach was only marginal. This situation changed decisively after it was realized that TSR experiments can be performed under conditions which indeed justify the use of simple theoretical approaches for the determination of trapping parameters. First, instead of dealing with a distribution of traps, only a limited number of trap levels is present. This is best achieved by rigorously purifying the material and subsequently back-doping it with known impurities. However, only for a limited number of semiconductors has this been accomplished [1.16, 36]. Secondly, the experimental conditions must eliminate certain transitions which complicate the kinetic equations. For example, retrapping may be neglected under high-field conditions in thermally stimulated current experiments [1.26], a situation that is characteristic for junction spectroscopy as well (see Chap. 3).

1.6.6 Nonsteady-State Relaxation

The establishment of a steady-state equilibrium between trapped and free carriers was the basis of the discussion in the preceding sections. It remains one of the key assumptions in the kinetic of thermoluminescence, thermally stimulated conductivity and exoelectron emission in insulators and semiconductors. The kinetics of these relaxation reactions yields differential equations [such as (1.33)] that prove to be intractable unless one resorts to drastic approximations. It was, therefore, one of the major steps toward practical utilization of thermally stimulated relaxation phenomena as tools for trap spectroscopy, when it was realized that under certain conditions nonequilibrium relaxation may greatly simplify kinetic equations [1.26, 32]. An experiment can be performed in two steps: a) the equilibrium is perturbed as usual and a nonequilibrium steady state is established (i.e., filling of electrons and/or hole traps) and b) the experimental conditions for the relaxation process are set up such that it proceeds only in one direction, i.e., thermal release of trapped carriers takes place without retrapping. A simple analogy of this situation is the monomolecular (first order) kinetic reactions one observes, e.g., during flash desorption of adsorbed inert gases from a substrate [1.37, 38].

Of interest here are experimental conditions that preclude retrapping: Reverse biased p–n junctions, thin insulating or semiconducting films with at least one blocking contact, Schottky barriers, etc. Typical thicknesses are $10 \, \mu m$ or less and the electric fields produced by a bias voltage of up to $100 \, V$ reaches up to $10^6 \, V \, cm^{-1}$. Any carriers, thermally released from traps, are immediately swept out of this thin layer, thus preventing the establishment of a quasi-equilibrium steady state between free and trapped carriers. For example, the rate of electron emission from a single trap level becomes

$$\frac{dn}{dt} = -\alpha n \tag{1.34}$$

where α is given by (1.21).

Various experimental techniques are known to study trapping states in solids on the basis of this simple concept. They are collectively referred to as junction spectroscopy or space-charge spectroscopy (Sect. 1.7.1 and Chap. 3).

1.7 Trap Level Spectroscopy by Thermally Stimulated Release of Trapped Carriers – A List of Experimental Methods

All thermally stimulated relaxation reactions discussed so far involve the release of trapped charge carriers into either the conduction band or valence band and their subsequent capture by recombination centers or recapture by other traps (retrapping). Their experimental investigation is undertaken with the goal to determine the characteristic properties of the traps: capture cross sections, thermal escape rates, activation energies, concentrations of traps and trapped carriers. None of the TSR techniques listed below is suitable to identify the microscopic structure and chemical nature of the centers involved. This information has to come from independent experiments.

1.7.1 Direct Methods

The reaction rate or thermal escape rate can be monitored directly and, thus, can be determined by measuring the concentration of trapped carriers as a function of time and/or temperature. This is accomplished by using the material to be studied in a capacitor configuration (e.g., a p–n junction, as a Schottky barrier or, in general, as thin film sandwiched between electrodes) and by recording the changes in capacitance during the relaxation process. The capacitance change may be measured isothermally at one or several fixed temperatures and the experimental techniques employed have become known as isothermal capacitance transients (ICAPT) [1.39–41], deep-level transient spectroscopy (DLTS) [1.42] and double-correlation deep-level transient spectroscopy (DDLTS) [1.43]. These methods differ in their experimental sophistication, convenience in use and sensitivity.

Nonisothermal capacitance methods employ thermal scans [heating programs $T(t)$] and are known as thermally stimulated capacitance (TSCAP) [1.44–47]. A comparison of these methods is given by *Miller* et al. [1.48].

All other experimental TSR techniques used in trap level spectroscopy in semiconductors and insulators are indirect methods for the determination of trapping parameters. The techniques involve the measurement of phenomena that are due to charge carriers emitted after thermal stimulation from the traps.

1.7.2 Indirect Methods

A carrier, thermally released from the trap into a transport band, may be either retrapped by the same species of traps or a different one and, under the influence of an electric field, may contribute to an externally measurable current or a Hall voltage. It may either be swept out of the region being probed or recombine with a recombination center. Some of the electrons may even overcome the work function barrier and leave the solid altogether. The traffic of these charge carriers from the traps to the recombination centers or out of the material can be monitored at various stages and, thus, information on the thermal emission rates obtained indirectly.

During the thermally stimulated relaxation process, the concentration of free holes and electrons in extrinsic semiconductors and insulators is determined by the balance between thermal emission from and the recapture by traps and the capture by recombination centers. In principle, integration of (1.22) or (1.33) yields $n_c(t, T)$ and $p(t, T)$ for both isothermal current transients or during irreversible temperature scans. It is obvious that the trapping parameters, listed above, together with the capture rates of carriers in recombination centers, determine these concentrations. Measurement of the current density $J = e(\mu_n n_c + \mu_p p)$ will, therefore, provide trap-spectroscopic information. The experimental techniques employed in an attempt to perform trap level spectroscopy on this basis are known as isothermal current transients (ICT) [1.49], thermally stimulated conductivity (TSC) [1.23, 50 56], and thermally stimulated Hall effect (TSH) [1.57].

An additional method is thermally stimulated capacitor discharge (TSCD) [1.58–63]. It involves filling the traps at some high temperature (e.g., room temperature) by the application of a high field and subsequent cooling to a lower temperature with the field applied. Thereafter, the field is removed and the sample heated in the usual manner. The current, measured during heating, consists of two components: a) the dielectric relaxation current (see Chap, 4), and b) the current due to carriers thermally released from the trap into the two upper bands. Thus, in order to utilize TSCD as a trap-spectroscopic tool, one has to subtract component a) from the total current [1.64]. Component b) is, of course, closely related to TSC.

Trapping states located in a surface layer of generally less than 50 Å thickness can be probed with the technique of thermally stimulated exoemission (TSEE) [1.65–69]. During a thermal scan and after perturbation of the thermal equilibrium via exposure of the surface to ionizing radiation, low energy electrons are emitted from some insulators, which are thought to originate from free electrons in the conduction band, and, thus, their measurement provides indirect information on spectroscopic parameters of traps near the surface (Chap. 5). Recently, evidence is growing that not all emitted negative particles are electrons [1.70]. Negative ions and/or neutral particles are also emitted in many cases and have to be discriminated against to obtain emission that is solely due to thermal release of trapped electrons.

One of the most commonly employed techniques is thermally stimulated luminescence (TSL), which monitors photons as a function of temperature during the thermal scan. These photons are the result of raditative transitions (luminescence) of free carriers, previously released from traps, to recombination centers (see Chap. 2).

1.8 Dielectric Depolarization Spectroscopy

Thermally stimulated dielectric relaxation of a solid in a polarized state (electret) was selected for this volume (Chap. 4) for two reasons:

I) Both experimental methods as well as the formal kinetics of the process are closely related to trap level spectroscopy by thermally stimulated release of trapped charge carriers.

II) Thermally stimulated discharging of electrets provides similar spectroscopic information on the involved species as trap level spectroscopy does, most importantly on their density and the activation energy required for the relaxation process to proceed.

The polarized state (nonequilibrium steady state) is created by applying a dc voltage at an elevated temperature and by subsequent cooling of the solid to a temperature that is sufficiently low that rapid relaxation is prevented. Thereafter, the dc bias is removed. The currents that can be measured during either isothermal or nonisothermal relaxation back to thermal equilibrium are used to monitor the relaxation processes involved.

A detailed discussion of the statistical thermodynamic aspects of thermally stimulated dielectric relaxation will not be given here. At this point it should suffice to merely state that the kinetics of most of the processes are again complicated and that the phenomenological kinetic theories used to describe the observed thermally stimulated currents make use of assumptions which, being necessary to simplify the formalism, may not always be justified. Just as in the general case of TSL and TSC, the spectroscopic information may in principle be available from the measurement of thermally stimulated depolarization currents. However, it is quite frequently not possible to extract it unambiguously from such experiments.

1.9 General Aspects of TSR During Irreversible Temperature Scans

In this section we emphasize common features in the formal phenomenological theory of thermally stimulated relaxation phenomena in solids. The discussion will be restricted to nonisothermal processes that occur in the sample *after* perturbation of its equilibrium occupational function of states and *during* a monotonous rise in its temperature according to a known heating program.

These perturbations are all "macroscopic" in nature, brought about externally by a variety of means, e.g., mechanical strain, electric fields, exposure to ionizing radiation, etc. The process of perturbation is generally called "excitation" and the sample is referred to as "excited". The excited system is characterized by a collective of "microscopic" quantum states which are in thermal contact with the host lattice but usually independent of each other. The nonequilibrium occupation function, in the excited state of the system, will change toward the equilibrium occupation function by individual microscopic processes which involve the absorption of activation energy – in form of single or multiphonon absorption – from the host lattice. Since not individual quantum transitions, but rather a large number of them occurring per unit time, are monitored during a TSR experiment, the mathematical description of the relaxation is necessarily statistical in nature. Thus, the probability for an individual state to relax back to its equilibrium occupation becomes the key element of the statistical phenomenological theory. The coupling of the state to the host lattice and the requirement of phonon absorption for the "activation" of the thermally stimulated relaxation process leads to a relaxation probability $\alpha(T)$ which is strongly temperature dependent. One specific example of this has been discussed previously in Sect. 1.6. Another important situation is the relaxation of the electret state in polar materials (Chap. 4).

1.9.1 Formal Structure

The evolution of the TSR process during irreversible temperature scans can be visualized with the aid of diagram which are quite similar to logic diagrams used in the design of analog circuits. To demonstrate this we first describe two simple examples and subsequently discuss a more general case. A formal mathematical structure will be discussed in Sect. 1.9.2.

Let us consider the simple case of thermal emission of electrons from traps into the conduction band and transition of these electrons to recombination centers (Fig. 1.12a). The traps of density N are occupied with a density n of electrons. The filling level or degree of occupation, n/N, depends on the initial conditions (excitation and the temperature history) of the sample. During the temperature scan n will change according to the heating program $T(t)$. We denote the temperature–time dependent density of trapped electron $n\{T(t)\}$. The rate at which electrons are released from the traps may then be represented as

$$r_\alpha = \frac{\Delta n}{\Delta t} = -\alpha(T)n\{T(t)\} , \qquad (1.35)$$

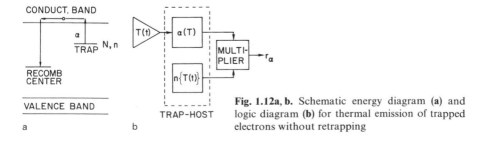

CONDUCT. BAND

RECOMB CENTER

VALENCE BAND

TRAP-HOST

Fig. 1.12a, b. Schematic energy diagram (**a**) and logic diagram (**b**) for thermal emission of trapped electrons without retrapping

a classical decay law that is the product of a trap-host system specific emission probability $\alpha(T)$ and a collective-statistical number density term. The logic "circuit" diagram of this situation is shown in Fig. 1.12b. A signal $T(t)$ is put into the trap-host system which produces the output r_α by multiplying $\alpha(T)$ and $n\{T(t)\}$.

As a second example, let us consider a dielectric that has externally been homogeneously polarized and subsequently cooled down to a temperature T_0 sufficiently low to prevent dipole relaxation. The internal polarization $P(T)$ will gradually vanish as the sample is warmed up according to a, e.g., linear heating program $T(T)=T_0+qt$ (q is the constant heating rate). Under short-circuit conditions, a depolarization current J_{TSD} will be measured that can be identified with the rate r_D at which the dipoles $n_D\{T(t)\}$ relax

$$J_{TSD}=\mu r_D=\mu n_D\{T(t)\}\alpha(T). \tag{1.36}$$

Here μ is the dipole moment and α is again the relaxation probability per unit time.

Since $J_{TSD}\equiv -dP(T)/dt$, we find with $P(T)=\mu n_D$ [1.65–70] and $dT=qdt$

$$P(T)=P(T_0)\exp\left[-\frac{1}{q}\int_{T_0}^{T}\alpha(T^*)dT^*\right], \tag{1.37}$$

$$J_{TSD}=P(T_0)\alpha(T)\exp\left[-\frac{1}{q}\int_{T_0}^{T}\alpha(T^*)dT^*\right], \tag{1.38}$$

and the density of dipoles, n'_d, that have vanished at time t is

$$n'_D(t)=\frac{1}{\mu}\int_{0}^{t}J_{TSD}dt=n_D(0)-n_D(t). \tag{1.39}$$

In the logic "circuit" diagram of this situation n_D and n'_D are "integrators" and $\alpha(T)$ may be considered a gain function (Fig. 1.13).

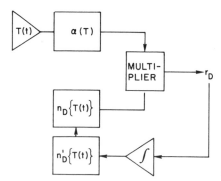

Fig. 1.13. Logic diagram for thermally stimulated depolarization according to (1.36) through (1.39)

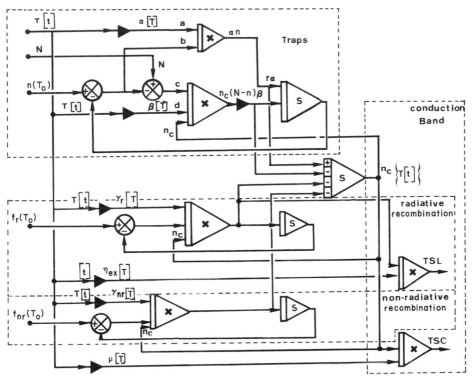

Fig. 1.14. Logic diagram for thermal release of trapped electrons n which subsequently can be either retrapped or recombined via a radiative and a nonradiative channel with recombination centers f_r and f_{nr}. The output functions that can be externally monitored are the TSL intensity and the thermally stimulated conductivity (TSC). The coefficients for radiative recombination are γ_r and γ_n, respectively. The temperature-dependent external detection efficiency for luminescence photons is η_{ex} (see Sect. 2.6) [1.73]

For both simple examples discussed here the logic diagram does not actually contribute anything to the understanding of the TSR process. However, most other thermally stimulated relaxation phenomena involve far more complicated patterns of kinetic relaxation as can be seen from a more realistic diagram of thermal release of trapped carriers (Fig. 1.14). In such a case the diagram is an instructive illustration of the complexities characterizing the kinetics and it helps to clarify the role of the various states and transitions involved in the rate processes.

The progress of the relaxation process can be monitored at various points in the diagram. The measured quantity may be called the "output" obtained from the system ("circuit") by putting in the "signal" $T(t)$. A list of these outputs or externally measurable phenomena was given in Sect. 1.7.

Of course, these logic schemes may also be helpful in analog simulations of relaxation processes if the involved probability functions $\alpha(T)$ are known. Very little work has been done on this subject up to this time [1.71].

1.9.2 Some Formal Mathematical Aspects

The mathematical description of TSR during a temperature scan reduces to solving a set of kinetic rate equations. These may take the form of (1.22) or (1.33), usually, however, simpler sets of coupled differential equations are chosen as the starting point of model calculations. In principle, these equations do not pose any problem. The methods to solve them numerically are well developed (Sect. 2.6). If the heating program, the transition rates, and the initial occupation function are known, the measurable outputs or physical phenomena that can be used to externally monitor the relaxation process can be calculated for a given model. Usually it is assumed in these calculations that a quasi-equilibrium exists between the state that relaxes (e.g., the occupied trap) and the state which it reaches by absorbing the activation energy (e.g., the transport band). The conditions for which this assumption is justified have been discussed in Sect. 1.6.2. Clearly, the quasi-equilibrium approximation is not valid in general. This, together with the irreversibility of the temperature scan, produces some remarkable properties of the reaction kinetics. The irreversibility is given by the fact that the occupational function of the relaxing system at a temperature T_1 is not reproduced by lowering the temperature back from a higher temperature T_2 to T_1. Consequently, all outputs are not simply functions of T or t, but should rather be called functionals, denoted by { }, e.g., $n\{T(t)\}$ [1.72].

The output functionals $X\{T(t)\}$ may be considered the result of processing the input signal $T(t)$ by a physical operator (sample) shown as a black box in Fig. 1.15. A particular given input signal $T(t)$ will then correspond to a particular representation, a so-called graph, $X_{T(t)}(t)$ of the general functional $X\{T(t)\}$. The operator $X\{\ \}$ applied to the signal $T(t)$ has some unusual properties [1.72]:

I) $X\{cT(t)\} \neq cX\{T(t)\};\quad c=\text{const.,}$

II) $X\{T(t+\tau)\} \neq X_{T(t)}(t+\tau),\quad$ and $\qquad\qquad\qquad$ (1.40)

III) $X\{T_1(t)+T_2(t)\}$ is meaningless.

The operation on $T(t)$ cannot be performed similar to a transfer function, spectral analysis or Volterra operator [1.73], nor by a variational calculation.

As stated before, a graph $X_{T(t)}(t)$ can be calculated from a set of coupled kinetic rate equations for given initial conditions and a heating program and partial derivatives with respect to time or temperature exist. The value of such a graph depends, however, on the exact temperature–time history of the sample;

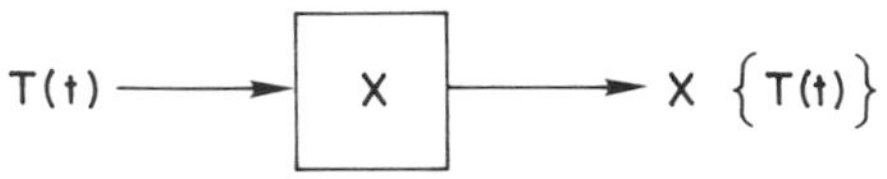

$$\text{T(t)} \longrightarrow \boxed{X} \longrightarrow X\ \{T(t)\}$$

Fig. 1.15. Schematic diagram of processing the input signal $T(t)$ by a physical operator X (trap system) to obtain the output functional $X\{T(t)\}$

that is to say, if one starts from a given situation of the system at T_0 and heats up to T_1 with a given heating program within a time $t_1 - t_0$, the value of the graph obtained with a different heating program at the same temperature and time will be different. This implies that partial derivatives of the general functional $X\{T(t)\}$ are difficult to conceive [1.74].

In conclusion it is noted that the general problem of thermally stimulated relaxation during irreversible temperature scans is mathematically complex because it involves functionals rather than functions of T and t. A way out of this situation may be sought by either designing the experiment in such a way that the kinetics becomes inherently simple (direct methods to obtain trap-spectroscopic information in TSR experiments discussed in Sect. 1.7.1 are examples) or to measure simultaneously several different outputs and thereby eliminate cumbersome functional dependencies on T or t [1.75].

References

1.1 S. Glasstone, K. J. Laidler, H. Eyring: *The Theory of Rate Processes* (McGraw-Hill, New York 1941)

1.2 R. M. Noyes: *Handbook of Physics*, ed. by E. U. Condon, H. Odishaw (McGraw-Hill, New York 1967)

1.3 M. Eigen, L. de Maeyer: "Investigation of Rates and Mechanisms of Reactions", in *Technique of Organic Chemistry*, Vol. 8, Pt. II, ed. by S. L. Friess, E. S. Lewis, E. Weissberger, 2nd ed. (Interscience, New York 1963)

1.4 G. Porter: Science **160**, 1299 (1968)

1.5 A. Rose: *Concepts in Photoconductivity and Allied Problems* (Interscience, New York 1963)

1.6 R. H. Bube: *Photoconductivity of Solids* (Wiley, New York 1960)

1.7 R. H. Bube: *Electronic Properties of Crystalline Solids* (Academic Press, New York, London 1974)

1.8 C. H. Henry, V. D. Lang: Phys. Rev. B**15**, 989 (1977)
R. Pässler: Phys. Status Solidi (b) **85**, 203 (1978)

1.9 R. Chen: J. Mat. Science **11**, 1521 (1976)

1.10 R. Chen: J. Electrochem. Soc., Solid State Sci. **116**, 1254 (1969)

1.11 W. Shockley, W. T. Read, Jr.: Phys. Rev. **87**, 835 (1952)

1.12 E. A. Moelwyn-Hughes: *The Kinetics of Reactions in Solutions* (Oxford University Press, London 1947)

1.13 L. S. Kassel: *Kinetics of Homogeneous Gas Reations* (Reinhold, New York 1932)

1.14 V. Maxia: Lett. Nuovo Chim. **20**, 443 (1977)

1.15 V. Maxia: Phys. Rev. B**17**, 3262 (1978)

1.16 A. G. Milnes: *Deep Impurities in Semiconductors* (Wiley, New York 1973)

1.17 N. W. Ashcroft, N. D. Mermin: *Solid State Physics* (Holt, Rinehart and Winston, New York 1976)

1.18 J. G. Simmons, G. W. Taylor: Phys. Rev. B**4**, 502 (1971)

1.19 S. C. Agarwal, H. Fritsche: Phys. Rev. B**10**, 4351 (1974)

1.20 B. T. Kolomiets, T. F. Mazets: J. Non-Cryst. Solids **3**, 46 (1970)

1.21 R. A. Street, A. D. Yoffe: Thin Solid Films **11**, 161 (1972)

1.22 P. Bräunlich: "Thermoluminescence and Thermally Stimulated Current – Tools for the Determination of Trapping Parameters", in *Thermoluminescence in Geological Materials*, ed. by D. J. McDougall (Academic Press, London 1968) Chap. 2.5

1.23 I. Broser, R. Broser-Warminsky: Ann. Phys. **6F**, 16, 361 (1955); Br. J. App. Phys. Suppl. **4**, 90 (1955)
1.24 W. Shockley: *Electrons and Holes in Semiconductors* (Van Nostrand, New York 1950)
1.25 J. S. Blakemore: *Semiconductor Statistics* (Pergamon Press, Oxford, New York 1962)
1.26 J. G. Simmons, G. W. Taylor: Phys. Rev. **B5**, 1619 (1971)
1.27 M. H. Cohen, H. Fritsche, S. Ovshinsky: Phys. Rev. Lett **22**, 1065 (1969)
1.28 G. A. Marlor: Phys. Rev. **159**, 540 (1959)
1.29 M. Lax: J. Phys. Chem. Solids **8**, 66 (1959)
1.30 M. Lax: Phys. Rev. **119**, 1502 (1960)
1.31 D. Curie: *Luminescence in Crystals* (Wiley, New York 1963)
1.32 J. T. Randall, M. H. F. Wilkins: Proc. R. Soc. London A**184**, 366 (1945)
1.33 G. F. J. Garlick, A. F. Gibson: Proc. R. Soc. London **60**, 574 (1948)
1.34 P. Kelly, M. J. Laubitz, P. Bräunlich: Phys. Rev. **B4**, 1960 (1971)
1.35 P. Kivits, H. J. L. Hagebeuk: J. Luminesc. **15**, 1 (1977)
1.36 P. J. Dean: "III–V Compound Semiconductors", in *Electroluminescence*, ed. by J. I. Pankove, Topics in Applied Physics, Vol. 17 (Springer, Berlin, Heidelberg, New York 1977)
1.37 D. Menzel: "Desorption Phenomena", in *Interactions on Metal Surfaces*, ed. by R. Gomer, Topics in Applied Physics, Vol. 4 (Springer, Berlin, Heidelberg, New York 1975) p. 102
1.38 L. D. Schmidt: "Adsorption Binding States on Single-Crystal Planes", in *Catalysis Reviews*, ed. by W. Heinemann, J. J. Carberry (M. Dekker, New York 1974) p. 115
1.39 R. Williams: J. Appl. Phys. **37**, 3411 (1966)
1.40 Y. Furukawa, Y. Ishibashi: Jpn. J. Appl. Phys. **5**, 837 (1966)
1.41 L. D. Yau, C. T. Sah: Appl. Phys. Lett. **21**, 157 (1972)
1.42 D. V. Lang: J. Appl. Phys. **45**, 3014, 3023 (1974)
1.43 H. Leferre, M. Schulz: IEEE Trans. ED-**7**, 973 (1977)
1.44 J. C. Carballes, J. Lebailly: Solid Stat. Commun. **6**, 167 (1968)
1.45 C. T. Sah, L. Forbes, L. L. Rosier, A. F. Tasch, Jr.: Solid-State Electron. **13**, 759 (1970)
1.46 C. T. Sah, W. W. Chan, H. S. Fu, J. W. Walker: Appl. Phys. Lett. **20**, 193 (1972)
1.47 C. T. Sah, J. W. Walker: Appl. Phys. Lett. **22**, 384 (1973)
1.48 G. L. Miller, D. V. Lang, L. C. Kimmerling: Ann. Rev. Mater. Sci. **7**, 377 (1977)
1.49 N. F. J. Mathews, P. J. Warter: Phys. Rev. **144**, 610 (1966)
1.50 C. J. Delbecq, P. Pringsheim, P. H. Yuster: Z. Phys. **138**, 266 (1954)
1.51 R. H. Bube: Phys. Rev. **83**, 393 (1951); **99**, 1105 (1955); **101**, 1668 (1956); **106**, 703 (1957)
1.52 R. R. Hearing, E. N. Adams: Phys. Rev. **117**, 451 (1960)
1.53 K. H. Nicholas, J. Woods: Brit. J. Appl. Phys. **15**, 783 (1964)
1.54 G. H. Dussel, R. H. Bube: Phys. Rev. **155**, 764 (1967)
1.55 I. J. Saunders: J. Phys. C**2**, 2181 (1969)
1.56 P. Bräunlich, P. Kelly: Phys. Rev. **B1**, 1596 (1970)
1.57 P. Kivits: J. Phys. C, Solid State Phys. **9**, 605 (1976)
1.58 M. C. Driver, G. T. Wright: Proc. Phys. Soc. London **81**, 141 (1963)
1.59 V. F. Zolotaryov, D. G. Semak, D. V. Chepur: Phys. Status Solidi **21**, 437 (1967)
1.60 A. G. Zhdan, V. B. Sandomirskii, A. D. Ozheredov: Solid-State Electron. **11**, 505 (1968)
1.61 A. Servini, A. K. Jonscher: Thin Solid Films **3**, 341 (1969)
1.62 H. A. Mar, J. G. Simmons: Phys. Rev. **B8**, 3865 (1973; Appl. Phys. Lett. **25**, 503 (1974)
1.63 B. T. Kolomiets, V. M. Lyubin: Phys. Status Solidi (a) **17**, 11 (1973)
1.64 S. C. Agarwal: Phys. Rev. B**10**, 4340 (1974)
1.65 J. Kramer: Z. Phys. **129**, 34 (1951; **133**, 629 (1952)
1.66 A. Bohun: Czech. J. Phys. **3**, 394 (1953); **4**, 91 (1954)
1.67 H. Nassenstein: Z. Naturforsch. **10a**, 944 (1955)
1.68 K. Becker: "Stimulated Exoelectron Emission from the Surface of Insulating Solids", in *Critical Reviews in Solid State Physics*, Vol. 3, ed. by D. E. Schuele, R. W. Hoffman (The Chemical Rubber Co. Press, Cleveland 1972) p. 39
1.69 V. Bichevin, H. Käämbre: Phys. Status Solidi (a) **4**, K 235 (1971)
1.70 I. V. Krylova: Usp. Khim. **45**, 2138 (1976), [English transl.: Russ. Chem. Rev. **45**, 1101 (1976)]
1.71 K. N. Razdan, W. G. Wiatrowski, W. D. Breiman: J. Appl. Phys. **44**, 5489 (1973)

1.72 A.Sparatu: *Theorie de la transmission de l'information* Part I (Masson, Paris 1970)
      J.P.Fillard, J.Gasiot: Phys. Status Solidi (a) **32**, K85 (1975)
      J.P.Fillard, J.Van Turnhout (ed.): *Thermally Stimulated Processes in Solids – New Prospects*,
      Proceedings of the International Workshop on Thermally Stimulated Processes, La Grande
      Motte, France 1976 (Elsevier 1977)
1.73 J.Gasiot: Thesis, University of Montpellier (1976)
      J.P.Fillard, J.Gasiot: J. Electrostatics **3**, 37 (1977)
1.74 B.K.P.Scaife: J. Phys. D**7**, L171 (1974)
      J.Van Turnhout: J. Phys. D**8**, L68 (1974)
      B.K.P.Scaife: J. Phys. D**8**, L72 (1975)
      B.Gross: J. Phys. D**8**, L127 (1975)
      J.Kratochvil, J.Sestak: Thermochim. Acta **7**, 330 (1973)
1.75 J.P.Fillard, J.Gasiot, M.De Murcia: J. Electrostatics **3**, 99 (1977)
      J.P.Fillard, J.Gasiot, J.C.Manifacier: Phys. Rev. B**18**, 4497 (1978)

2. Thermally Stimulated Luminescence and Conductivity

P. Bräunlich, P. Kelly, and J.-P. Fillard

With 16 Figures

Observations of thermally stimulated luminescence (TSL) have been reported as early as the 17th century, but *Urbach* [2.1] is generally credited with proposing it as a potentially useful experimental technique for trap level spectroscopy. However, only after the publication of the work of *Randall* and *Wilkins* [2.2] in 1945 did TSL receive much attention [2.3]. First measurements of thermally stimulated conductivity (TSC) alone or simultaneously with TSL were performed by *Bube* [2.4], *Delbecq* et al. [2.5], and *Broser* and *Broser-Warminsky* [2.6].

Both TSL and TSC are observed in a broad variety of materials, e.g., fish scales, bones, dental enamel, plastics, and other organic solids [2.7], archaeological artifacts, minerals and ceramics as well as in amorphous, polycrystalline and single crystalline semiconductors and insulators.

The occurrence of TSL and TSC during a thermal scan of a previously excited ("perturbed") material is probably the most direct evidence we have for the existence of electronic trap levels in these materials. A TSL/TSC spectrum (for historical reasons frequently referred to as "glow curve") usually consists of a number of more or less resolved peaks in luminescence intensity or electric conductivity vs temperature (Fig. 2.1) which, in most cases, may be attributed to a species of traps. Its appearance is a direct representation of the fact that

   I) the escape probability of trapped carriers is a sharply increasing function with temperature, and

   II) the supply of trapped carriers is limited to start with and decreases with their ongoing thermal release from the traps.

Since the escape probability of carriers from trapping sites is proportional to $\exp(-E/kT)$ [2.2], the location of a glow peak on the temperature scale provides encoded information on the value of the thermal activation energy E. Hence, a glow curve represents a spectrum of energies which are required to free carriers from the various species of traps in the material.

The procedures used to decode the glow spectrum and retrieve the desired trap-spectroscopic data appear, on first sight, obvious and straightforward: A measured curve is analyzed to obtain such characteristics as location of the peak on the temperature scale, its widths, initial rise, etc. These data are then utilized to compute trapping parameters via an *appropriate* model for the reaction kinetic processes that occur during the temperature scan. However, exact knowledge of the proper kinetics is mandatory for this analysis to yield quantitative values.

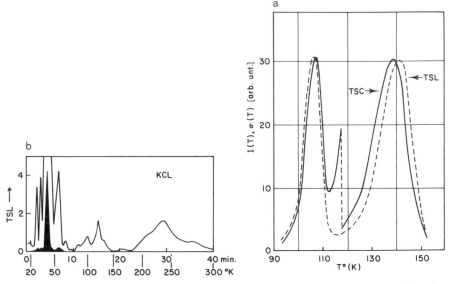

Fig. 2.1. (a) Example of a glow spectrum. The thermally stimulated luminescence intensity, measured with a nearly constant quantum efficiency between 900 and 5000 Å, is plotted vs temperature for a KCl single crystal that was exposed to X-rays at 10 K and subsequently heated to room temperature at a rate of approximately $0.1 \ \mathrm{K \ s^{-1}}$. The shaded curve represents a reduction of the solid curve by a factor of 20 [2.56]. **(b)** Thermoluminescence (———) and thermally stimulated conductivity (– – –) glow spectra of a nominally pure LiF single crystal after its exposure to X-rays at 85 K and subsequent heating at a rate $q = 0.05 \ \mathrm{K \ s^{-1}}$. This glow spectrum is an example for the frequently observed correlation between TSL and TSC peaks. Typically, a temperature shift between the location of the luminescence and conductivity peaks is observed in experiments of this kind. Of interest here is the occurrence of the first TSC peak at a lower temperature as compared to the TSL peak, an observation that cannot be explained with the so-called single trap model (Sect. 2.6.1) unless one assumes an exponential temperature dependence of the carrier mobility [2.279]

The most simple of reaction kinetics which actually yields TSL and TSC peaks is the so-called single trap model [2.2–17] described in Sect. 2.6.1. The field of TSL and TSC developed along two lines. The first one merely made use of the capability of deep traps in certain insulating materials to "store" charge carriers at or below room temperature for a long time – sometimes thousands of years – without being much concerned about the mechanism of this information storage and its eventual retrieval in a thermal scan in the form of thermoluminescence or (less frequently) thermally stimulated conductivity. Very successful applications of these phenomena in dosimetry, geology, archaeology, etc., were the result (see Chap. 6). The other approach concentrated on quantitative trap level spectroscopy; employing curve-fitting techniques on the basis of the single trap model [2.15, 17] and on efforts to completely understand the detailed features of TSL and TSC curves calculated within the framework of this model [2.10–13, 16, 17]. For a number of years, the often rather limited success to measure trapping parameters by this approach was no deterrent for ever continuing attempts along this line. A poor fit of experimental and computed glow curves was rarely related to inadequacy of the single trap model but rather to the fact that only approximate solutions of its rate equations were available. Exact solutions of this set of "stiff" differential equations were possible only after the development of powerful numerical computation techniques. The work of *Kelly* et al. in 1971 [2.11–13] revealed the starting variety of different thermal emission curves that could be obtained from a physically meaningful range of trapping parameters.

It then became immediately evident that it is extremely difficult to correlate, by curve fitting alone, theory and experiment with any degree of confidence. Any measured and well-resolved glow peak that may reasonably be expected to be due to a single type of traps can be fitted with a solution of the single trap model by appropriately adjusting several out of a set of many model parameters. Unfortunately, such a fit is not unique, since a number of different simple model descriptions, some of them derived from (1.22) in Chap. 1 as well, are conceivable in addition to the single trap model. Very little effort has been extended to investigate them [2.18–20]. However, they pose no principal mathematical problems any longer.

The origin for this lack of uniqueness has to be traced to the fact that both thermally stimulated conductivity and thermoluminescence are only *indirect* trap-spectroscopic methods (Sect. 1.7.2). In contrast to thermally stimulated capacitance techniques (Chap. 3), the thermal release from traps or the capture of charge carriers in traps is not measured directly. Rather, the transient traffic of thermally released carriers toward available levels, which will eventually be populated according to the thermal equilibrium occupational function, is monitored at various points along the way. Naturally, as long as the course of this traffic is not precisely known, unambiguous information on the type and the origin (traps) of the carriers cannot be readily obtained by solely studying its dynamics at isolated points along one of its several possible routes.

The full potential of TSL and TSC as quantitative spectroscopic tools has not yet been realized. In their present state of development they give a quick and very sensitive survey of the number and relative concentration of traps without being able to discriminate between hole and electron traps. Only estimates of activation energies are possible. TSL and TSC measurements provide at best a test for the validity of a kinetics mechanism if most of the pertinent trap level parameters are known from independent measurement techniques. However, if one starts from a point of complete ignorance, the general complexity of the kinetics will usually prevent the extraction of quantitative trap-spectroscopic data by solely employing presently available TSL and TSC techniques.

2.1 Background

This chapter is written to critically appraise the current state of TSL and TSC techniques as tools in trap level spectroscopy and hopefully to provide a stepping-stone for future advancement. Progress toward the goal to develop the

full potential of nonisothermal relaxation phenomena such as thermoluminescence and thermally stimulated conductivity as investigative tools of trap
properties will most likely come by

I) continuing careful studies of defect levels in nonmetallic solids with such
independent techniques as electron spin resonance [e.g., 2.21–23], optical
absorption and emission spectra [e.g., 2.24–27], and other trap-influenced
phenomena such as thermally stimulated capacitance, photocapacitance,
photoconductivity, carrier injection and space charge limited conduction,
luminescence, Hall effect, diffusion, impurity conduction, electric field impact
ionization, etc. [2.28], and correlating the obtained results whenever possible
with TSL and TSC investigations;

II) devising novel experimental TSL and TSC techniques that are designed
to direct the recombination traffic of carriers into one well-understood
direction or channel and, thus, eliminate much of the complexity otherwise
associated with the kinetic rate equations.

The latter approach appears to be particularly promising. Decisive steps in
this direction have been taken by *Simmons* and *Taylor* [2.29], and the
developers of thermally stimulated current and capacitance techniques in
semiconductor junctions (*Driver* and *Wright* [2.30], *Carballes* and *Lebailly*
[2.31], *Sah* and co-workers [2.32, 33], and *Lang* [2.34, 35]). These workers
realized that, by choosing an appropriate experimental arrangement, e.g., a thin
layer of the material to be studied either sandwiched between two blocking or
one blocking and one Ohmic contact or a *p-n* junction, a high field can be
applied during the thermal scan. Carriers, thermally released from traps, are
rapidly swept out of the layer through the contacts which has the immediate
effect of channelling all the carrier traffic along a one-way street. Retrapping
and recombination transitions can be neglected resulting in simplified reaction
kinetic rate equations and eliminating many, if not all the ambiguities otherwise
so characteristic of TSL and TSC experiments. However, even here field effects
may influence the results obtained by these techniques and therefore have to be
carefully considered (Chap. 4).

These new approaches toward quantitative trap level spectroscopy have
been implemented in thermally stimulated capacitance techniques and culmi-
nated in the development of deep-level transient spectroscopy (DLTS). Chapter
3 gives an account of this major advancement in the utilization of thermally
stimulated relaxation phenomena as powerful tools for the investigation of
nonradiative transitions between defect levels and the transport bands in
nonmetallic solids.

Surprisingly, the high-field TSC techniques proposed by *Simmons* and
Taylor have not yet been employed to any extent. Perhaps the more difficult
preparation of the sample together with the often occurring background
polarization and depolarization currents have been a deterrent up to this point.
However, since the DLTS method can equally well be employed for current
measurements, considerable progress can be expected in the near future,
particularly in nonpolar materials. The observation of thermoluminescence

under high-field conditions from a thin layer is far more complicated. In addition to elaborate sample preparation, one has to find optimal field conditions which forbid retrapping but also allow some of the free carriers to recombine with recombination centers in the thin layer. Clearly, new experimental techniques are needed to measure luminescence in this situation. Thermostimulated exoemission (TSEE) of charged or neutral particles (see Chap. 5) may not easily lend itself to high-field measurements. Perhaps thin film materials, deposited onto metallic substrates, may be used in external electric fields. In this connection it appears intriguing to use the rate-window concept (see Chap. 3) in conjunction with pulsed excitation (e.g., a pulse electron source) and, thus, take advantage in TSEE experiments of the entire range of experimental improvements introduced with the DLTS technique.

In the following we will first give a brief description of shallow and deep defect levels in nonmetallic solids and present a survey of measurement techniques other than TSL and TSC which not only reveal the existence of traps in these materials but may also be used in conjunction with thermal stimulation to determine trap parameters. Following the introduction of the primary trap-spectroscopic parameters that can be experimentally determined with the methods presented in this chapter, the theory of nonradiative capture and release of charge carriers from traps is reviewed (Sect. 2.5). The recombination kinetics of TSL and TSC is developed in Sect. 2.6, followed by a detailed description of experimental techniques.

2.2 Defect States in Insulators and Semiconductors

Even though radiative transitions that occur during the thermal scan are measured in TSL, the primary objects of investigation in both TSL and TSC experiments are nonradiative transitions between the ground level or excited level of the defect (trap) and the conduction band and/or valence band. In certain cases, namely when the two different impurities – a donor and an acceptor – are close enough that their wave functions overlap, a tunneling process may take place from an excited level of the donor to the acceptor. In this case the excited level can be occupied from the ground state via a nonradiative transition and, after tunneling to an excited state of the acceptor, light may be emitted by the subsequent decay to its ground state (thermoluminescence) [2.36–40]. If the entire recombination traffic proceeds via tunneling between donor – acceptor pairs, no TSC will be observed (Sect. 2.6.3).

There are other thermally stimulated luminescence and electron transport phenomena which are associated with electronic transitions between defect levels and the states of the solid, which can be exploited to obtain information on the concentration of these defect and transition rates. Defect levels are localized electronic states in the solid due to a variety of causes but all leading to a loss of translational symmetry of the crystal lattice: substitutional or interstitial impurities, vacancies, dislocations as well as the termination of the lattice at the surface. The properties of these levels are studied with the conventional methods of solid-state spectroscopy (optical absorption, lumines-cence, photoconductivity, electron spin resonance, X-ray photoelectron spec-troscopy, etc.) [2.39, 41–47].

The theory of electronic defect states and *optical* transitions was developed on the basis of the adiabatic approximation. The many-particle system, consisting of electrons and nuclei and their mutual interactions, is thereby separated into a nuclear part and an electronic part. The coupling between the two is neglected. The resulting many-electron Schrödinger equation is fre-quently reduced further to a one-electron equation which describes the behavior of a single representative electron in the periodic potential V of the perfect lattice which is modified by an additional potential U centered around the location of the point defect [2.42, 48]. This "adiabatic" theory has reached a high level of sophistication in two limiting cases: deep and shallow defect states.

2.2.1 Tightly Bound (Deep) Defect States

If the wave functions of the defect states are localized in the region where U dominates, the lattice potential may only act as a small perturbation and the electronic defect states can be calculated with well-known methods. Successfully treated problems include calculations of the splitting of the free atom or ion levels as the impurity is imbedded in the matrix of the host crystal [2.49, 50] and of the energy required to remove one or more electrons from the impurity (ionization energy). The agreement between calculated and observed values of the ionization energy for a number of impurities in such semicon-ductors as GaAs, Si and Ge is remarkable (see, e.g., the extensive tables given by *Milnes* [2.28]). The bulk of our knowledge of deep impurity levels in semiconductors stems, however, from experimental investigations. Optical techniques (absorption and photoluminescence) have been used extensively since they have the advantage of easy sample preparation and straightforward data analysis. Transport phenomena (including TSC and DLTS) often require relatively tedious analysis, measurement over a wide range of temperatures as well as more complicated sample preparation [2.28, 39]. However, their use appears essential whenever the recombination is predominantly nonradiative or when strong lattice coupling produces broad photon distributions which do not permit a clear identification of the involved transition by optical means.

2.2.2 Weakly Bound (Shallow) Defect States

Electronic states that are only weakly bound to a defect are characterized by eigenfunctions which extend over a large area of the lattice. In this case the defect potential U acts as a small perturbation on the periodic lattice potential

V [2.46, 51–55]. Classic examples are phosphorous, arsenic, and antimony as donor impurities in silicon and germanium and aluminum, gallium and indium as acceptor impurities. The ionization energies of these impurities are very small (typically 0.01 eV in Ge and 0.05 eV in Si). They are conveniently determined by infrared absorption techniques or low temperature Hall measurements [2.28]. Thermoluminescence and thermally stimulated conductivity have only occasionally been performed at sufficiently low temperatures to observe glow peaks which are produced by thermal release and traps with activation energies less than 50 meV [2.56]. Quantitative TSL and TSC measurements of such small activation energies have not been reported. However, they should be possible if the model for recombination kinetics is known.

2.2.3 Intermediately Bound Defect States

The theory of electronic defect levels in solids where neither of the previous approximations is valid is still in a rather incomplete state and in most cases only experimental investigations provide accurate values of transition probabilities and the associated energics. The defect is considerably localized but the electronic levels are often not just perturbed by the surrounding lattice but grossly distorted. Absorption and luminescence spectra are usually broad even at low temperatures. Certain color centers in wide band gap materials are examples. A wealth of experimental information on the optical properties of defects in these ionic crystals can be found in the literature [2.57–60]. A detailed list of impurity and other defect spectra in the visible and uv in alkali halides, alkaline earth halides and some metal oxides is given by *Sparks* [2.61].

The theoretical study of these defects in ionic insulators is also difficult because of strong polarization effects which produce severe distortions of the ions surrounding the impurity. These distortions are quite different for a defect ground state as compared to excited states, a situation which is visually represented by the Seitz coordinate diagram [2.62–64], discussed in Sect. 2.5.3. It should be noted here that color centers and many other defect levels in wide gap materials are usually deep levels as far as their position in the forbidden gap is concerned [2.41–44] and that the aforementioned theoretical difficulties are the sole reason for including them in this section.

The bulk of the thermoluminescence literature and a considerable fraction of the literature on thermally stimulated conductivity is concerned with these defect levels in insulators. The study of trap parameters with TSL and TSC techniques in wide gap materials is, however, limited to those deep defect levels whose activation energies for the release of trapped carriers do not exceed about 1.5 eV. Any larger value will place the glow peak at temperatures so high that it becomes difficult to discriminate against the black body background and ionic conductivity.

2.3 Traps and Thermal Transport Properties

Defect levels (used here as a generic term for all defect and impurity states in the forbidden gap) determine the density of carriers in the transport bands of extrinsic semiconductors and wide band gap materials. Because of the strong temperature dependence of thermal emission rates and, frequently, also of nonradiative capture rates, the study of thermal transport properties in a range of temperatures provides a natural means to obtain information on these trapping parameters.

Thermal transport properties in this connotation refer only to those related to charge carrier transport in a rather wide sense, but not to thermal conductivity, impurity diffusion, self-diffusion, etc., which involve the transport of phonons or atomic species.

Thermal emission rates (1.18, 20) are in most cases not measured directly. A notable exception is the thermally stimulated capacitance technique (Chap. 3). A plot vs temperature of some transport property that is influenced or dominated by thermal carrier release from traps permits the determination of thermal ionization energies which are either equal or related to trap depths (see Sect. 2.5.3). Interpretation of measured data in terms of a physical model for the reaction kinetics will often lead to the determination of the carrier density and the frequency factors, defined with (1.21) as $v(T) \equiv \alpha/\exp[-(E_c - E_t)/kT]$ and $v(T)^* \equiv \alpha^*/\exp[-(E_t - E_v)/kT]$.

In general, various different types of traps and recombination centers may be present and their involvement in the reaction kinetic process will greatly change with temperature. The temperature range in which a specific trap dominates must therefore be identified. This is most conveniently achieved with the aid of nonisothermal temperature scans (Sect. 2.3.3) during which thermally stimulated luminescence (TSL) and conductivity (TSC) are monitored. In semiconductors such novel isothermal techniques as deep-level transient spectroscopy may be used as well (Chap. 3). Of course, the microscopic physical and chemical nature of traps cannot be determined with these methods.

All trap-spectroscopic techniques which are based on thermal transport properties have in common that the interpretation of empirical data is often ambiguous because it requires knowledge of the underlying reaction kinetic model. Consequently, a large number of published trapping parameters – with the possible exception of thermal ionization energies in semiconductors – is uncertain or not very accurate. Data obtained with TSC and TSL techniques, particularly when applied to insulators and photoconductors, are no exception [2.4, 14, 65–69].

In order to place TSL and TSC techniques in perspective relative to other thermal transport methods, it may be instructive to briefly list the most important representatives of the latter group. Not all of them can be classified as being strictly thermal transport techniques. However, lifetimes and recombination rates obtained are often considerably influenced by nonradiative transitions between the transport bands and trap levels and, therefore, can yield valuable information on trapping parameters. Rather extensive lists of methods suitable for lifetime and capture cross section measurements in semiconductors were given by *Bullis* [2.70], and *Milnes* [2.28].

The temperature dependence of various thermal transport phenomena can be measured isothermally at a number of different temperatures in which the sample is either in thermal equilibrium, in a steady-state equilibrium in the presence of some external excitations or decays after a pulsed excitation in a transient fashion. In contrast to this, TSL and TSC experiments are noniso-thermal and are observed only during a programmed change in sample temperature.

The very large number of publications on characterizing trap levels with thermal transport measurements preclude a thorough review in this chapter. As a rule only monographs or classic papers in the field are referred to together with recent papers dating back no further than 1976. This should help the reader to find the standard, but older texts as well as the most up-to-date publications in the field.

2.3.1 Thermal Equilibrium and Steady-State Phenomena

The most direct and, in semiconductors, the easiest way to determine carrier densities and thermal activation energies is to measure ohmic conduction. The current density is given by

$$J = eF(n_c\mu_n + p\mu_p) .$$
(2.1)

The method requires nonblocking ("ohmic") contacts, the knowledge of the carrier mobilities, μ_n or μ_p, and the type of the majority carriers. F is the electric field strength. The thermal activation energy, obtained by a semilogarithmic plot of the ohmic current vs $1/T$, is usually interpreted as either half the bandgap in intrinsic semiconductors or the trap depth of the dominant trap in extrinsic semiconductors. In compensated materials, sometimes also called nonextrinsic impurity conductors, such simple interpretation is not necessarily possible. Particularly in wide-gap materials which need be only weakly compensated, *Schmidlin* and *Roberts* [2.71] have shown that the empirical activation energy of ohmic conduction is the arithmetic average of the depths of the dominant electron trap and the dominant hole trap. Their analysis excludes, however, the case when the Fermi level or quasi-Fermi level falls within a few times kT of a dominant trap level. Here again the activation energy is equal to the thermal ionization energy of the trap. Precisely this situation is encountered in most TSC experiments that are performed under ohmic conduction conditions.

Ohmic conduction is present only as long as the carrier density is not influenced by the electric field F. At sufficiently high F, applied to the sample that is sandwiched between contacts (one of which must be "ohmic" or an injecting contact [2.71]), carriers are injected into the material and, as a result, the current becomes space-charge limited [2.28, 72–74]. If only one type of carrier is injected, e.g., electrons into a n-type material, the current density, J_s, increases with F^2 [2.71]

$$J_s = \frac{9}{8} \frac{F^2}{L} \varepsilon\mu \frac{N_c}{N} \exp[-(E_c - E_t)/kT] . \qquad (2.2)$$

Here ε is the dielectric constant, L the length of the sample and N the density of the dominant electron traps of depth $E_c - E_t$. Again, measuring J_s as a function of T permits the determination of the trap depth and, with the aid of simple model description, the trap density as well as the mobility [2.71]. Interesting effects are observed when both types of carriers are injected [2.28, 75, 76]. However, their discussion goes well beyond the confines of this chapter. Recent theoretical developments include the treatment of single carrier space-charge controlled currents in the presence of trap distributions [2.77, 78] as well as neutral traps and traps possessing attractive Coulomb potentials [2.79]. However, exact expressions for the $J - V$ characteristic cannot be given in explicit form even for monoenergetic traps uniformly distributed in space [2.80]. Electric field profiles and spacial variations of carrier concentrations during minority-carrier injection into semiconductors containing traps have been calculated as well [2.81]. Experimental investigations have recently been performed with SiC [2.82], CdTe [2.83], sapphire [2.84], chalcongenide glass [2.85], bismuth germanate films [2.86], chlorophyll [2.87], and other organic solids [2.88, 89]. Methods for the determination of trapping parameters are reviewed in [2.90–92]. Trapping effects have also been studied by measuring non-ohmic $J - V$ characteristics in various semiconductor–insulator devices such as MOS structures [2.93, 94], FETs [2.95], CCDs [2.96]. Double-carrier injection in the presence of traps has recently been measured in $p - i - n$ diodes [2.97], diodes with S-shaped switching characteristics [2.98, 99], and n-type GaAs : O [2.100]. Hot carrier injection and trapping effects in various semiconductor devices are reported in [2.101–105]. Transient space-charge controlled currents are mentioned in the following section.

Very useful for providing information on traps is the study of steady-state luminescence, optical absorption, photoconductivity, Hall effect, photo-Hall effect, generation–recombination noise spectra, and the Dember and photomagnetoelectric effects, all measured at a range of different temperatures. All of these techniques have been described in detail in the monographs by *Milnes* [2.28], *Bube* [2.75], and *Rose* [2.72], and, therefore, will only be mentioned here.

Empirical investigations frequently make use of a combination of different steady-state and transient techniques involving these phenomena together with the measurement of optically and thermally stimulated luminescence or carrier density variations. In general, such combinations of experimental methods are advantageous as they often reduce the uncertainties associated with the reaction kinetic models used for the interpretation of empirical data.

Again, we will restrict ourselves to a brief list of very recently published investigations and remind the reader that only those phenomena are selected which reveal information on defect states (traps) by measuring them in a range of different temperatures. Steady-state methods usually do not yield non-radiative transition rates (thermal release rates or capture rates) but they are excellent means to measure activation (thermal ionization) energies and, frequently, densities of traps.

The roles of hole or electron traps in the photographic process in halides [2.106, 107] and in photochromic darkening [2.108] have recently been investigated by optical absorption measurements. The temperature dependence of luminescence of doped ZnS phosphors [2.109–111] and CdS [2.112] was measured to determine the depth of trapping levels in these materials.

Densities and/or depths of various trapping states in CdS [2.113], CdTe [2.114], ZnS [2.115], GaAs [2.116–118], AgBr [2.119] were determined with photoconductivity measurements. The Dember effect (a potential difference caused by different electron and hole mobilities and measured in the direction of strongly absorbed radiation) was exploited to determine hole trap depths in AgBr : I, LiH and LiD [2.120]. Using a combination of steady-state and transient photoelectric techniques, trap densities and depths were studied also in MOS structures [2.121], barium titanate [2.122] and stannic oxide [2.123]. The photo-Hall effect was employed for these trap measurements in [2.116] and photocurrent-voltage characteristics in [2.121].

The photoelectromagnetic effect (Dember voltage in the presence of a magnetic field that is applied perpendicular to the direction of the radiation) has been previously employed to deduce capture cross sections of traps (for references see *Milnes* [2.28]), however, no work has been published since 1977.

Hall effect studies have confirmed that the nitrogen isoelectronic trap forms a bound state in $GaAs_{1-x}P_x$ [2.124] and that γ radiation creates electron traps in CdTe [2.125]. The temperature dependence of the trap-controlled drift mobility was measured in CdS [2.126] and photoexcited KCl [2.127]. It should be mentioned that in wide-gap materials all methods that require a conveniently measurable density of free carriers (e.g., dark current, Hall effect, etc.) are not immediately applicable. External excitation for carrier generation is usually necessary (e.g., photo-Hall effect).

The free carrier density of semiconductors fluctuates according to the statistical nature of the generation (thermal emission) and recombination (capture) processes. Studies of generation–recombination noise (or carrier fluctuation noise) may therefore be another effective way to determine trapping parameters. This has occasionally been attempted, resulting in the measurement of capture cross sections of traps in a number of semiconductors. The frequency spectrum of the noise is measured at various temperatures and interpreted with theoretical calculations. The noise is usually characterized by certain frequencies which correspond to capture lifetimes of the carriers [2.28, 72]. Recent measurements of noise spectra in GaAs [2.117], MOS transistors [2.128, 129] and MOSFET devices [2.130] demonstrate both the potential usefulness of this technique in characterizing traps in these devices as well as the considerable complexity of interpreting the empirical results.

2.3.2 Isothermal Transient Phenomena

Measurements of the growth and/or the decay of luminescence, photoconductivity as well as transient space-charge controlled currents and capacitance changes of junctions are very valuable methods for the determination of thermal emission and capture rates and, thus, lifetimes of excess free carriers and capture cross sections of traps.

Again, empirical data have to be interpreted in terms of a reaction kinetic model whose validity must be established to obtain unambiguous results. A major advancement in this area was the development of deep-level transient spectroscopy (DLTS). A forward biased Schottky barrier or similar junction is employed to inject excess carriers, some of which are trapped. Monitoring the change of the capacitance under reverse bias permits the direct determination of the thermal emission rate of these carriers from the traps (Chap. 3).

Again, it seems inappropriate to repeat here the detailed descriptions of these techniques given in the literature [2.28, 45, 72, 73, 131].

Recent advances in the theory of transient space-charge controlled currents in dielectrics containing traps were reported in [2.132–134]. Transient absorption or luminescence phenomena were analyzed in terms of trapping effects in SiO_2 : Al [2.135], alkali halides [2.136, 137] and silver halides [2.115, 117–120, 122, 123], $BaCaU_6$ [2.138] and anthracene [2.139]. Recent theoretical treatments of photocurrent transients in the presence of traps are given in [2.140–146]. Trapping parameters were determined with photocurrent transients in anthracene [2.147], dimethylnaphthalene [2.148], Cu_2O [2.149], CdTe [2.150], ZnS [2.151], $CdIn_2S_4$ [2.152], Si [2.153], alkali halides [2.154] and vitreous As_2S_3 [2.155]. No recent work is known to the authors on using diffusion length and drift techniques for the determination of trap dominated lifetimes [2.28].

2.3.3 Nonisothermal Phenomena

We now return to the subject of this chapter, namely thermally stimulated luminescence and conductivity measured during a temperature scan. If one were to judge the importance of TSL and TSC relative to other thermal transport methods for trap characterization by the number of recently published papers (about 400 between the summers of 1977 and 1978 alone), these techniques appear by far to outweigh all others combined, even though about one third of the publications are concerned with applications in dosimetry, geology, archaeology and related TSL instrumentation. However, the large number of articles on TSL and TSC does not necessarily indicate any advantage in their usefulness as trap-spectroscopy tools over the other methods. What can safely be concluded is that nonisothermal thermally stimulated relaxation is presently still a very active field of research. Its attraction can be traced to the relative ease with which a TSL or TSC glow spectrum can be measured and its extraordinary sensitivity for detecting traps (perhaps as low as 10^7 cm^{-3}). However, as mentioned before, the unequivocal interpretation of the data is rather difficult and unambiguous quantitative results are in general obtained *only* by performing elaborate experiments involving independent techniques for trap characterization such as optical absorption, photoluminescence, electron spin resonance, etc., in conjunction with TSL and/or TSC.

Let us now briefly consider the various steps involved in a typical TSL or TSC experiment in a wide band gap material such as an alkali halide. We choose electromagnetic radiation as a means of excitation (perturbation of the statistical thermodynamic equilibrium). The interaction of this radiation with the solid leads to a number of electronic phenomena, many of which are not yet clearly understood [2.156]. They include the production of new defects (radiation damage) as well as filling of traplevels with electrons or holes:

I) High energy electromagnetic radiation (X, γ, uv) produces in the crystal hot electrons which may multiply by impact ionization and subsequently quickly thermalize so that one is left with free carriers and excitons (Fig. 2.2). A fraction of the incoming energy gives rise to radiation damage effects [2.156]. Recently it has been shown that some of these defects can be produced via multiphoton absorption by intense visible light as well [2.157–160].

II) The excess free carriers and excitons do not represent stable excited states of the solids. A fraction of them recombine directly after thermalization either radiatively or by multiphonon emission (Fig. 2.3). In most materials nonradiative transitions to defect states in the gap are the dominant mode of decay. The lifetime of free carriers

$$\tau = 1/f\langle v \rangle S$$

is determined by the density, f, of recombination centers, their thermal velocity $\langle v \rangle$, and the capture cross section S and may span the range 10 to 10^{-14} s [2.72]. The methods used to measure τ have been discussed in Sects. 2.3.1, 2. Electrons, captured by states above the demarcation level E_d^n for electrons and holes captured by states below the hole demarcation level E_d^p may get trapped. The condition for trapping is given when the occupied electron trap has a very small cross section for recombining with a free hole. The trapping process has, until recently, not been understood well. Theoretical calculations of non-radiative multiphonon transitions cannot be based on the adiabatic approximation. The work of *Henry* and *Lang* [2.161], and *Pässler* [2.162, 165] represents

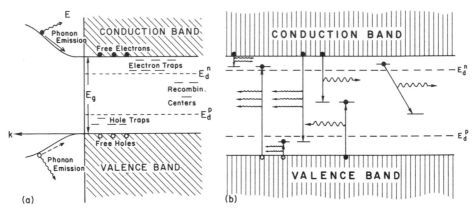

Fig. 2.2. (a) Schematic representation of the energy states of free electrons and valence electrons as a function of the crystal momentum k. Hot electrons and holes, generated during the excitation of the crystal, dissipate their excess energy as phonons, reaching the edges of the transport bands (left). On the right the electron states of a crystal containing defects are schematically represented as a function of a not specified crystal coordinate. E_d^n is the demarcation level for electrons and E_d^p that for holes. **(b)** Diagram of possible electron and hole transitions. Some of these are assumed to be radiative (photon emission, $\leadsto$), others nonradiative (phonon emission, $\leadsto$)

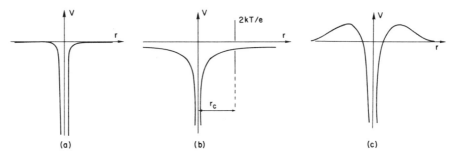

Fig. 2.3a–c. Schematic potential distributions of defect levels in insulators and semiconductors: Neutral trap **(a)**, Coulomb-attractive trap **(b)**, and Coulomb-repulsive trap **(c)**. The effective radius r_c is that radius at which the Coulomb potential in **(b)** has reached a value that is smaller by $2kT/e$ as compared to the continuum (e.g., lower edge of the conduction band)

the first genuine progress in this area even though their theories are still semiempirical (see Sect. 2.5).

III) After the decay of the excess free carriers due to recombination and trapping transitions, the solid is in the so-called excited state which is characterized by the perturbation of the statistical equilibrium. The concentration of the remaining free carriers is now determined by the balance between thermal emission of carriers from the traps, retrapping transitions and capture by recombination centers.

If the excitation occurred at a low temperature such that the thermal emission rate of carriers from traps is very small, the perturbed equilibrium will exist for a long time and only upon an appropriate increase of the sample temperature can the relaxation process proceed at a rate that permits one to monitor it by measuring the conductivity $\sigma(T) = e(n_c\mu_n + p\mu_p)$ of the sample (TSC) or the luminescence (TSL) emitted by radiative recombination of the carriers thermally released from the traps.

We should not fail to mention two additional thermally stimulated transport phenomena which are measured during an irreversible temperature scan. These are thermally stimulated Hall effect (Hall voltage measured during a TSC experiment) [2.27] and thermally stimulated surface potential, described by *Yamashita* et al. [2.166]. The former may be used to determine the temperature dependence of the Hall mobility and the latter to study interface states, e.g., in MIS structures.

2.4 Simple Trap Models

A defect state can either act as a trap or a recombination center, depending on its location with respect to the demarcation levels E_d^n for electrons and E_d^p for holes (Sect. 1.6.3). The concept of demarcation levels is based entirely on the ratio of the probability for thermal escape of an electron (or hole) to the probability of capture of a hole (or free electron). Since the former is largely determined by the Boltzmann factor $\exp(-E/kT)$ (here E is the thermal activation energy for release of a trapped carrier), defect states around the center of the forbidden gap tend to be recombination centers, those in the upper part electron traps and those in the lower part hole traps (Fig. 1.6). The capture cross section of a trap is, in a rough approximation, largely determined by its charge state. Values reported in the literature span the range from 10^{-15} to 10^{-12} cm^2 for Coulomb-attractive centers, 10^{-17} to 10^{-15} cm^2 for neutral centers down to 10^{-22} cm^2 for Coulomb-repulsive centers.

In principle, any defect state in the forbidden gap has a finite cross section for capture of free holes *and* electrons. Obviously, however, a trap being Coulomb-attractive for free electrons may be neutral or even repulsive for holes and, thus, have a vastly smaller cross section for hole capture as compared to that for electron capture.

In the familiar Shockley–Read process [2.167] the charge Z (in units of e of a defect center that is effective for hole capture differs from that effective for electron capture processes by unity. A given defect state can be considered attractive or neutral always for one of both types of carriers. If it is singly attractive ($|Z| = 1$) for one type it will be neutral for the other and if it is doubly ($|Z| = 2$) or triply ($|Z| = 3$) attractive for one, it will be singly or doubly repulsive for the other [2.162–165].

Rose [2.72] has given an estimate of capture cross sections for Coulomb-attractive and repulsive traps on the basis of simple model potentials (Fig.

2.3a, b). Assuming that the free carrier can be regarded as a diffusing particle having a mean free path for energy loss that is much smaller than the effective capture radius r_c of the trap, he postulates that the probability $p(r)$ for finding it at a distance r from the trap is proportional to the Boltzmann factor $\exp[eV(r)/kT]$ and the area $2\pi r^2$. Calculating the minimum in $p(r)$, the capture radius $r_c = e^2/2\varepsilon kT$ is obtained for a Coulomb potential $V(r) = -e^2/\varepsilon r^2$ (Fig. 2.3b) and the capture cross section $S \equiv \pi r_c^2$ is seen to be proportional to T^{-2}.

Coulomb-attractive defect centers can capture more than one charge carrier and thus turn neutral or even Coulomb-repulsive. The latter type of traps is assumed to be surrounded by a potential barrier of height ΔV. Capture of a free carrier requires either tunneling through or going over this barrier. For example, the cross section for electron capture may be taken as S_n, the cross section for a Coulomb-attractive trap, times the probability to find the electron at an energy $e\Delta V$ above $E_c = eV_\infty$ in Fig. 2.3c [2.168]

$$(S_n)_{repul} = S_n \frac{\int\limits_{e\Delta V}^{\infty} f(E)N(E)\,dE}{\int\limits_{E_c}^{\infty} f(E)N(E)\,dE}. \tag{2.3}$$

This expression is difficult to evaluate for most distribution functions $f(E)$. An estimate of the range in which Coulomb-repulsive cross sections are expected to fall may be obtained by assuming the probability to find an electron at $e\Delta V$ is smaller by $\exp[-e\Delta V/kT]$ as compared to finding it at E_c [2.72]. Thus barrier heights of only a few tenths of an electron volt may reduce S_n by many orders of magnitude. Further, the cross section for Coulomb-repulsive trap centers is expected to be very temperature dependent.

For neutral trap centers, an estimate of the capture cross section may be obtained in a similar way as for a Coulomb-attractive center. The Coulomb potential must be replaced by a moderately long-range r^{-4} potential [2.169] that results from the interaction between the charge carrier and the polarization of the trap and its immediate host lattice surrounding. The resulting cross sections are expected to be smaller than those for long-range r^{-2} potentials because of a reduced capture radius r_c.

2.5 Nonradiative Transitions

Capture of carriers in traps or recombination can be nonradiative and thermal emission of carriers from traps is a nonradiative transition. Nonradiative recombination processes are difficult to identify because their occurrence can usually only be inferred from a low luminescence emission efficiency η. In an elegant experiment with GaAs p-n junctions, *Narayanamurti* et al. [2.170] have recently been able for the first time to directly observe phonons generated by nonradiative capture of free carriers injected into the junction. Their work is of considerable importance for it provides direct proof that phonons are involved in dissipating the energy in these transitions. If the radiative and nonradiative recombination processes have lifetimes τ_r and τ_{nr}, respectively,

$$\eta = (1/\tau_r)/(1/\tau_r + 1/\tau_{nr}). \tag{2.4}$$

The luminous efficiency η can be temperature dependent, decreasing with increasing T (thermal quenching) [2.44]. A number of simple models have been employed to explain this experimental observation. When radiative and nonradiative transitions compete within a luminescence center, *Mott* and *Gurney* [2.171], and *Seitz* [2.62] found, on the basis of a configural coordinate theory, $\eta = 1/[1 + c \exp(-E^*/kT)]$, where c is a constant and E^* an activation energy. A similar expression was obtained by *Klasens* [2.172], and *Schön* [2.173], however, they postulated the existence of centers which thermally emit holes into the valence band, thus, reducing the capture cross section of these centers for radiative recombination with an electron from the conduction band.

The theory of nonradiative capture has been the subject of numerous papers and reviews [2.161–165, 174, 175]. Three plausible mechanisms of nonradiative capture have evolved:

I) The Auger effect; the energy lost by the captured carrier excites another nearby carrier in the crystal [2.174, 176].

II) Cascade capture; the electron loses energy by dropping down a ladder of closely spaced excited levels of the defect, emitting one single phonon in each step [2.28, 169, 177, 178].

III) Multiphonon capture; the energy of the electron is dissipated by multiphonon emission [2.161–166, 179–185].

Thermal emission rates have, up to this point in time, eluded quantitative theoretical calculations. *Marlor* [2.186] has touched upon this problem in his field formulation of rate processes. The "radiationless" transition from the ground state of the trap to an excited state from which recombination is possible is considered occurring in a "phonon field" that is characteristic of the lattice and depends on T. An estimate of the probability that a captured electron is *not* reemitted was presented by *Henry* and *Lang* [2.161] within the framework of their semiempirical theory of multiphonon capture. *Pässler* [2.162–165] has developed a more complete semiempirical theory of multiphonon capture. He was able to show that the relations between thermal emission rates and capture rates, previously only obtained from detailed balance considerations (Sect. 1.6.1), indeed reduce to the form

$$\frac{\alpha_i(T)}{\beta_i(T)} = g_{c,t}^{-1} N_c(T) \exp\left[-(E_c - E_{t,i})/kT\right] \tag{2.5}$$

$$\frac{\alpha_i^*(T)}{\beta_i^*(T)} = g_{v,t}^{-1} N_v(T) \exp\left[-(E_{t,i} - E_v)/kT\right]. \tag{2.6}$$

Here it is assumed that the system contains i discrete and different traplevels at energies $E_{t,i}$. The capture coefficients for electrons are α_i and for holes α_i^* and the coefficients for thermal release are β and β^*, respectively. N_c and N_v are the density of states in the respective transport bands and $g_{c,t}$ and $g_{v,t}$ are the spin degeneracies of the trap level i that are effective for electron capture and hole capture, respectively. In kinetic theories of TSL and TSC, degeneracies are usually neglected or "hidden" in the so-called frequency factors $v_i \equiv g_{c,t}^{-1} \beta_i N_c$ and $v_i^* = g_{v,t}^{-1} N_v$.

As can be seen from (2.5, 6), measurements or calculations of rates, cross section or coefficients for capture immediately yield the associated rates, etc., for thermal emission and vice versa, if the density of states N_c and N_v and the trap depths are known.

2.5.1 Auger Capture

The role of Auger transitions involving traps in TSL and TSC experiments is not altogether clear and experimental information on this radiationless trapping or retrapping mechanism is rather scarce.

Examples of possible Auger processes are given in Fig. 2.4. It should be noted that the energy exchange is not necessarily confined to carriers in similar states.

While band–band Auger recombination has been investigated theoretically [2.187–190] as well as experimentally (e.g., *Auston* et al. [2.191] measure the Auger-rate constant $\gamma = n_c^{-3} dn_c/dt$ to be 1×10^{-31} cm^6 s^{-1} in Ge using single-photon carrier generation with the aid of psec Nd laser pulses), free-to-bound transitions have not been clearly demonstrated experimentally and may only be inferred from a number of experiments performed in semiconductors and insulators. Probably the best evidence for these transitions was provided by the measurement of the minority carrier lifetime τ_h in heavily doped Ge by *Karpova* and *Kalashnikov* [2.192] which was found to be a function of the majority carrier concentration n_c. The Auger process involved is of the type (b) in Fig. 2.4 and its contribution to the hole decay is $dp = -\gamma p n_c n dt$, where again γ is the Auger coefficient and n the density of electron-occupied traps (or, which is the same, recombination centers for holes). Curve fitting of the empirical data $\tau_h(n_c)$ yielded $\gamma = 10^{-26}$ cm^6 s^{-1}. Other experimental evidence was presented by *Jayson* et al. [2.193].

The only manifestation of an Auger process in a thermally stimulated relaxation experiment may be the observation of exoelectrons during thermal destruction of self-trapped holes (V_k centers) in a number of alkali halide crystals by *Bichevin* and *Käämbre* [2.194]. Since thermal ejection of self-trapped holes into the valence band cannot directly result in electron emission from the sample into the vacuum, it is very likely that the free hole recombines with one of the impurity centers and the released energy leads to the ionization of another center. If this process occurs close to the crystal surface it may result

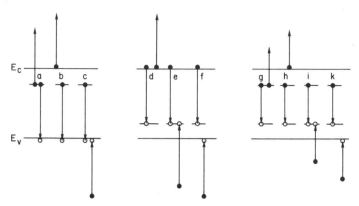

Fig. 2.4. Examples of possible Auger recombination processes involving defect states in insulators or semiconductors. Processes (a), (b), (d), (g), and (h) could conceivably lead to exoemission of electrons

in the observed emission of electrons. *Henry* and *Lang* [2.161] have studied capture rates for electrons in Cr centers and in the so-called unidentified B centers in GaAs and for holes in the oxygen center in GaP as a function of carrier concentration. They found *no* evidence for Auger capture. It is probably safe to state that direct contributions of Auger transitions to the carrier traffic in TSL and TSC experiments in most semiconductors and in insulators are negligible. Their rate coefficients are small and considerable densities of free and trapped carriers are required for Auger transitions to affect other dominant radiative or multiphonon capture and emission rates.

2.5.2 Cascade Capture

In an attempt to account for large cross sections measured for Coulomb-attractive and neutral centers, *Lax* [2.177] has treated the problem of nonradiative capture as a cascade process. If the trap center has a dense set of excited states available, the electron can occupy one of these levels with the emission of a phonon. As a result of subsequent single-phonon transitions it will either escape again or move closer toward the ground state. The final step, from the first excited state to the ground state, may be a multiphonon or radiative transition. The probability for an electron, after occupying one of the higher excited states, to reach the ground state without leaving it again is called the "sticking probability." In shallow traps, having ionization energies around kT, this sticking probability is very small because the electron can easily be emitted back to the continuum by single-phonon absorption. If the ionization energy is large, it is more favorable for the electron to reach the ground state via a cascade of single phonon transitions. A key element in the theory is its prediction of the temperature dependence of the capture cross section. As in *Rose*'s simple Coulombic trap model, lowering the temperature permits contributions from states with increasing radius if one assumes the sticking probability to approach sizable values for all states located at energies larger than kT below the continuum level. Thus, the capture radius r_c of the trapping center increases with T. Indeed, the relation $S \propto T^{-n}$ (with $n = 2$ to 4) has been confirmed, the classical example being the work of *Ascarelli* and *Rodriques* [2.195]. Calculated cross sections are usually larger than empirical values. Improvements of the theory have included a more detailed treatment of the sticking probability [2.196] as well as the basic formalism [2.169]. However, the theory is not universally accepted. *Henry* and *Lang* [2.161] rejected it outright as an explanation for capture cross section of nine defect states they measured in GaAs and of four in GaP. None of these deep levels have excited states that are more than 50 meV away from the band edges. In addition, it is uncertain whether neutral centers with a r^{-4} potential have sufficient, if any, excited states to allow the cascade process to occur [2.197]. Even though some of the capture cross sections measured by these authors [2.161] indeed decrease with increasing T, there is no apparent need to invoke the cascade mechanism whose only confirmation so far appears to be its explanation of large cross sections of hydrogen-type shallow donors such as As and Sb in Ge, measured at very low temperatures [e.g., 2.195].

2.5.3 Multiphonon Capture

Multiphonon transitions are favored whenever there is a very strong electron–phonon coupling which is associated with large changes in the equilibrium lattice coordinates as well as in the energy of the defect state during capture or ejection of the electronic charge carrier. This situation is not only characteristic for many defects (color centers) in ionic lattices such as the alkali halides but also, as has been confirmed in the last 4 years mostly by DLTS measurements, in a large fraction of deep levels in such semiconductors as Ge, Si, GaAs, GaP etc.

In the following we will attempt to discuss the basic physical processes that occur during nonradiative capture and the simplifications and approximations made in the semiempirical theory of multiphonon transitions. In doing so we will be guided by the elegant exposés by *Henry* and *Lang* [2.161], and *Pässler* [2.162–165] who presented intuitively satisfying descriptions of this complex subject which cannot be found in such clarity elsewhere in the literature.

Let us begin with a discussion of Fig. 2.5 [2.161] which illustrates, as a function of a single "effective" lattice coordinate, the electron–lattice interactions as they affect the "electronic" energy level of a trap i before, immediately after, and long after capture of an electron. Lattice vibrations cause the energy level $E_{t,i}$ to move up and down the energy gap. Sufficiently large vibration will cause it to cross into the free electron levels (conduction band) where it may capture an electron. The probability for this to take place at

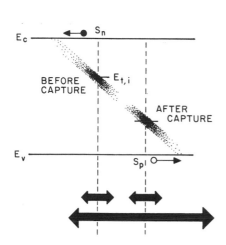

Fig. 2.5. Schematic illustration of nonradiative energy states of the defect are shown as a function of the lattice coordinate. The dashed lines indicate the equilibrium positions of the coordinates prior to and after capture. The shaded regions within the gap indicate how the energy of the defect (rap) levels changes as the lattice vibrates. For sufficiently large vibration, the energy level of the trap can cross into the conduction band and capture an electron with a capture cross section S_n. Similarly, the energy level of the occupied trap may cross into the valence band and capture a hole with a cross section S_p. The small arrows represent the amplitudes of the lattice vibrations before and after electron capture; the large arrow demonstrates the violent lattice vibrations immediately after capture. The vibrations are rapidly damped by phonon propagation away from the location of the defect, justifying this process being called nonradiative capture by multiphonon emission [2161].

high temperatures (no tunneling) is proportional to $\exp(-U_{c,i}/kT)$. Here $U_{c,i}$ is the lattice energy necessary for the shift of $E_{t,i}$ up to or very close to the free electron states. Immediately after capture the lattice near the defect has absorbed the available energy and vibrates violently (large arrow in Fig. 2.5) about the new equilibrium position of the lattice coordinate. These vibrations damp out rapidly as the energy propagates away from the defect in form of lattice phonons until a new equilibrium position of the energy state in the gap is reached as well. The cross section for capture turns out ot be proportional to the product of the probability for such "crossing-inducing" vibrations to occur, the probability for capture during crossing and the probability that the electron will not be reemitted after capture (note the similarity between the latter and the "sticking probability," introduced in the theory of cascade capture). An immediate consequence of this concept is that the capture cross section is temperature dependent according to $\exp(-U_{c,i}/kT)$ and $U_{c,i}$ becomes the activation energy for nonradiative multiphonon (NMP) electron capture.

Figure 2.5 depicts only the energy of the electron in the conduction band, trap and valence band as a function of the lattice coordinate Q. For these three energy states the potential well (or its energy equivalent for the lattice, namely "electronic" energy plus elastic energy of the lattice) is again conveniently displayed in a single-coordinate Seitz-type diagram [2.161, 162] (Fig. 2.6). *Pässler* chose the lattice coordinate Q to be dimensionless. The values $Q = 1/2$ are the identical equilibrium positions when the electron is in the conduction band or in the valence band (this corresponds to the occupational state of the trap "empty") and $Q = -1/2$ is the position of the potential when the electron is trapped.

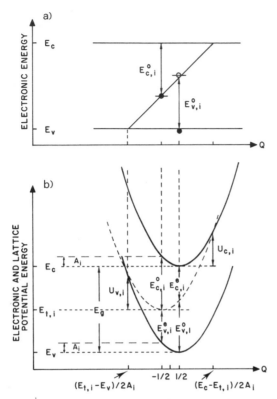

Fig. 2.6a, b. Schematic diagram of the electronic energy (**a**) and of parabolic lattice potentials (**b**) as a function of an effective lattice coordinate for trap levels with strong electron-lattice coupling. This situation is representative for deep levels in semiconductors and insulators (e.g., color centers). The dashed parabola is the potential energy for an occupied trap i at energy $E_{t,i}$. The solid parabolas correspond to the trap conditions "empty". The diagram is valid both for electron as well as hole capture. Capture of an electron requires an activation energy $U_{c,i}$. If the trap captures a hole, the activation energy is $U_{v,i}$. The thermal ionization energy for electron ejection from the trap is $(E_c - E_{t,i}) + U_{c,i}$ and that for thermal hole emission is $(E_{t,i} - E_v) + U_{v,i}$. Optical transitions occur without immediate lattice readjustment (vertical transitions). Optical ionization of the trap by absorption of photons will yield absorption bands centered around $E_{c,i}^\circ$ (for electron emission from the trap) or $E_{v,i}^\circ$ (for hole ejection). Radiative capture transitions of free electrons to the trap would result in luminescence bands centered around $E_{c,i}^e$ and, similarly, around $E_{v,i}^e$ for radiative hole capture. The effective lattice coordinates for empty traps are assumed to be $Q = 1/2$. A_i is the lattice adjustment energy [2.162–165]

The lattice potentials are assumed to be parabolas of the form

$$E_c + (Q - 1/2)^2 A_i, \quad E_v + (Q - 1/2)^2 A_i \quad \text{and} \quad E_{t,i} + (Q + 1/2)^2 A_i,$$

respectively.

The so-called lattice adjustment energy A_i is a result of the different equilibrium positions of the energy parabolas for the empty as compared to the occupied trap levels. It represents the change in the energy separation of the minima that is associated with radiative (vertical) transitions.

Thus, the characteristic differences between optical and thermal ionization energies, given by the Franck–Condon principle, are featured in Fig. 2.6 as well. The centers of gravity of optical absorption bands are $E^o_{c,i}$ which are *larger* by A_i than the respective trap depths $(E_c - E_{t,i})$ and $(E_{t,i} - E_v)$. Similarly, the luminescence bands are centered around $(E_c - E_{t,i}) - A_i$ and $(E_{t,i} - E_v) - A_i$, so that $2A_i$ is the familiar Stokes shift.

Pässler [2.162] considers nonradiative electron–hole recombination as a two-step process consisting of electron capture and subsequent hole capture at one and the same trap i associated with an empirical trap level position $E_{t,i}$. The available energies $E_c - E_{t,i}$ from the first step and $E_{t,i} - E_v$ from the second are equal to $E_c - E_v \equiv E_g$, the energy of the gap. The thermal activation energy for NMP electron capture is $U_{c,i}$. The activation energies for thermal ejection of charge carriers are $(E_c - E_{t,i}) + U_{c,i}$ and $(E_{t,i} - E_v) + U_{v,i}$. As a consequence, measured activation energies for thermal emission are *not* equal to the trap depths $E_c - E_{t,i}$ or $E_{t,i} - E_v$ whenever the capture processes are thermally activated (i.e., $U_{c,i} \neq 0$, $U_{v,i} \neq 0$). Therefore, *the activation energy for carrier capture must be subtracted from the empirical activation energy for thermal ejection to obtain the true trap level depths.*

Early theories of NMP relaxation were based on an adiabatic electron–lattice coupling scheme [2.198–201], resulting in values for capture cross section that were orders of magnitude smaller than measured ones. Only after *Helmis* [2.202], and later *Kovarskiy* and *Sinyavskiy* [2.203], *Henry* and *Lang* [2.161], and *Pässler* [2.262–265] considered so-called static coupling schemes has the NMP relaxation mechanism been accepted as a viable explanation for the large cross section of deep traps measured in many materials. In this scheme the quasi-stationary states of charge carriers in the solid are affected by an effective average lattice potential and the vibrations of the ion cores of the lattice together with the oscillations of the electron–lattice interaction potential are considered as transition-inducing perturbations. *Henry* and *Lang* [2.161] demonstrated that the adiabatic approximation breaks down precisely near the level crossings at $Q = (E_v - E_{c,i})/2A_i$ and $Q = (E_c - E_{t,i})/2A_i$. It is, further, an inherent property of the static coupling scheme that the effective lattice potential is still parabolic at these crossing points.

Pässler [2.162] states:

"In terms of this configurational coordinate scheme, a NMP carrier capture or ejection process can be visualized as an energy-conserving (i.e. "horizontal") transition undergone by the total system, whereby the decrease or increase of the energy of the electronic subsystem (by an amount of $E_c - E_{t,i}$ or $E_{t,i} - E_v$) is compensated by a corresponding increase or decrease, respectively, of the energy of heat motion (i.e. the sum of kinetic and potential energy) of the lattice. At this, for nonvanishing $U_{v,i}$ or $U_{c,i}$, the transitions between the corresponding quasi-stationary states of the lattice subsystem may be interpreted

I) at high temperatures to be the consequence of thermal activation of the lattice into highly excited states that correspond to classical motions of the lattice oscillators on trajectories encompassing the point of intersection (= the unique point where a change of electronic energy to potential energy of the lattice, i.e. without affecting the kinetic energy of the lattice, can occur), and

II) at low temperatures to proceed via quantum-mechanical tunneling of the lattice oscillators, since such classically allowed trajectories through the point of intersection are then thermally inaccessible.

This distinction gives yet some qualitative explanation for the NMP carrier-capture cross sections to possess relatively large magnitudes associated with activation-type dependences $\exp(-U_{c,i}/kT)$, or $\exp(-U_{v,i}/kT)$, at high temperatures, as contrasted to the smaller magnitudes associated with rather weak temperature dependences, at low temperatures."

Henry and *Lang* [2.161] calculated capture cross sections of neutral trapping centers for electrons, S_n, and for holes, S_p, in semiconductors and found them to be given by

$$\left.\begin{array}{c} S_n \\ S_p \end{array}\right\} = S_\infty \exp(-E_s/kT), \tag{2.7}$$

where $E_s \approx U_{c,i}$ (or $U_{v,i}$) was experimentally found to be in the range from 0 to 0.56 eV. Calculations yield $S_\infty \approx 10^{-14}$ or 10^{-15} cm^2. This is of the same order of magnitude observed for a number of traps in these materials (see Chap. 3, Fig. 3.21).

Pässler [2.162–165] extended these calculations to Coulomb-attractive and repulsive centers and presented a detailed discussion of the temperature dependence of electron and hole capture cross sections (Table 2.1). His work confirms the results obtained by *Henry* and *Lang*.

Both of these theories assume lattice anharmonicities as well as nonlinear carrier–lattice interactions to be negligible. However, *Henry* and *Lang* found evidence that nonlinear behavior of the bound-state energy E_t with lattice displacement affects the magnitude of NMP cross sections considerably. They observed that such nonlinearities tend to reduce S_n and increase S_p.

Further simplifications include the neglect of phonon dispersion and of temperature variations of the band gap. Despite these approximations, the

Table 2.1. Low- and high-temperature limits of the temperature dependence of the cross sections S_n or S_p for multiphonon capture of electrons or holes by Coulomb-attractive, Coulomb-repulsive and neutral traps. The choice of T^0 or T in the parentheses must be made on the basis of whether $(E_{b,i} - E_{t,i})^2/2kTA_i$ is larger or smaller compared to certain dimensionless trap-volume parameters. The index b stands for c in case of electron capture and v in the case of hole capture (see Fig. 2.6). $\Theta_b^{|Z|}$ are charge-state specific temperature parameters [2.165]

	T	Coulomb-attractive	Coulomb-repulsive	Neutral
S_n	High	$\propto (T^0 \text{ or } T)T^{-3/2}$ $\cdot \exp\left(-\dfrac{U_{b,i}}{kT}\right)$	$\propto (T^0 \text{ or } T)T^{-5/3}$ $\cdot \exp\left\{-\dfrac{U_{b,i}}{kT} - 3\left[\dfrac{\Theta_b^Z}{T}\left(\dfrac{|E_{b,i}-E_{b,i}|}{2A_i} + \dfrac{1}{2}\right)\right]^{1/3}\right\}$	$\propto (T^0 \text{ or } T)T^{-1}$ $\cdot \exp\left(-\dfrac{U_{b,i}}{kT}\right)$
or				
S_p	Low	$\propto T^{-1}$	$\propto T^{-7/6}\exp\left[-3\left(\dfrac{\Theta_b^Z}{T}\right)^{1/3}\right]$	$\propto T^{-1/2}$

semiempirical theory of NMP capture of charge carriers into deep trapping states that are strongly coupled to the surrounding lattice yields not only considerable insight in the process of carrier trapping but also a set of analytical expressions for trapping parameters that are experimentally accessible. Thus, direct comparison of empirical and theoretical NMP capture data has now become possible.

In this connection it is of great value that *Pässler* was able to confirm the validity of (2.5, 6), originally obtained from detailed balance considerations for the case of thermal equilibrium. Requirements for (2.5, 6) to evolve from the theory of NMP capture are, in addition to the previously mentioned assumptions on linearity of the electron–lattice interaction and lattice harmonicity, that the lattice is in thermal equilibrium prior to capture, the free carrier distribution is a Boltzmann distribution at the lattice temperature T and that the transport bands are parabolic in the electron wave vector at least close to the edges E_c and E_v.

A recent calculation of multiphonon capture coefficients of Coulomb-attractive centers for electrons was presented by *El-Waheidy* [2.204]. He also studied the effect of external electric fields F. S_n is found to increase with F but not rapid enough to produce negative differential photoconductivity.

In conclusion we hasten to point out that much experimental evidence of carrier capture into deep trapping states still cannot be explained as NMP, Auger or cascade capture [2.161].

2.6 TSL and TSC Kinetics

In this section we develop electron kinetic models for thermoluminescence and thermally stimulated conductivity, based on the statistical considerations described in Chap. 1, and then present solutions for several special cases.

Assume an arbitrary distribution $N(E)$ of defect levels in the forbidden gap of the nonmetallic solid and an occupation function $f(E)$ that has been removed from statistical thermodynamic equilibrium by generation of free electrons and holes and subsequent trapping. This situation is shown in Fig. 1.11 for the special case $N(E) = $ const. It represents the initial condition at $t = t_0$ for the relaxation kinetics under study. Equations (1.22) are now

$$dn_c/dt = - \int_{E_v}^{E_c} c_n(E) N(E) [1 - f(E)] dE + \int_{E_v}^{E_c} \alpha(E) N(E) f(E) dE$$

and   (2.8)

$$dp/dt = - \int_{E_v}^{E_c} c_p(E) N(E) f(E) dE + \int_{E_v}^{E_c} \alpha^*(E) N(E) [1 - f(E)] dE,$$

where $c_n = \beta n_c$ and $c_p = \beta^* p$, β and β^* being the capture coefficients for electrons and holes, and n_c and p the concentration of free electrons and holes, respectively.

We assume further that at t_0 the sample is at a uniform temperature, T_0, low enough to prevent thermal emission of both holes and electrons from their respective traps. Increasing the temperature according to a heating program $T(t)$ will increase the thermal emission coefficients α and α^* according to (1.21) and, thus, eventually lead to the redistribution of the trapped carriers over available defect states until, finally, thermodynamic equilibrium is reached again at some higher temperature. Knowledge of $N(E)$, c_n, c_p, α and α^* at $T = T_0$ is in principle sufficient for the complete characterization of the relaxation kinetics of a given solid as a function of time or temperature. The initial occupational function at T_0 is given by (1.23) and its behavior with temperature during the irreversible scan $T(t)$ is fully described by (2.8). Such solutions have rarely been attempted for several reasons [2.29], the most important of which is probably the fact that no simple method is known to measure $f(E, T)$ directly. Phenomenological theories of nonisothermal reaction kinetics have therefore concentrated on solutions of (2.8) that can be directly related to such experimentally accessible reaction-kinetic quantities as TSL and TSC.

In the general form of (2.8) the kinetics is, of course, mathematically intractable. Therefore, numerical or approximate solutions can be attempted at best. However, little is gained by this procedure, because knowledge of all the rate coefficients and the trap distributions is a prerequisite. Indeed, the problem one faces in an attempt to characterize the kinetics of a material is of a different nature: The rate coefficients for a given solid are unknown and must be determined experimentally. Such indirect measurement methods as TSL and TSC are readily performed, but must somehow be interpreted to yield quantitative information on the trapping parameters. The commonly employed procedure to do this is to model the relaxation kinetics based on simplified versions of (2.8). TSL and TSC curves are calculated for wide variations of the rate coefficients, initial conditions, etc., and compared with measured ones. In the following we give a brief review of these phenomenological theories of thermally stimulated nonisothermal relaxations.

Energy states of the electrons in defect levels i are frequently centered around a narrow energy distribution $g(E - E_{t,i})$ or are even discrete. In the former case the volume density of the states i is given by

$$N_i = \int_{E_v}^{E_c} N(E)g(E - E_{t,i})dE \quad \text{with} \quad \int_{E_v}^{E_c} g(E - E_{t,i})dE = 1. \tag{2.9}$$

Discrete levels have a δ-function distribution on the energy scale. If we have j different discrete states in the forbidden gap, (2.8) may then be written in the form

$$dn_c/dt = \sum_{i=1}^{j} \int_{E_v}^{E_c} N(E)\delta(E - E_{t,i}) \left[(\alpha + n_c\beta_i)f(E) - n_c\beta_i\right] dE$$

$$= \sum_{i=1}^{j} \alpha_i n_i - \beta_i(N_i - n_i)n_c \tag{2.10}$$

$$dp/dt = \sum_{i=1}^{j} \alpha_i^*(N_i - n_i) - \beta_i^* n_i p .$$

Here α_i and α_i^* are the thermal emission coefficients for electrons and holes, β_i and β_i^*, the capture coefficients, and n_i the volume densities of electrons in the states i. Thus, depending on the position of state i in the gap, $N_i - n_i$ may be interpreted as the density of trapped holes, empty electron traps or empty recombination centers for electrons. Similarly, n_i may also be looked upon as the density of empty recombination centers for holes or occupied electron traps.

Equations (2.10) are coupled by the condition

$$dn_c/dt - dp/dt = - \sum dn_i/dt = + \sum d(N_i - n_i)/dt . \tag{2.11}$$

In order to proceed with model calculations of TSC or TSL glow curves, we must now specify, for a given solid, which of the levels N_i at $E_{t,i}$ are traps or recombination centers. Several special cases shall be discussed [2.17].

1) One or several states N_i are located above the demarcation level E_d^n for electrons and one or several states A_i are located between E_d^p and E_d^n. We assume that the shifts of E_d^n and E_d^p during the thermal scan are small enough so that none of the levels changes its nature as trap or recombination level. After excitation (perturbation of the statistical equilibrium), the levels above E_d^n are partially or completely filled with electrons and the recombination levels are unoccupied so that at $t = t_0$ all electrons deposited in the traps are missing from the recombination levels (Fig. 2.7a, b). Note that we now use the symbol N_i only for the volume density of electron traps and A_i was introduced for that of recombination centers.

An equivalent case for hole traps below E_d^p can be constructed. It is formally described by the same phenomenological theory and will therefore not be discussed in detail (Fig. 2.7c).

II) Electron traps N_i above E_d^n are present together with hole traps $N_{h,i}$ below E_d^p (Fig. 2.7d). Only one case, namely when there are no additional states present between E_d^p and E_d^h, has been treated [2.19, 40], however, modern numerical methods are available to readily solve the problem when these states exist (see Sect. 2.6.1).

III) The presence of trap-recombination center (or donor–acceptor) pairs constitutes a special case. The defect centers are not randomly distributed throughout the solid. Instead, donors and acceptors are associated in pairs with

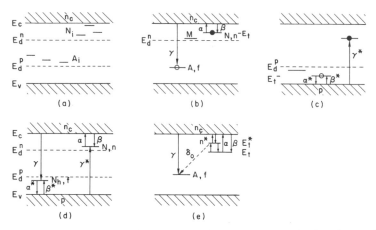

Fig. 2.7a–e. Reaction kinetic models involving discrete trap levels and recombination centers. The energy diagram (**a**) is a somewhat general case. The solid contains different discrete electron traps i of densities N_i located above the demarcation level for electrons E_d^n. Different discrete recombination centers of densities A_i are located between E_d^n and the demarcation level E_d^p for holes. From the situation (**a**) the so-called single electron trap model has evolved (**b**). Only one kind of discrete traps of density N is retained. These are thermally connected via the transitions α and β to the conduction band. Traps M are thermally disconnected. They are assumed to neither capture free electrons n_c nor eject electrons trapped by them. The recombination centers of density A are characterized by an energy-independent recombination coefficient γ. The densities of filled traps is n and that of empty activation centers f. The single-hole trap model (**c**) is the hole-trap analogy to the single-electron trap model (**b**). The energy diagram of a solid containing both electron traps (which can also act as recombination centers for free holes) and hole traps (which can be considered recombination centers for free electrons) is shown in (**d**). Trap-recombination center pairs (**e**) may involve charge exchange transition δ_0 from an excited trap level E_t^* in addition to thermal emission and capture transitions α, β, and γ

varying spatial separation. These configurations are well known from luminescence investigations [2.39, 42, 43, 48]. However, they have been discussed only hypothetically in conjunction with TSL and TSC [2.20] and there is, aside from some curve-fitting agreement, no hard evidence available of "pair-type" glow curves. Because of the spatially close association of trap and recombination center, carriers may recombine without being thermally ejected into a transport band (Fig. 2.7e).

2.6.1 TSL and TSC Due to Electron Traps

This is case I) of the previous section (Fig. 2.7a). We start with a solid that contains only one type of electron traps of volume density N at the discrete level E_t and, in addition, a set of occupied, but otherwise unspecified deeper electron traps of density M (Fig. 2.7b). The experiment is performed in a temperature range in which traps N empty but traps M remain "thermally disconnected" and act only as an untapped reservoir of trapped electrons. The density of recombination centers is unspecified, but a density f of them are empty. At $T = T_0$ a concentration of f_0 empty recombination centers exists due to excitation. Charge neutrality of the sample requires

$$f = n_c + n + M. \tag{2.12}$$

Again n_c is the density of free electrons and n that of occupied traps of type N. Equations (2.10) simplify to

$$dn_c/dt = \alpha n - \beta n_c(N - n) - \gamma n_c(n_c + n + M). \tag{2.13}$$

Here we denote the capture coefficient $\beta_i \equiv c_n/n_c$ for electron traps by β and that for recombination by γ. With (2.12) we obtain

$$df/dt = dn_c/dt + dn/dt = -\gamma n_c(n_c + n + M). \tag{2.14}$$

The recombination coefficient γ is the sum of γ_r and γ_{nr}, the coefficients for radiative and nonradiative recombination, respectively. The TSL intensity per unit volume is $I(T) = n_c \tau_r^{-1} = \gamma_r n_c f$. With the aid of the luminous efficiency $\eta = \tau_r^{-1}/(\tau_r^{-1} + \tau_{nr}^{-1}) = \gamma_r/(\gamma_r + \gamma_{nr})$, we can write $I(T) = \eta \gamma n_c f$. The temperature dependence of n_c and f has to be calculated from (2.13, 14) with $T(t)$. However, only $\eta_{ex} I(T)$ is experimentally accessible, where $\eta_{ex} < 1$ accounts for absorption losses and internal reflection on the sample surface. This external efficiency η_{ex} is usually less than a few percent. Since η_{ex} is in general not measured in TSL experiments, absolute quantum yields or quantitative determinations of recombination coefficients are not reported in the literature.

Numerical solutions of (2.13, 14) have been obtained first by *Kelly* et al. [2.12]. Approximate analytic solutions have a long history [2.2, 9] and are possible if

$$n_c \ll n \quad \text{and} \quad dn_c/dt \ll dn/dt, \tag{2.15}$$

yielding immediately

$$-dn/dt = n(M + n)\alpha/[(1 - R)n + M + RN] \tag{2.16}$$

and

$$n_c = \alpha n/\gamma[(1 - R)n + M + RN], \tag{2.17}$$

where $R \equiv \beta/\gamma$ is the retrapping coefficient.

Approximate numerical solutions for thermally stimulated conductivity $\sigma = e\mu_n n_c$ and/or thermoluminescence $I = \eta \gamma(n_c + n + M)$ have been presented by *Kelly* and *Bräunlich* [2.11, 12], *Bräunlich* [2.40, 205], *Böer* et al. [2.206],

Saunders [2.16], Kivits [2.17], and Dussel and Bube [2.10], Böhm and Scharmann [2.207], Kemmey et al. [2.208], and Land [2.209].

The time dependence of the rate equations (2.16, 17) is replaced by the temperature dependence via the heating program, which is customarily taken to be linear such that $dT=qdt$, where q is the heating rate. Occasionally quadratic heating programs $(dT=\alpha T^2 dt; \alpha=const)$ have been employed because they simplify certain integrals which appear in the approximate expressions obtained from (2.16, 17) for $\sigma(T)$ and $I(T)$ [2.210–212].

It is, of course, not really necessary to ignore n_c relative to n in order to decouple (2.13, 14). It suffices to assume $dn_c/dt \ll dn/dt$, which then leads to

$$n_c = (B/2\gamma)[(1+4\alpha\gamma n/B^2)^{1/2} - 1], \tag{2.18}$$

where

$$B = \beta(N-n) + \gamma(n+M).$$

Approximate solutions of this type have a somewhat longer temperature range of validity, for, generally, $dn_c/dn < n_c/n$ [2.13]. Broadly speaking, it has been shown [2.13] that the approximate solutions are valid for only a part of the range of physically plausible trapping parameters. The exact solutions exhibit an even greater variety of shapes, peak positions and magnitudes. Hence, the effect of thermally disconnected traps or of various initial filling ratios on TSL and TSC was found to be different from that observed in the approximate solutions. Kelly et al. [2.13] concluded that any analysis of TSL and TSC in the absence of extensive other information is highly unlikely to lead to unambiguous values for the trapping parameters.

Exact numerical solutions were first obtained using a Runge–Kutta–Gill fourth-order process [2.13]. Since then Hagebeuk and Kivits [2.213] have developed a stable method which greatly increases the speed of computation, and Haridoss [2.214] has successfully employed Milne's fifth-order predictor–corrector method. However, we have found that the system of stiff equations (2.13) and (2.14) is most readily solved by the method of Gear [2.215].

We have timed solutions by this latter method for a wide range of the trapping parameters, on an IBM 360/67 time-shared system and found that the central processor unit time per unit step in temperature is less than 0.01 s for $N \le 10^{15}\, cm^{-3}$; less than 0.04 s for $N \le 10^{17}\, cm^{-3}$ and less than 0.4 s for $N = 10^{18}\, cm^{-3}$.

This is about 10 to 100 times faster than the estimate of time as given by Hagebeuk and Kivits who used a Burroughs 6700 machine [2.213].

The variety of glow curve shapes which one can obtain using these techniques has been fully described in the literature and we refer the reader to those works [2.13]. Some examples are shown in Figs. 2.8, 9.

The classic solution for the TSL intensity $I(T)$ by Randall and Wilkins [2.2] is obtained for $R=0$ (no retrapping), $M=0$ (no thermally disconnected traps),

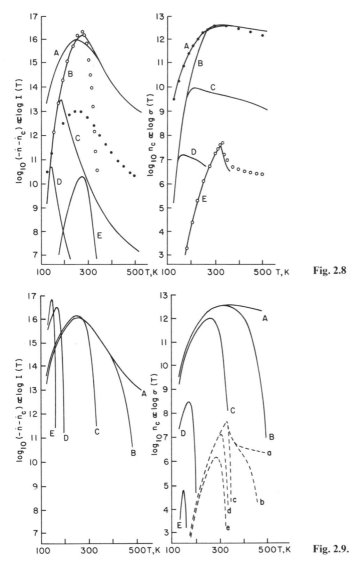

Fig. 2.8. Numerical solutions of (2.13, 14) for the case $M=0$ (no thermally disconnected traps). The ratio $R \equiv \beta/\gamma$ determines the basic shapes of the TSL and TSC curves $I(T)$ and $\sigma(T)$. $R=10^6$ for A and dots; 1 for B–D; 10^{-6} for E and circles. The ratio of the frequency factor v to the heating rate q determines the peak position for a given shape: $v/q=10^{11}$ K^{-1} for A, D, and dots; 10^8 K^{-1} for C; and 10^5 K^{-1} for B, E and circles. The density of traps N determines the magnitude of $I(T)$: $N=10^{18}$ cm^{-3} for A, B, and circles; 10^{15} cm^{-3} for C and dots; and 10^{12} cm^{-3} for D and E. The recombination coefficient γ divided by the heating rate q determines the magnitude of $\sigma(T)$: $\gamma/q=10^{-14}$ cm^3 K^{-1} for A, B and dots $\gamma/q=10^{-11}$ cm^{-3} K^{-1} for C; 10^{-8} cm^{-3} K^{-1} for D, E, and circles [2.13]. The thermal trap ionization energy is $E/k=4000$ K in all cases

Fig. 2.9. Numerical solutions of (2.13, 14) for various concentrations of thermally disconnected traps M. Solid curves: $R \equiv \beta/\gamma=10^6$, $v/q=10^{11}$ K^{-1}; curves A–E correspond to $M/N=0$, 10^{-2}, 1, 10^4, and 10^8; E may be physically unrealistic and is included only to show the trend. Dashed lines: $R=10^{-6}$, $v/q=10^5$ K^{-1}; curves a–e correspond to $M/N=0$, 10^{-8}, 10^{-6}, 10^{-2}, and 1, respectively; all these cases produce only one single $I(T)$ curve which is shown by circles in Fig. 2.8. The thermal trap ionization energy is $E/k=4000$ K in all cases

$dn_c \ll dn$ and $n_c \ll n$

$$I(T) = \eta n(T_0) v \exp\left[-\frac{E}{kT} - \frac{v}{q}\int_{T_0}^{T} \exp(-E/kT)dT\right]. \tag{2.19}$$

Here η again is the luminous efficiency of the transition of free electrons to the recombination centers; $v \equiv g^{-1}\langle v \rangle N_c S_n$, is the so-called frequency factor (here taken as temperature independent), g the trap degeneracy (usually assumed to be unity), and E is the thermal activation energy. As discussed in Sect. 2.5.3, the capture cross section S_n can be thermally activated and may be expressed as $S_\infty \exp(-U_c/kT)$. Note that in this case $v = g^{-1}\langle v \rangle N_c S_\infty$ and the trap depth $E_c - E_t$ is *not* identical with the thermal activation energy $E \equiv (E_c - E_t) + U_c$. The temperature dependence of frequency factors associated with multiphonon capture has been discussed in detail by *Pässler* [2.162].

Similarly, *Garlick* and *Gibson*'s [2.9] solution is obtained for $R = 1$ (equal chance for free electrons to be retrapped or to recombine), $M = 0$ and, again, T-independent v and η

$$I = \eta N n(T_0)^2 v \exp(-E/kT)\left[N + n(T_0)\frac{v}{q}\int_{T_0}^{T} \exp(-E/kT)dT\right]^{-2}. \tag{2.20}$$

The validity of the so-called hypothesis of a stationary electron distribution in the conduction band [2.216], that is $dn_c/dt \ll dn/dt$, has been justified by *Maxia* [2.217] on the basis of *Prigogine*'s theory of nonequilibrium thermodynamics [2.218]. It was found to be a good approximation for a wide range of physically reasonable trapping parameters except for the special case $M = 0$ and $N < 10^{15}$ cm^{-3}. In this situation, calculated TSC curves have no maximum for $R = 0$ and tend to be much broader than many experimental curves [2.12, 40, 65, 205, 206].

Approximate analytic solutions for $\sigma(T) = e\mu_n n_c$ have been obtained under the assumption that the recombination lifetime $\tau = 1/\gamma(n_c + n + M)$ is constant, which is reasonable only if $f = n + n_c + M \gg n$. For this condition to hold, M has to be much larger than N. *Bube* [2.119], and *Kulshreshtha* and *Goryunov* [2.220] obtained a solution for σ, that is, except for factors, identical to *Randall–Wilkins*-type shapes [2.221].

It is obvious that, under equal conditions, the temperature at the glow maxima for TSL and TSC are equal in this constant-lifetime approximation for any value of R. However, there is strong experimental evidence that this is not the general case [2.6, 7, 65].

Further analytic solutions for $\sigma(T)$ were reported by *Simmons* and *Taylor* [2.29] for the case that retrapping can be neglected in a thin sample at high electric fields. They also considered the presence of several trap levels of density N_i and demonstrated the superposition of the individual glow peaks when the thermal ionization energies of these levels are very similar (Fig. 2.10). The development of the trap occupancy function $f(E, T)$ during the thermal scan is

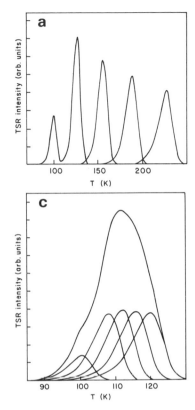

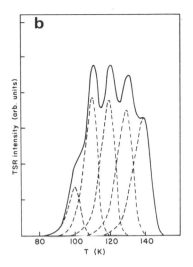

Fig. 2.10a–c. Examples of glow spectra resulting from several different discrete trap levels which are (a) widely separated, resulting in clearly resolved glow peaks; (b) not sufficiently separated to be clearly resolved in the glow spectrum; and (c) overlapping to produce a single, unresolved glow peak [2.29]

discussed by these authors as well. *Bosacchi* et al. [2.222] showed how these approximate analytic solutions of (2.13, 14) (with $R-0$) can be extended to continuous trap distributions $N(E)$, thus, their results are approximate solutions of (2.8) for negligible retrapping.

2.6.2 TSL and TSC Due to Electron and Hole Traps

The simple "single trap model" is inappropriate whenever electrons as well as holes are simultaneously released from their respective traps. Naturally, the phenomenological rate theory of TSL and TSC becomes very complex. Detailed solutions have not been discussed in the literature so far. They should pose, however, no principal problems. Approximate solutions have been described [2.19, 40] for a simplified case in which occupied hole traps are the only available recombination centers for free electrons and, similarly, filled electron traps are the only recombination centers for free holes (Fig. 2.7c). The conductivity is now

$$\sigma = e(\mu_n n_c + \mu_p p) \tag{2.21}$$

and, assuming that the transitions γ and γ^* are both radiative,

$$I = \gamma n_c f + \gamma^* p n . \tag{2.22}$$

Again, γ denotes the recombination coefficient for free electrons and γ^* that for free holes. For simplicity, the luminescence efficiency is assumed to be unity. Similarly, α is the coefficient for thermal release of trapped electrons, α^* that for holes and the respective capture coefficients are β and β^*. We define the retrapping coefficients as $R = \beta/\gamma$ and $R^* = \beta^*/\gamma^*$. Equations (2.5, 6) are valid for both types of traps. The recombination kinetics is now characterized by two trap depths, $E_c - E_{t,n}$ and $E_{t,p} - E_v$ and two thermal activation energies

$$E = (E_c - E_{t,n}) + U_c \quad \text{and} \quad E_A \equiv (E_{t,p} - E_v) + U_v .$$

U_c and U_v are defined in Sect. 2.5. They are nonzero for multiphonon capture. Here $E_{t,n}$ and $E_{t,p}$ are the energies of the electron traps and hole traps, respectively.

Approximate analytic solutions have been obtained for stationary free carrier distributions in the transport bands for the cases I) $R \approx 0$, $R^* \approx 0$; II) $R \gg 1$, $R^* \gg 1$; III) $R \approx 0$, $R^* \gg 1$; and IV) $R \gg 1$, $R^* \approx 0$ [2.40]. The curves so obtained are simply the sum of two *Randall–Wilkins* or *Garlick–Gibson*-type curves and their value lies solely in demonstrating the complexity added to the problem by the additional set of hole-trapping parameters.

Simultaneous release of electrons and holes from their respective traps can also lead to negative thermally stimulated conductivity, predicted in 1967 [2.223] and recently observed in CdTe [2.224].

2.6.3 TSL and TSC Due to Trap-Recombination Center Pairs

The acceptor–hole pair concept of *Williams* [2.225] has proven to be a physically realistic situation in many solid systems. TSL and TSC curves can be obtained for the case that, in addition to thermal release into ,the transport bands, recombination of trapped carriers may also occur simultaneously by thermally activated nonradiative transitions from an excited trap level. The activation energy E_A for those transitions is assumed to be smaller than E (Fig. 2.7d). Analytic solutions were obtained again for the familiar condition $dn_c/dt \ll dn/dt$ and $n_c \ll n$ [2.20] for the case of $R > 1$ and $n(T_0)/N < 1$. As expected, the glow curves are characterized by the two activation energies E and E_A.

If thermal ejection of carriers from traps occurs exclusively via such transitions within the pair, no TSC curve can be measured and a TSL curve is obtained only, when the transition δ_0 is radiative [2.226]. Little experimental work has been done on thermally stimulated relaxation of trap-recombination center pairs [2.20].

2.6.4 General Order Kinetics

In direct analogy to "formal order" chemical reaction kinetics a formal phenomenological theory of TSL and TSC can be constructed on the basis of the rate equation [2.227, 228]

$$dn/dt = -vn^l \exp(-E/kT),\tag{2.23}$$

where again v is a frequency factor (assumed to be temperature independent), E a thermal activation energy and n the density of trapped carriers. The "order" of the process is given by the constant l which is not necessarily an integer. The case $l=1$ corresponds to (2.19) and $l=2$ to (2.20). For $n>2$ no simple physical model correspondence can be identified and hence this approach to relaxation kinetics is essentially classificatory in nature, giving no insight into the physics of the process or correlations with other observable phenomena.

2.6.5 Methods of TSL/TSC Trap Level Spectroscopy

The principle goal of TSL/TSC trap level spectroscopy is to experimentally determine, by comparison of model-glow curve (or parts of them) with measured ones, the characteristic parameters which govern the nonisothermal relaxation kinetics of the solid. During the long history of trap level spectroscopy, a number of principal difficulties with this approach have been established, the discussion of which is the purpose of this section.

We start with the "ideal" situation of a thermally stimulated relaxation process that somehow is known to be exactly described by the single trap level model (2.13, 14). The relaxation process is completely determined by the initial conditions T_0 and $n(T_0)$, the heating program and five temperature-independent trapping parameters: the thermal ionization energy E, the frequency factor v, the recombination coefficient γ, the retrapping ratio $R = \beta/\gamma$, and the ratio of the densities of "thermally connected" traps N to that of "thermally disconnected" traps M. Numerical solutions of (2.13, 14) show that a given physically reasonable combination of these parameters and knowledge of $n(T_0)/N$ will yield a unique pair of TSL and TSC glow curves which are part of the established curves that have been shown to be solutions for a wide range of these parameters [2.13]. Further, any given glow curve pair out of this set of possible solutions can be uniquely associated with one set of these five trapping parameters. That is, a given curve permits one, in principle, to determine these parameters unambiguously [2.213] if the validity of the single trap model is established and the initial condition $n(T_0)/N$ is known. These facts are taken as the physical basis for trap level spectroscopy by thermally stimulated luminescence and conductivity. However, the experimental realization of this method is a different matter. What looked obvious turned out to be exceedingly difficult.

TSL/TSC trap level spectroscopy consists of the following procedure:

I) Establish the validity of the kinetic model. This must usually be done by independent experiments.

II) Measure the shape of the TSL/TSC glow curves as accurately as possible for various initial conditions and/or heating rates.

III) Correct for experimental artifacts.

IV) Analyze the shape (maybe in normalized form) in such a way as to determine the five key trapping parameters.

V) Calculate the glow curves with these parameters and compare with the experimental curves for consistency.

Step I) is far from being trivial and is therefore usually not taken by experimentalists, thus, throwing doubt on the obtained empirical data. Tests for the validity of the single trap model have been established and should be performed as a minimum precaution (see Sect. 2.6.6). Steps II) and III) involve details concerning the experimental facility. Step IV) requires further discussion.

The principal trap parameters are the thermal activation energy E and the frequency factor. They are related to the thermal emission rate via $\alpha = v \exp(-E/kT)$ and the capture cross section via $v = g^{-1} \langle v \rangle S_n N_c$. Numerous analytical methods have been derived from approximate solutions of (2.13, 14) (assuming certain values for the remaining three parameters) which allow one to determine E from experimental glow curves. Detailed reviews of these methods are given in [2.14, 65]. Procedures for the determination of v have been discussed in [2.65] and are not repeated here. Recent publications on this subject are listed in the Appendix of this volume.

Chiefly, the methods for the determination of the thermal activation energy or the frequency factor can be divided into three categories: those making use of heating rate variations, those based on geometrical approximations of the glow peak and various others, most important among them *Garlick*'s and *Gibson*'s initial rise method [2.9] according to which the slope of the initial rise of a glow peak, plotted semilogarithmically vs $1/T$, is equal to the thermal activation energy E.

The accuracy of all the known methods to extract $\dot{E}$ from an analysis of experimental glow curves was tested by *Hagebeuk* and *Kivits* [2.14] who applied them to computer generated glow curves, assuming as initial condition $n(T_0)/N \equiv 1$. Thus, this test evaluates *only* the validity of the assumptions made to derive the methods and does not take into account experimental artifacts. Further, it requires that the single trap model in the form of (2.13, 14) with the given initial condition is an *exact* description of the relaxation kinetics. If this can be shown to be the case for a given glow peak, TSL and TSC are indeed helpful tools for determining trapping parameters.

Since the shape of TSL and TSC glow curves depends strongly on the reaction kinetics (e.g., trap distribution instead of discrete levels, presence of hole traps in addition to electron traps, excited trap levels, trap-recombination center pairs, etc.), kinetics-specific analysis methods have to be derived for each situation. This can be done [2.19, 20, 40, 65], but again the validity of a given model kinetics has to be established prior to application of these methods for trap-spectroscopic analyses of empirical data.

It is further not difficult, in principle, to correct all these methods for the temperature dependence of the luminous efficiency for recombination, $\eta(T)$, or the carrier mobilities, $\mu_n(T)$ and $\mu_p(T)$, etc., provided these are known [2.17]. Thus, the potential of TSL and TSC techniques for trap level spectroscopy is still a real one, even though most of the work done to date in this field is based on kinetic models that are either too simplistic or not applicable outright. To make use of this largely untapped potential, experimental techniques have to be used or newly established which permit one to determine the exact model of the kinetics. Since TSL and TSC are indirect trap-spectroscopic tools (see Sects. 1.7 and 2.1), a most promising approach is to devise experiments that reduce the complexity of the recombination traffic.

2.6.6 Tests for the Validity of the Single Trap Model

Establishing the applicability of the single trap model to a particular thermally stimulated relaxation process is a complex procedure in many insulating materials. Semiconductors are often known well enough through independent studies to warrant its cautious application.

Experimentally, only currents and not the conductivity are measured and it is mandatory to check for uniform F-field distribution and negligible injection effects (see Sect. 2.7.4). Ideally, the carrier composition (holes vis-à-vis electrons) should be established as well and independent measurements of $\mu_n(T)$ and $\mu_p(T)$ should be made. Temperature gradients in the sample must be carefully avoided [2.228].

Spectral analysis of the thermoluminescence output will establish the involvement of one or several recombination centers or even a broad distribution. The temperature dependence of the luminous efficiency can only be inferred indirectly from measurements of photoluminescence transients, etc., however, knowledge of its absolute value is not necessary for trap level spectroscopy.

Correlated measurements of $I(T)$ and $\sigma(T)$ are recommended whenever possible, however, they do not necessarily permit detailed conclusions concerning the validity of the kinetic model [2.12, 229].

Numerous independent control experiments are usually required to establish the applicability of a particular kinetic model. Two examples are discussed below. However, no single or simple test is available to date that will prove the validity of a given kinetic model.

We have seen that, in general, the TSR process is irreversible, not because of the physical properties of the individual defect states but rather because of the fundamental thermodynamic properties associated with the occupation function of the perturbed equilibrium of the system defect states – host lattice.

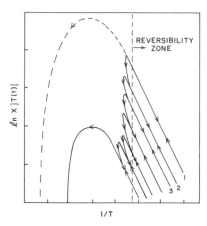

Fig. 2.11. Schematic semilogarithmic plot of the intensity $X\{T(t)\}$ of a TSR process vs $1/T$. The figure illustrates the linear initial rise of the intensity at low temperatures as well as the fact that, in first approximation, the process is reversible as long as T is not increased beyond the so-called reversibility zone. During the initial rise only a very small fraction of the total number of trapped charge carriers is released and, therefore, the TSR glow curve can be retracted. Heating to temperatures past the reversibility boundary and subsequent cooling yield again a linear curve but with reduced intensity. Heating cycles that produce a succession of curves 1, 2, 3... (solid line) are used in the technique of fractional trap emptying. The thermal ionization energy of the trap can be determined in this manner because the slope of these linear sections is equal to $-[(E_c-E_{t,\,i})+U_{c,\,i}]/k$

Consequently, if one were able to measure in a TSR experiment only the properties of the individual defects (traps) without changing the occupation function, the inherent aspect of irreversibility could be eliminated.

Experimentally this can be achieved by increasing the temperature, starting from a low temperature steady-state situation, just enough to obtain a very small but measurable output so that, *in first approximation*, the occupation function has not been affected. Any output will then be proportional to the thermal emission probability $\alpha(T)$, and as a result, information on it can be obtained in such a way. This situation is exploited in the so-called initial rise method for the measurement of activation energies [2.9] or in fractional emptying experiments (Sect. 2.7.1). Under these conditions the functional $X\{T(t)\}$, discussed in Sect. 1.9, is now independent of the heating program as long as the increase in temperature remains sufficiently small. It should be noted that such an initial rise of $X(T)$ may be measured repeatedly even after T had been increased above the reversibility domain. Cooling afterwards will reestablish a new initial condition for the occupation function and a subsequently repeated initial rise experiment will again yield information on $\alpha(T)$ (Fig. 2.11).

One approach to process TSR data so as to test for the validity of the kinetic model is to measure two or more output variables simultaneously during one temperature scan and to attempt to remove complex functionals or replace them with functionals that are easier to handle. We will demonstrate this possibility in the special case of thermoluminescence and thermally stimulated conductivity of an n-type material which are given, respectively, by

$$I(T)=\eta n_c\{T(t)\}/\tau\{T(t)\} \quad \text{and} \quad \sigma(T)=\mu_n e n_c\{T(t)\}.$$

Here μ_n is the carrier mobility, e the electron charge, F the electric field, τ the lifetime of free carriers due to transitions to recombination centers and η is the luminous efficiency. The lifetime, being a function not only of T and t but also of the concentration f of unoccupied recombination centers, is again a functional. Forming the ratio

$$r^* \equiv \frac{\sigma}{I} = \frac{\mu_n e}{\eta} \tau\{T(t)\} \tag{2.24}$$

one obtains, with $\tau\{T(t)\} = 1/\gamma f\{T(t)\}$

$$r^* = \frac{\mu_n e}{\eta\gamma} \frac{1}{f\{T(t)\}}. \tag{2.25}$$

The ratio r^* is easily accessible experimentally and may be analyzed either in a reversible experiment such as the "initial rise" or "fractional emptying" experiment or in an irreversible temperature scan [2.230]. The reversible T scan permits one to determine the temperature dependence of $\mu_n/\eta\gamma$ because of the condition $df \approx 0$. With its knowledge the temperature dependence of $f\{(T)\}$ can then be determined by measuring $r^*(T)$ in a complete reversible T scan for a given initial condition and heating program. Forming the derivative of the so obtained empirical function $f\{(T)\}$ with respect to T permits one, in a further step, to check the commonly made assumption $I(T) \propto -df(T)/dT$. In the case that the measured TSL glow curve $I(T)$ is proportional to the curve $-df(T)/dT$ obtained with this procedure, it can safely be concluded that the radiative recombination measured via the TSL curve is indeed the only mechanism for recombination of free electrons with recombination centers. If this proportionality cannot be confirmed, recombination transitions other than those monitored with the TSL glow curve must contribute to the decay of the free carriers. Thus, valuable information on the validity of a reaction kinetic model can be obtained in this manner. The test hinges on the assumption that only one type of free carrier is generated by thermal ejection from traps. Its application to ZnS, ZnSe, SnO_2 and AgBr strongly indicates that, at least in these systems, the simple model is not valid [2.230, 231].

A second, less rigorous test was proposed by *Fields* and *Moran* [2.232]. These authors assumed, in first approximation, the ratio μ_n/γ to be T-independent. If this assumption, which is usually made without further considerations, is indeed correct, a linear correlation should exist within the framework of the single trap model between the integrated TSL signal, $\int_{T_0}^{T} I(T')dT'$ and the ratio of $I(T)/\sigma(T)$. Application to several peaks in LiF did not confirm that μ_n/γ is temperature independent and/or that the single trap model is applicable [2.232].

A method to estimate yet another parameter, namely the degeneracy factor g of a trap level, was suggested by *Landsberg* [2.233].

The development of new experimental tests to unambiguously establish the validity of a kinetic model is required in order to elevate TSL and TSC techniques to quantitative trap-spectroscopic tools.

2.7 Experimental Techniques

The main attraction of TSL and TSC as experimental methods for the study of defects in nonmetallic solids was, for many years, their apparent simplicity. The excited sample merely had to be placed onto a heater in front of an optical detector and/or, after attachment of two metallic contacts for voltage biasing, connected to a sensitive current meter. Work at low initial temperatures required the experiment to be performed inside a vacuum chamber in an inert gas atmosphere.

As the field evolved, more complicated heating programs were found to be advantageous in certain situations and the need for spectral analysis of thermoluminescence arose. However, the basic experimental arrangement was never considered one of great complexity.

Yet, as we have stressed in Sect. 2.6, measurements of TSL and TSC alone, performed independently or simultaneously, with whatever resolution, sensitivity, heating program, or degree of excitation of the sample, will in general not yield all the trapping parameters of a single defect state that enter even a simple set of rate equations in the phenomenological kinetic theory unless the validity of the model is established independently [2.213]. The sole exceptions known to date are cases where the thermal ejection from and capture of carriers in traps are the only transitions to be considered. Since these special experimental conditions are not given or cannot be realized in many actual measurement situations (e.g., in thermoluminescence or, in general, whenever a sufficiently thin sample cannot be prepared), trap level spectroscopy requires the use of auxiliary measurement techniques in addition to TSL and TSC. Which of these should be selected depends on the individual material and type of defect under study. The identification of the chemical nature of a defect may be possible with electron spin resonance, optical absorption, photoluminescence and/or intentional back-doping of a previously cleaned material. Numerous examples are reported in the literature where TSL and TSC have been used successfully in conjunction with these independent techniques to shed light on the detailed properties of trap levels, recombination centers and transition rates for charge carriers involved in the recombination traffic during the irreversible temperature scan. Here we will describe only experimental methods pertaining directly to TSL and TSC measurements.

2.7.1 Heating Programs

For most experiments on nonisothermal thermally stimulated relaxation simple cooling of the sample to the desired initial temperature and a linear

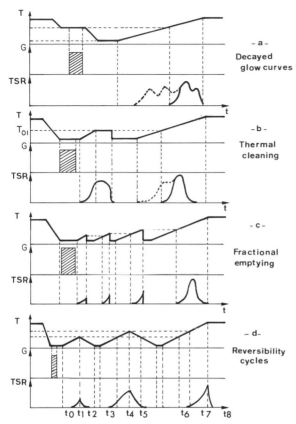

Fig. 2.12a–d. Excitation (G) and heating cycles employed in TSR experiments. To remove disturbing glow peaks at the low temperature side of the peak under study, the sample is excited at a temperature chosen high enough to not fill traps responsible for the low temperature peaks (**a**). The sample may also be excited at the low temperature but preheated prior to an appropriate higher temperature to measure the TSR peak under investigation (**b**). The technique of fractional emptying (**c**) is based on the existence of temperature zones in which the TSR process is approximately reversible (**d**)

increase in T after excitation is sufficient to obtain TSL and TSC glow curves. Some techniques require more elaborate heating cycles, the details of which depend on the relaxation mechanism under study and on whether it is necessary to discriminate between simultaneously occurring processes, e.g., thermally stimulated depolarization and TSC (see Chap. 4) [2.234–236].

Heating programs can also be designed to partially overcome one of the major problems in the measurement of thermally stimulated relaxation, namely the occurrence of unresolved peaks. Several procedures have been devised to this end: the decayed glow curve technique, thermal cleaning and fractional emptying (Fig. 2.12). The first consists of selecting an initial temperature T_0 that is high enough not to fill those traps during excitation which produce the low temperature peak in Fig. 2.12a. Thermal cleaning achieves this same result by preheating the sample, excited at T_0, to T_{01} (Fig. 2.12b).

Fractional emptying consists of measuring the initial rise of glow peaks (from which the thermal activation energy can be determined) during successive T ramps of gradually increasing upper temperature limits. A histogram of the slopes, obtained from a semilogarithmic plot of the glow curve intensity vs $1/T$, reveals the spectrum of the activation energies involved in the otherwise unresolved glow curve [2.237]. If these temperature ramps include cooling at the same rate as used during the heating ramp (Fig. 2.12d), the measured relaxation phenomenon will be reversible in certain temperature zones (Figs. 2.11, 12d). Reversibility of this type is indicative of not significantly changing the density of electrons in the traps during the T ramp.

The rates q in a linear heating program $dT=qdt$ should be carefully considered. In general, a compromise between fast heating (to improve the signal-to-noise ratio) and uniform heating of the sample is chosen. *Bonfiglioli* [2.238] showed that the temperature difference between the back (heated) and front surface of a flat sample for a given heating rate q may be estimated from

$$\Delta T = q d^2 c \xi / \kappa. \tag{2.26}$$

Here c is the specific heat, κ the thermal conductivity, ξ the sample density and d its thickness. For a typical ionic material of $d=1$ mm, ΔT may reach more than 1 K with $q=1\,\mathrm{K\,s^{-1}}$. Temperature gradients across the sample may lead to hysteresis effects during fractional emptying experiments [2.237]. For certain applications (e.g., TSL measurements of geological or archaeological samples in the form of very small quartz grains [2.8, 239, 240] or in dosimetry [2.241]) rates $q>1\,\mathrm{K\,s^{-1}}$ may be justified. Comparisons between the TSL output of different samples requires simply that the heating program is *reproducible* without it being linear or heating the specimen uniformly. Heating with photon beams (lasers) [2.242, 243] or "hot anvil" systems [2.244] are used to achieve extreme rates.

TSL and TSC can be observed from liquid helium temperature up to several hundred degrees K. Three temperature regions have been used most often according to the availability of experimental facilities or required application of the glow curve technique:

I) $-77 < T < 450\,\mathrm{K}$: Thermal activation energies in the range from 0.1 to about 1 eV may be studied. The sample holder is cooled by liquid nitrogen and is either inside a vacuum chamber or immersed in a dry gas to prevent condensation of moisture [2.245, 246] or electrical leakage [2.247]. The gas may also be used to cool the sample to the initial temperature [2.248, 249]. In a vacuum the sample holder is usually a copper tip in contact with an external reservoir of liquid N_2. The sample is either clamped directly or attached to it with the aid of a contact agent of high thermal conductivity such as silicon grease, silver paint or indium [2.250–252]. If electrical isolation is required, thin platelets of teflon, mica, beryllium oxide or boron nitride ceramics or even epoxy [2.234] have been used.

II) $T > 450$ K: The natural limitation in temperature is given for TSL by black body radiation or, for TSC, by intrinsic or ionic conductivity at elevated temperatures. No vacuum is required. However, in many cases immersion of the sample in dry N_2 is recommended to avoid spurious emission (see Chap. 6). High temperature operation is mandatory for dosimetry of ionizing radiation [2.241, 252] and in archeological dating [2.8, 239]. Commercial TSL readers for dosimetry are available [2.241] as well as for dating [2.239]. Integrated equipment for material analysis is marketed as well (Unirelax, USA [2.253]) which makes use of TSL, TSC in conjunction with other optical, mechanical and thermodynamic analysis methods. Large automated TSL readers, used in the monitoring of environmental radiation hazards, are able to evaluate up to 0.5 million dosimeters per year [2.244].

III) $T < 77$ K: This temperature range requires the use of liquid helium or hydrogen cryostats. Very few experiments on TSL and TSC have been performed at these temperatures [2.56, 240, 242, 254, 255]. The sample is again attached to the cryotip or cooled by immersion into a stream of precooled gas.

The experimental realization of a particular heating program usually starts by simply removing the contact of the sample holder with the coolant reservoir and/or, if one starts at room temperature, by heating with a constant heating current. The temperature–time function so obtained has a characteristic S–shape which can be linearized by appropriately modifying the heating current. The established time variations of the heating current may then be reproduced by mechanical means (e.g., a cam drive) [2.256]. Saw-tooth heating programs for fractional emptying are more difficult to implement particularly if active cooling is a requirement. Modern electronic feedback systems using microprocessors are now becoming available with which any desired heating cycle that is commensurate with the thermal response of the sample holder [2.257–259] can be preprogrammed.

2.7.2 Cryostats and Heaters

Standard optical metal or metal–glass cryostats are commonly used in low temperature studies of TSL and TSC after installation of a heater system [2.240, 260]. Less expensive facilities for liquid N_2 temperatures have been designed [2.237, 242] and closed cycle cryotips may be employed as well. Since thermally stimulated relaxation requires dynamic temperature cycles, it is generally advantageous to design the sample holder–dewar arrangement as small as possible to enable quick turnaround times [2.261]. Most designs require emptying the coolant reservoir before the temperature rise. This can be avoided by using removable thermal connections between the sample holder and the reservoir either in the form of a regulated gas flow [2.262] or a calibrated copper wire [2.252].

The heaters commonly consist of a resistive wire or a co-axial heating cable with an insulated heating wire inside a bendable metal tubing. These are usually powered by dc supplies or special electronic control units [2.250, 258]. "Optical" heating using light pipes has been explored also [2.262].

Transparent windows are required for TSL studies or whenever the sample is excited by optical means. The window material is chosen according to the transparency requirements: glass (0.33–2 µm); quartz (0.18 µm to visible); LiF (0.11 µm to infrared); alkali halides, Ge or Si (infrared); thin foils of polyethelene, mylar, beryllium or aluminum (X- and γ-rays).

Electrical leads and feed throughs are designed for minimal leakage currents and stray capacitances.

2.7.3 Sample Excitation

The statistical equilibrium of the sample can be perturbed in many ways. TSL and TSC can be produced after mechanical deformation, however, ionizing radiation or carrier injection are used most often.

For excitation in the visible range of the electromagnetic spectrum lasers or incandescent and gas discharge lamps together with filters or monochromators have been used. The excitation of the sample may involve band–band transitions with subsequent trapping of the carriers or lead to internal conversion of localized defects [2.263]. Such secondary effects as optical quenching or sample heating during exposure to radiation should be minimized or controlled. Strongly absorbed light in conjunction with an appropriate bias voltage leads to photoinjection of carriers [2.245, 264]. Exposure to infrared light may enhance or quench glow peaks [2.265]. X-rays [2.254] and γ-rays [2.266] are frequently used as well. Beams of β^-, β^+ [2.267], α [2.268], neutrons [2.250, 269, 270] or even π_0 mesons [2.271] have been employed for the purpose of sample excitation as have beams of electrons up to several hundred keV [2.272, 273]. High energy radiation does, of course, not only fill traps but produces new defects in most samples (radiation damage) [2.156].

2.7.4 Electrical Measurements

TSC experiments are customarily analyzed assuming the sample behaves "ohmic", that is the contacts do not introduce an inhomogeneity in the distribution of the electric field or carrier density and a uniform bulk density of carriers extends through the entire sample. Contact barriers are neglected. Recent work by *Henisch* and *Popescu* [2.274–276] has contributed much toward a more realistic appraisal of the situation. They analyze carrier and field distributions under isothermal and stationary conditions and find that, in the limiting case of zero current flow, a convenient measure of the field inhomogeneity near the contacts is given by the Debye length

$$L_D = \left(\frac{\varepsilon k T}{2e^2 n_c} \right)^{1/2} . \tag{2.27}$$

It turns out that L_D is rather small for so-called lifetime semiconductors (free carrier lifetime τ is greater than the dielectric relaxation time τ_D) and that it can be appreciable in comparison to the sample length in relaxation semiconductors ($\tau_D > \tau$).

Stationary current flow (carrier injection from the contacts into the sample) results in more complex field and carrier distributions along the sample, which depend largely on the injection or current composition ratio. For example, if one considers minority carrier injection, the ratio of the current carried by minority carriers to the total current assumes a certain value γ_0 far away from the contacts; it may be larger (injection) or smaller (exclusion) than γ_0 at the minority carrier injecting contact end, again, larger (accumulation) or smaller (extraction) at the other contact. The presence of traps has been considered as well, but nonstationary and nonisothermal situations have so far not been dealt with [2.277]. Without going into further details we point out only that the measurement of the carrier density via TSC experiments is not a simple matter and inhomogeneities in both the carrier and the electric field distributions along the sample may seriously affect the validity of the results obtained [2.276]. The traditional TSC theory yields expressions for $\sigma(T) = e(n_c\mu_n + p\mu_n)$, but experiments measure currents only and not the conductivity. Inhomogeneities are always associated with diffusion currents as well, a fact that is usually neglected in the theory of TSC. Experiments should always be carried out in such a way as to minimize injection effects.

Contact materials and configurations used in TSC experiments vary widely and depend on the particular application. Metal electrodes can be attached to the sample by evaporation, ultrasonic soldering (e.g., indium or indium–gallium alloys and gold), by application of conductive pastes (silver paint or epoxy) or metallo–organic compounds (noble metals precipitate upon heating the sample and the organic residue evaporates). Semiconductor devices rely, of course, on proven techniques developed by the semiconductor industry. Typical contact configurations are shown in Fig. 2.13.

Biasing the sample is in general a simple matter. Stable dc power supplies, floating or with one grounded outlet, or dry cell batteries are convenient to use because the currents through the sample are in almost all cases extremely small and require the use of dc picoammeters with either linear or logarithmic output signal (Fig. 2.14). Lock-in amplifiers can be considered whenever the signal can be modulated. However, response and decay times in insulators can be large and, therefore, the modulation frequencies have to be quite low. Vibrating-reed current meters are best whenever currents smaller than 10^{-13} A have to be measured [2.20, 40, 281, 282].

In order to eliminate at least the uncertainties associated with transient current and field inhomogeneities along the sample during TSC experiments, it has been suggested to perform experiments under constant-current conditions and to compare with the usual constant voltage results. Observed differences will be symptomatic of the complications discussed above [2.276]. Constant-current sources for this type of test are commercially available.

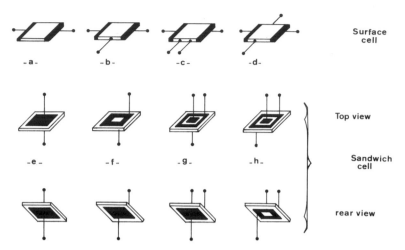

Fig. 2.13a–h. Typical electrode configurations used for the measurement of thermally stimulated currents. The simple surface cell (**a**) may be augmented by additional contacts (**b**)–(**d**) to monitor potential distribution along the sample. The sandwich cell (**e**)–(**h**) is ideally suited for the use of guard rings of either rectangular or circular shapes [2.279, 280]. Simultaneous TSL measurements are best performed with configurations (**a**)–(**d**) or by using transparaent NESA or indium tin oxide films [2.278] in configuration (**e**)–(**h**)

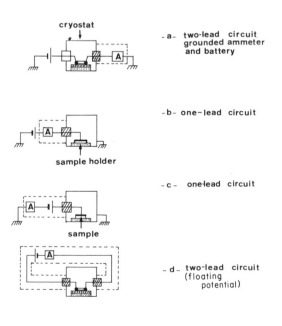

insulator
ammeter
shielded and insulated zone

Fig. 2.14. Examples of experimental arrangements suitable for the measurement of thermally stimulated conductivity

Quite difficult are measurements of the transient Hall effect during TSC in wide band gap materials, because the carrier densities are small even in the TSC peak [2.283]. *Kivits* described a technique that exploits an alternating magnetic field for the determination of the Hall signal and the carrier mobility [2.27].

2.7.5 Optical Measurements

The measurement of TSC entails the detection of usually weak light emission from the sample which is a function of $T(t)$ and $\lambda(t)$. The TSL intensity depends on the number of radiative recombination transitions per unit time and on the luminous efficiency $\eta(\lambda, T)$. The spectral composition of the light may change as the occupational function for recombination centers changes during the temperature scan.

We have not discussed in this chapter the physics of radiative transitions which is adequately described in the literature [2.41–44]. The strength of TSL techniques is not the investigation of the optical properties of recombination centers. This information can be obtained, in most instances, from reversible isothermal optical measurements such as absorption and photoluminescence spectroscopy. Emphasis in this volume is placed on radiative transitions as monitors for the thermally stimulated recombination traffic of carriers ejected from traps and, consequently, optical TSL measurement techniques are discussed only in view of how to achieve this. For example, since η may change with temperature as well as the shape of the thermoluminescence emission band, the TSL glow peak, recorded with a photodetector having a wavelength-dependent quantum efficiency, may be distorted and not accurately represent the recombination kinetics. In addition, through the knowledge of the thermo-luminescence spectrum, the nature of the recombination center may be identified by comparison with photoluminescence spectra. This has been successfully done in many cases where the electron capture leads to tightly bound defect states [2.27].

Popular detectors for TSL are photomultipliers which may be cooled to reduce the dark current. They can be used in an analog or single-event counting mode. Their quantum efficiency becomes usually quite small in the near IR; if this fact is not sufficient to reduce black body background radiation, IR cut-off filters [2.239, 241] or novel differential thermoluminescence techniques [2.284, 285] may be used.

Recently silicon diodes or even vidicons have been employed [2.257, 286]. Their advantage is a broad and flat response between 0.35 and 1.2 µm.

When the emission spectrum is to be measured as a function of $T(t)$, special facilities are required. Since the TSC intensity is in most cases very small, spectral resolution is often sacrificed for the sake of fast time response. Rotating discs with interference filters, oscillating dispersive elements [2.278, 287–290] or oscillating exit slits [2.266] have been tried. The use of color transparency film is not recommended [2.291].

An elegant solution of the problem of measuring time-resolved luminescence spectra is a modern optical multichannel analyzer (OMA, Princeton Applied Research, Princeton, New Jersey) or any other linear or two-dimensional array of photodetectors with associated scanning electronics at the output of a mono- or polychromator [2.257]. The output of these detectors (e.g., vidicon) can be displayed on a cathode ray tube (CRT). Sweep times shorter than ms are available. Electronic processing permits automatic background subtraction, gating, monochannel selection, cycle compilation, etc., and interfacing with digital data processors [2.287, 292].

Finally, the photographic plate should be mentioned as an integrating spatial detector. It is widely used for TSC application in geology and archaeology and may serve conveniently to record the two-dimensional output of microchannel plate image intensifiers.

Because of the low light levels of TSL, efficient light collecting optics is often required. Mirrors, imaging optics or simply close proximity of sample and detector will improve the collection efficiency. Light pipes [2.241, 293] or fiberoptic cables have been used to this end as well. Very small samples, typical in geological applications of TSL, require observation and/or detection through a microscope [2.294], where the phototube is directly attached to the ocular.

2.7.6 Data Recording and Processing

Experiments aimed at the extraction of trap-spectroscopic information – be it by curve-fitting techniques or by performing critical experiments aimed at testing the applicability of a given kinetic model to a given material – require acquisition, recording, and processing of empirical data.

The simplest and often adequate way to obtain a record or measure of TSL and TSC intensities is a multiple-pen chart recorder which displays as a function of time, $I(T, \lambda)$, $\sigma(T)$ and the temperature. $X - Y$ recorders are convenient for this purpose as well. They may serve to directly plot, e.g., $I(T)$ vs T or, if initial rise techniques are used for the measurement of thermal ionization energies, $\ln[I(T)]$ vs $1/T$. If recorder response times pose a problem, CRT displays are advantageous. Single-photon counting techniques in TSL lend themselves to the use of multichannel analyzers. The thermal TSL spectrum is obtained by counting, for a short time interval, the number of emitted photons, displaying them in one of the channels, then advance toward the next channel and repeating the procedure. Interfacing with computers for data analysis is easily done with modern multichannel analyzers. The usefulness of computer data processing for apparent complete characterization of the traps in KCl:Tl has been convincingly demonstrated by *Mattern* et al. [2.287]. Their facility consisted of an electronically controlled sample heater, a grating spectrometer scanned at a rate of 1 scan/s between 2500 and 5500 Å and a calibrated optical detector. Computer processing involved signal averaging of 32 scans for each temperature interval, normalizing the data and correcting for

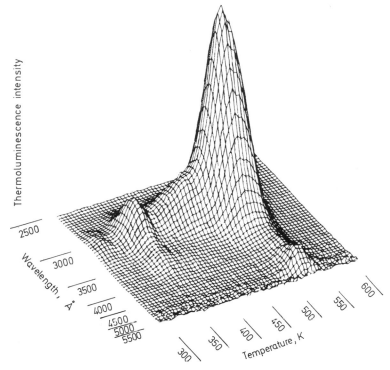

Fig. 2.15. Thermoluminescence of KCl:Tl between 300 and 600 K obtained by simultaneous measurement of the spectral distribution and the intensity of the emitted luminescence $I(T, \lambda)$ [2.287]

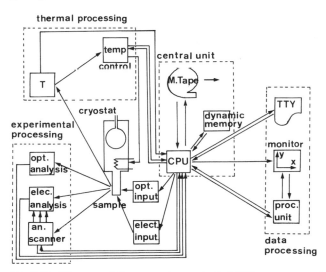

Fig. 2.16. Diagram of the computer-controlled TSR measurement facility "Spartacus" at the University of Montpellier. The central processing unit (CPU) controls the heating program, performs the analysis of the optical and electrical TSR signals and processes the raw experimental data. Statistical analysis, necessary manipulations for trap level spectroscopy, etc., are performed automatically [2.286]. an. scanner: analog scanner (multiplexer) for sequential interrogation of different analog signal channels, T: temperature monitor, TTY: teletype unit

calibration, providing a three-dimensional plot of $I(T, \lambda)$ (Fig. 2.15) and, finally, analyzing the data on the basis of a simple kinetic model and configuration-coordinate trap model.

A similar system was described by *Castagne* [2.286]. Instead of scanning the dispersive element of the spectrometer, a linear array of photodetectors is placed in the exit plane and the computation and control function are now handled by a minicomputer or microprocessors (Fig. 2.16). Experimental advances of these kinds are expected to dramatically improve the value of TSL and TSC techniques for trap level spectroscopy. As the cost of digital equipment decreases, automated data recording and processing facilities will become more common and this development will certainly contribute towards full realization of the potential of TSL and TSC techniques as tools for monitoring thermally stimulated relaxation in solids and for complete characterization of all relevant trapping parameters.

2.8 Concluding Remarks

We have stressed in this chapter the potential of TSL/TSC techniques in trap level spectroscopy as well as the problems associated with their application. Little space was devoted to the large number of excellent papers published in which these techniques were employed only as one of several investigative tools with the aim of identifying the nature of deep defect levels. Classic examples are the studies of optical color center bleaching and annealing in relation to the enhancement or destruction of certain prominent TSL or TSC glow peaks, especially in alkali halide lattices [e.g., 2.25, 26, 295–300], thus unambiguously relating some of these peaks to well-known defect levels. Charge transfer from traps to impurity states (rare-earth ions) in CaF_2 was studied in a series of eloquent publications by *Merz* and *Pershan* [2.27], and *Schlesinger* and *Whippey* [2.27] who applied absorption and thermoluminescence emission spectroscopy in conjunction with conventional glow curve measurements. Similarly fruitful was the combination of electron spin resonance measurements with TSL glow curve techniques pioneered by *Mason* et al. [2.21], and continued by *Scharmann* and co-workers [e.g., 2.22, 23, 301, 302].

Space considerations precluded a detailed discussion of this literature. The Addendum in which the most recent papers are collected is an attempt to compensate somewhat for this incompleteness.

References

2.1   F. Urbach: Wien. Berichte (II, A) **139**, 363 (1930)
2.2   J. T. Randall, M. H. F. Wilkins: Proc. R. Soc. (London) A **184**, 366 (1945)
2.3   F. Daniels, C. A. Boyd, D. F. Saunders: Science **117**, 343 (1953)

2.4   R.H.Bube: Phys. Rev. **83**, 393 (1951); **99**, 1105 (1955); **101**, 1668 (1956); **106**, 703 (1957); J. Chem. Phys. **23**, 18 (1955)

2.5   C.J.Delbecq, P.Pringsheim, P.H.Yuster: Z. Phys. **138**, 266 (1954)

2.6   I.Broser, R.Broser-Warminsky: Brit. J. Appl. Phys. Suppl. **4**, 90 (1955); Ann. Phys. Leipzig **6F**, **16**, 361 (1955)

2.7   F.Pliquett, M.K.Solncev: *Thermolumineszenz Biologischer Objekte*, Fortschritte der Experimentellen und Theoretischen Biophysik, Vol. 22 (Georg Thieme, Leipzig 1978)

2.8   D.J.McDougall (ed.): *Thermoluminescence of Geological Materials* (Academic Press, London, New York 1968)

2.9   G.F.J.Garlick, A.F.Gibson: Proc. R. Soc.(London) **60**, 574 (1948)

2.10 G.A.Dussel, R.H.Bube: Phys. Rev. **155**, 764 (1967)

2.11 P.Kelly, P.Bräunlich: Phys. Rev. B**1**, 1587 (1970)

2.12 P.Bräunlich, P.Kelly: Phys. Rev. B**1**, 1596 (1970)

2.13 P.Kelly, M.Laubitz, P.Bräunlich: Phys. Rev. B**4**, 1960 (1971)

2.14 H.J.L.Hagebeuk, P.Kivits: Physica **83**B, 289 (1976)

2.15 R.Chen: J. Mat. Science **11**, 1521 (1976)

2.16 I.J.Saunders: J. Phys. C**2**, 2181 (1969)

2.17 P.Kivits, H.J.L.Hagebeuk: J. Luminesc. **15**, 1 (1971)

2.18 P.Kivits: J. Luminesc. **16**, 119 (1978)

2.19 P.Bräunlich, D.Schäfer, A.Scharmann: Proc. Intern. Conf. on Luminescence Dosimetry, ed. by F.H.Attix (U.S. National Bureau of Standards, CONF-650637, Springfield, Virginia 1967) pp. 57–73

2.20 P.Bräunlich: Z. Phys. **177**, 320 (1964)

2.21 D.R.Mason, H.A.Koehler, C.Kikuchi: Phys. Rev. Lett. **20**, 451 (1968)

2.22 G.K.Born, R.J.Grasser, A.O.Scharmann: Phys. Status Solidi **28**, 583 (1968)

2.23 M.Böhm, B.Cord, A.Hofstaetter, A.Scharmann, P.Parrot: J. Luminesc. **17**, 291 (1978)

2.24 E.Sonder: Phys. Rev. B**5**, 3259 (1972)

2.25 C.C.Klick, E.W.Claffy, S.G.Gorbics, F.H.Attix, J.H.Schulman, J.A.Allard: J. Appl. Phys. **38**, 3867 (1967)

2.26 V.Ausin, J.L.Alvarez Rivas: J. Phys. C**5**, 82 (1972)

2.27 J.L.Merz, P.S.Pershan: Phys. Rev. **162**, 217 (1967); **162**, 235 (1967) M.Schlesinger, P.W.Whippey: Phys. Rev. **162**, 286 (1967); **177**, 563 (1969)

2.28 A.G.Milnes: *Deep Impurities in Semiconductors* (Wiley, New York 1973)

2.29 J.G.Simmons, G.W.Taylor: Phys. Rev. B**5**, 1619 (1971)

2.30 M.C.Driver, G.T.Wright: Proc. Phys. Soc. London **81**, 141 (1963)

2.31 J.C.Carballes, J.Lebailly: Solid State Commun. **6**, 167 (1968)

2.32 C.T.Sah, L.Forbes, L.L.Rosier, A.F.Tasch: Solid-State Electron. **13**, 759 (1970)

2.33 C.T.Sah, W.W.Chan, H.S.Fu, J.W.Walker: Appl. Phys. Lett. **20**, 193 (1972)

2.34 D.V.Lang: J. Appl. Phys. **45**, 3014 (1974)

2.35 D.V.Lang: J. Appl. Phys. **45**, 3023 (1974)

2.36 J.S.Prener, F.E.Williams: Phys. Rev. **101**, 1427 (1956)

2.37 F.E.Williams: J. Phys. Chem. Sol. **12**, 265 (1960)

2.38 J.I.Pankove (ed.): *Electroluminescence*, Topics in Applied Physics, Vol. 17 (Springer, Berlin, Heidelberg, New York 1977) pp. 1–39

2.39 P.J.Dean: "III–V Compound Semiconductors", in *Electroluminescence*, ed. by J.I.Pankove, Topics in Applied Physics, Vol. 17 (Springer, Berlin, Heidelberg, New York 1977) pp. 63–134

2.40 P.Bräunlich: Ann. Phys. (Leipzig) **7F**, **12**, 262 (1963)

2.41 J.I.Pankove: *Optical Processes in Semiconductors* (Prentice Hall, Englewood Cliffs, New Jersey, USA 1971)

2.42 F.Bassani, G.Pastori Parravicini: *Electronic States and Optical Transitions in Solids*, Science of the Solid State, Vol. 8 (Pergamon Press, Oxford 1975)

2.43 B.I.Stepanov, V.P.Gribkovskii: *Theory of Luminescence* (Iliffe Books, London 1968)

2.44 D.Curie: *Luminescence in Solids* (Methuen, London, and Wiley, New York 1963)

2.45 R.H.Bube: *Photoconductivity in Solids* (Wiley, New York, London 1960)

2.46 R.H.Bube: *Electronic Properties of Crystalline Solids* (Academic Press, New York 1974)

2.47   P.Goldberg (ed.): *Luminescence of Inorganic Solids* (Academic Press, New York 1966)
2.48   F.E.Williams: "Theoretical Basis for Solid-State Luminescence" in *Luminescence of Inorganic Solids*, ed. by P.Goldberg (Academic Press, New York 1966) pp. 2–52
2.49   D.S.McClure: "Electronic Spectra of Molecules and Ions" in *Solid State Physics*, Vol. 9, ed. by H.Ehrenreich, F.Seitz, D.Turnbull (Academic Press, New York 1969) p. 399
2.50   C.J.Ballhausen: *Introduction to Ligand Field Theory* (McGraw-Hill, New York 1962)
2.51   W.Kohn: "Shallow Impurities in Silicon and Germanium" in *Solid State Physics*, Vol. 5, ed. by H.Ehrenreich, F.Seitz, D.Turnbull (Academic Press, New York 1957) p. 257
2.52   J.H.Reuszer, P.Fischer: Phys. Rev. **135**, A 1125 (1964)
2.53   B.T.Alburn, A.K.Ramdas: Phys. Rev. **167**, 717 (1968)
2.54   P.J.Dean, C.J.Frosh, C.H.Henry: J. Appl. Phys. **39**, 563 (1968)
2.55   B.T.Alburn, A.K.Ramdas: Phys. Rev. **187**, 932 (1969)
2.56   F.Fischer: Z. Phys. **163**, 401 (1961)
2.57   W.D.Compton, J.H.Schulman: *Colour Centers in Solids* (Pergamon Press, New York 1968)
2.58   W.B.Fowler (ed.): *Physics of Colour Centers* (Academic Press, New York 1968)
2.59   J.J.Markham: *F-Centers in Alkali Halides* (Academic Press, New York 1966)
2.60   P.D.Townsend, J.C.Kelly: *Colour Centres and Imperfections in Insulators and Semiconductors* (Chatto and Windus, Brighton, London 1973)
2.61   M.Sparks: "Tabulation of Impurity Absorption Spectra – Ultraviolet and Visible", Vol. I and II, Tech. Rpt. No. 9 and No. 10, DARPA Contract No. DAHC 15-73-C-0127, Washington, D.C. (1977/78)
2.62   F.Seitz: Trans. Faraday Soc. **35**, 79 (1939)
2.63   F.E.Williams: J. Chem. Phys. **19**, 457 (1951)
2.64   D.L.Dexter: "Theory of Optical Imperfections in Nonmetals", in *Solid State Physics*, Vol. 5, ed. by H.Ehrenreich, F.Seitz, D.Turnbull (Academic Press, New York 1958)
2.65   P.Bräunlich: "Thermoluminescence and Thermally Stimulated Current – Tools for the Determination of Trapping Parameters" in *Thermoluminescence of Geological Materials*, ed. by D.J.McDougall (Academic Press, New York 1968) pp. 61–88
2.66   K.H.Nicholas, J.Woods: Brit. J. Appl. Phys. **15**, 783 (1964)
2.67   H.J.Dittfeld, J.Voigt: Phys. Status Solidi **3**, 1941 (1963)
2.68   K.Unger: Phys. Status Solidi **2**, 1279 (1962)
2.69   P.Kivits, J.Reulen, J.Hendricx, F. van Empel, J. van Kleef: J. Luminesc. **16**, 145 (1978)
2.70   W.M.Bullis: "Measurement of Carrier Lifetime in Semiconductors", Document AD 674627, National Tech. Inform. Service, Springfield, Virginia (1968)
2.71   F.W.Schmidlin, G.G.Roberts: Phys. Rev. Lett. **20**, 1173 (1968)
2.72   A.Rose: *Concepts in Photoconductivity and Allied Problems* (Interscience, New York 1963)
2.73   M.A.Lampert, P.Mark: *Current Injection in Solids* (Academic Press, New York 1970)
2.74   M.A.Lampert: Rep. Prog. Phys. **27**, 329 (1964)
2.75   K.L.Ashley, A.G.Milnes: J. Appl. Phys. **35**, 369 (1964)
2.76   P.N.Keating: J. Phys. Chem. Sol. **24**, 1101 (1963)
2.77   A.Rizzo, G.Micocci, A.Tepore: J. Appl. Phys. **48**, 3415 (1977)
2.78   J.Godlewski, J.Kalinowski: Solid State Commun. **25**, 473 (1978)
2.79   B.L.Timan, V.M.Fesenko, G.M.Gulevich: Izv. Vuz Fiz. **5**, 11 (1977)
2.80   C.Schnittler: Phys. Status Solidi (A) **45**, K 179 (1978)
2.81   J.-C.Manifacier, H.K.Henisch: Phys. Rev. B **17**, 2648 (1978)
2.82   V.A.Brodovoi, A.C.Gozak, G.P.Peka: Fiz. Tekh. Poluprovodn. **11**, 1022 (1977) [English transl.: Sov. Phys.–Semicond. **11**, 605 (1977)]
2.83   A.Zoul, E.Klier: Czech. J. Phys. Sect. B **27**, 789 (1977)
2.84   R.Y.Loo, C.R.Viswanathan: Extended Abstracts of the Electrochemical Society Fall Meeting, Las Vegas, USA, Oct. 1976 (unpublished)
2.85   N.T.Gurin, D.G.Semak, V.V.Fedak: Poluprovodn. Tekh. Mikroelektron. **26**, 21 (1977)
2.86   O.A.Gudaev, E.G.Kostsov, V.K.Malinovskii: Avtometriya **1**, 96 (1978)
2.87   G.A.Corker, I.Lundstrom: J. Appl. Phys. **49**, 686 (1978)
2.88   K.O.Lee, T.T.Gan: Phys. Status Solidi (A) **43**, 565 (1977)
2.89   S.Nespurek, J.Obrda, J.Sworakowski: Phys. Status Solidi (A) **46**, 273 (1978)

2.90 S. Nespurek, J. Sworakowski: Phys. Status Solidi (A) **41**, 619 (1977)

2.91 B. L. Timan, V. M. Fesenko, G. M. Gulevich: Fiz. Tekh. Poluprovodn. **11**, 1195 (1977) [English transl.: Sov. Phys.–Semicond. **11**, 707 (1977)]

2.92 A. N. Zyuganov, S. V. Svechnikov, E. P. Shul'Ga: Ukr. Fiz. Zh. **23**, 291 (1978)

2.93 J. B. Anthony, S. R. Butler, F. J. Feigl: Extended Abstracts of the Electrochemical Society Spring Meeting, Philadelphia, May 1977 (unpublished) p. 221

2.94 R. J. Powell: Appl. Phys. Lett. **31**, 290 (1977)

2.95 M. L. Lonky, A. P. Turley, F. J. Kub: Extended Abstracts of the Electrochemical Society Fall Meeting, Las Vegas, 1976 (unpublished) p. 802

2.96 N. S. Saks: IEEE Trans. NS-**24**, 2153 (1977)

2.97 A. Meisel, A. Zehe: Krist. Tech. **12**, 481 (1977)

2.98 V. A. Brodovoi, G. P. Peka, A. N. Smolyar: Fiz. Tekh. Poluprovodn. **11**, 280 (1977) [English transl.: Sov. Phys.–Semicond. **11**, 162 (1977)]

2.99 V. A. Brodovoi, G. P. Peka, A. N. Smolyar: Ukr. Fiz. Zh. **22**, 1835 (1977)

2.100 M. F. Leach, B. K. Ridley: J. Phys. C **11**, 2265 (1978)

2.101 J. M. Aitken, D. R. Young: IEEE Trans. NS-**24**, 2128 (1977)

2.102 R. Amantea, R. S. Muller: Jpn. J. Appl. Phys. **16**, 205 (1977)

2.103 P. K. Chaudhari: 15th Annual Proc. Reliability Physics, **5** (1977)

2.104 D. J. Coe: IEEE J. Solid-State and Electron Devices **2**, 57 (1978)

2.105 T. H. Ning, C. M. Osburn, H. N. Yu: J. Electron. Mater. **6**, 65 (1977)

2.106 T. Tani: J. Photogr. Sci. **24**, 133 (1976)

2.107 S. Sakuragi, H. Kanzaki: Phys. Rev. Lett. **38**, 1302 (1977)

2.108 C. L. Marquardt, J. F. Giuliani, G. Gliemeroth: J. Appl. Phys. **48**, 3669 (1977)

2.109 M. Tabei, S. Shionoya: J. Luminesc. **16**, 161 (1978)

2.110 K.-S. Rebane: Toim. Eesti Nsv Tead. Akad. Fuus. Mat. **26**, 404 (1977)

2.111 V. F. Tunitskaya, L. S. Lepnev: Zh. Prikl. Spektrosk. **26**, 706 (1977)

2.112 I. B. Ermolovich, V. V. Gorbunov, I. D. Konozenko: Fiz. Tekh. Poluprovodn. **11**, 1812 (1977) [English transl.: Sov. Phys.–Semicond. **11**, 1061 (1977)]

2.113 V. A. Shlapak, V. V. Serdyuk: Izv. Vuz Fiz. No. **2**, 138 (1978)

2.114 C. Lhermitte, D. Carles, C. Vautier: Rev. Phys. Appl. **12**, 273 (1977)

2.115 G. Jacobsen: Physica **92** B and C, 300 (1977)

2.116 W. Plesiewicz: J. Phys. and Chem. Sol. **38**, 1079 (1977)

2.117 A. F. Kravchenko, B. S. Lisenker, Yu. E. Maronchuk: Phys. Status Solidi (A) **42**, 647 (1977)

2.118 M. F. Leach, B. K. Ridley: J. Phys. C **11**, 2249 (1978)

2.119 J. D. Bierlein: Photogr. Sci. Eng. **21**, 241 (1977)

2.120 A. M. Kahan: Photogr. Sci. Eng. **21**, 237 (1977)
        C. B. Lushchik, V. G. Plekhanov, G. S. Zaht, I. L. Kuusmann: Eesti Nsv Tead. Akad. Fuus. Inst. Uurim. **47**, 7 (1977)

2.121 D. J. Dimaria, Z. A. Weinberg, J. M. Aitken, D. R. Young: J. Electron. Mater. **6**, 207 (1977)

2.122 G. Godefroy, P. Jullien: Ferroelectrics **18**, 55 (1978)

2.123 M. De Murcia, J. Bonnafe, J. C. Manifacier, J. P. Fillard: J. Appl. Phys. **49**, 1177 (1978)

2.124 L. J. Van Ruyven, H. J. A. Bluyssen, R. W. Van der Heijden: Appl. Phys. Lett. **31**, 685 (1977)

2.125 T. Taguchi, J. Shirafuji, Y. Inuishi: Nucl. Instrum. Methods **150**, 43 (1978)

2.126 V. V. Bezverkhii, G. I. Bulakh, I. V. Ostrovskii: Fiz. Tverdogo Tela **19**, 1456 (1977) [English transl.: Sov. Phys.–Solid State **19**, 847 (1977)]

2.127 A. W. Thompson, A. B. Scott: J. Chem. Phys. **66**, 5252 (1977)

2.128 A. Touboul, G. Pelloux, G. Lecoy, A. A. Choujaa, P. Gentil: Rev. Phys. Appl. **13**, 227 (1978)

2.129 A. Owczarek: Electron Technol. **9**, 43 (1976)

2.130 R. P. Jindal, A. Van der Ziel: Solid-State Electron. **21**, 901 (1978)
        S. T. Hsu: RCA Rev. **38**, 226 (1977)

2.131 A. Many, G. Rakavy: Phys. Rev. **126**, 1980 (1962)

2.132 E. G. Kostsov: Avtometriya **1**, 85 (1978)

2.133 G. Sh. Gil'Denblat, A. Ya. Karachentsev, Yu. N. Potashev: Radiotekh. and Elektron. **22**, 801 (1977) [English transl.: Radio Eng. Electron. Phys. **22**, 98 (1977)]

2.134 V. I. Arkhipov, A. I. Rudenko: Fiz. Tekh. Pluprovodn. **11**, 1380 (1977) [English transl.: Sov. Phys.–Semicond. **11**, 811 (1977)]
2.135 D. R. Young, D. J. Dimaria, W. R. Hunter: J. Electron. Mater. **6**, 569 (1977)
2.136 I. Schneider: Solid State Commun. **25**, 1027 (1978)
2.137 K. D. Nierzewski, T. Todorov, M. Georgiev: Phys. Status Solidi (B) **86**, 697 (1978)
2.138 D. M. Krol, G. Blasse: J. Chem. Phys. **69**, 3124 (1978)
2.139 A. Schmillen, H. Wolff: Z. Naturforsch. A **32** A, 798 (1977)
2.140 V. I. Arkhipov, A. I. Rudenko: Zh. Nauchnoi and Prikl. Fotogr. Kinematogr. **22**, 195 (1977)
2.141 V. I. Arkhipov, A. I. Rudenko: Fiz. Tekh. Poluprovodn. **11**, 1527 (1977) [English transl.: Sov. Phys.–Semicond. **11**, 897 (1977)]
2.142 J. Noolandi: Solid State Commun. **24**, 477 (1977)
2.143 J. A. Kalade: Litov. Fiz. Sb. **16**, 279 (1976) [English transl.: Sov. Phys.–Collect. **16**, 68 (1976)]
2.144 J. Noolandi: Phys. Rev. B **16**, 4466 (1977)
2.145 A. Nussbaum, E. L. Cook, D. M. Korn: Solid-State Electron. **20**, 583 (1977)
2.146 F. W. Schmidlin: Phys. Rev. B **16**, 2362 (1977)
2.147 G. Casalini, T. Corazzari, T. Garofano: Nuovo Cimento B **41** B, 273 (1977)
2.148 W.-W. Falter: Z. Naturforsch. A **31** A, 251 (1976)
2.149 J. P. Zielinger, C. Noguet, M. Tapiero: Phys. Status Solidi (A) **42**, 91 (1977)
2.150 G. H. Talat, M. Tomasek: Czech. J. Phys. B **28**, 331 (1978)
2.151 R. K. Agarwal: Indian J. Pure Appl. Phys. **14**, 627 (1976)
2.152 A. Anedda, F. Raga, A. Serpi: 3rd Intern. Conf. on Ternary Compounds, Edinburgh, Scotland, April 1977, p. 203
2.153 M. Abe, S. Okubo, H. Kuwano: Trans. Inst. Electron. Commun. Eng. Jpn. E **60**, 57 (1977)
2.154 G. T. Jenkin, D. W. Stacey, J. G. Crowder, J. W. Hodey: J. Phys. C **11**, 1841 (1978)
2.155 A. M. Andriesh, M. S. Iovu, E. P. Kolomeyko, M. R. Chernii: Proc. Intern. Conf. on Amorphous Semiconductors, Balatonfured, Hungary, Sept. 1976, p. 301
2.156 C. Lehmann: *Interaction of Radiation with Solids and Elementary Defect Production* (North-Holland, Amsterdam 1977)
2.157 R. T. Williams, J. N. Bradford, W. L. Faust: Phys. Rev. B **18**, 7038 (1978)
2.158 A. Schmid, P. Bräunlich, P. K. Rol: Phys. Rev. Lett. **35**, 1382 (1975)
2.159 R. T. Williams: "Photochemistry of F-Center Formation in Halide Crystals" in *Semiconductors and Insulators*, Vol. 3 (Gordon Breach, New York 1978) p. 251
2.160 M. N. Kabler, R. T. Williams: Phys. Rev. B **18**, 1948 (1978)
2.161 C. H. Henry, D. V. Lang: Phys. Rev. B **15**, 989 (1977)
2.162 R. Pässler: Czech. J. Phys. B **24**, 322 (1974)
2.163 R. Pässler: Czech. J. Phys. B **25**, 219 (1974)
2.164 R. Pässler: Phys. Status Solidi (B) **78**, 625 (1976)
2.165 R. Pässler: Phys. Status Solidi (B) **85**, 203 (1978)
2.166 K. Yamashita, M. Iwamoto, T. Hino: J. Appl. Phys. **49**, 2866 (1978)
2.167 W. Shockley, W. T. Read, Jr.: Phys. Rev. **87**, 835 (1952)
2.168 A. Barraud: "Current Instabilities in High Resistivity Gallium Arsenide Produced by Strong Electric Fields"; Ph.D. Thesis, Document N69-18810-3/26, National Technical Information Service, Springfield, Virginia (1967)
2.169 E. F. Smith, P. T. Landsberg: J. Phys. Chem. Solids **27**, 1727 (1966)
       A. F. Polupanov: Fiz. Tekh. Poluprovodn. **11**, 2044 (1977)
       R. M. Gibb, G. J. Rees, B. W. Thomas, B. L. H. Wilson: Philos. Mag. **36**, 1021 (1977)
2.170 V. Narayanamurti, R. A. Logan, M. A. Chin: Phys. Rev. Lett. **40**, 63 (1978)
2.171 N. F. Mott, R. W. Gurney: *Electronic Processes in Ionic Crystals*, 2nd ed. (Dover, New York 1964)
2.172 H. A. Klasens: J. Phys. Chem. Sol. **7**, 175 (1958)
2.173 M. Schön: Tech. Wiss. Abh. Osram-Ges. **6**, 49 (1953)
2.174 P. T. Landsberg: Phys. Status Solidi **41**, 457 (1970)
2.175 A. Haug: Festkörperprobleme **XIII**, 411 (1972)
2.176 P. T. Landsberg: J. Luminesc. **7**, 3 (1973)
2.177 M. Lax: Phys. Rev. **119**, 1502 (1960)

2.178 H.I.Ralph, F.D.Hughes: Solid State Commun. **9**, 1477 (1971)
2.179 H.Stumpf: *Quantentheorie der Ionenkristalle* (Springer, Berlin, Göttingen, Heidelberg 1961)
2.180 V.L.Bonch-Bruevich, E.G.Landsberg: Phys. Status Solidi **29**, 9 (1968)
2.181 A.Haug: *Theoretische Festkörperphysik II* (Deuticke, Wien 1970)
2.182 E.W.Schlag, E.Schneider, S.F.Fischer: Ann. Rev. Phys. Chem. **22**, 465 (1971)
2.183 V.A.Kovarskiy, I.A.Chaykovskiy, E.P.Sinyavskiy: Fiz.Tverd. Tela **6**, 2137 (1964)
2.184 D.W.Howgate: Phys. Rev. **177**, 1358 (1969)
2.185 E.P.Sinyavshiy, V.A.Kovarskiy: Fiz. Tverd. Tela **9**, 1464 (1966)
2.186 G.A.Marlor: Phys. Rev. **159**, 540 (1967)
2.187 R.Beattie, P.T.Landsberg: Proc. R. Soc. London, Ser. A **249**, 16 (1959)
2.188 L.Huldt: Phys. Status Solidi (a) **24**, 221 (1974)
2.189 L.R.Weisberg: J. Appl. Phys. **39**, 6096 (1968)
2.190 W.Rosenthal: Solid State Commun. **13**, 1215 (1973)
2.191 D.H.Auston, C.V.Shank, P.LeFur: Phys. Rev. Lett. **35**, 1022 (1975)
2.192 I.V.Karpova, S.G.Kalashnikov: Proc. Intern. Conf. Semicond. Exeter (1962), Inst. of Phys. and Phys. Soc., London (1962) p. 880
2.193 J.S.Jayson, R.N.Bhargava, R.W.Dixon: J. Appl. Phys. **41**, 4972 (1970)
2.194 V.Bichevin, H.Käämbre: Phys. Status Solidi (a) **4**, K 235 (1971)
2.195 G.Ascarelli, S.Rodriques: Phys. Rev. **124**, 1325 (1961)
2.196 D.R.Haman, A.L.McWhorter: Phys. Rev. **134**, A 250 (1964)
2.197 V.L.Bonch-Bruevich, V.B.Glasko: Fiz. Tverd. Tela **4**, 510 (1962) [English transl.: Sov. Phys.–Solid State **4**, 371 (1962)]
2.198 K.Huang, A.Rhys: Proc. R. Soc. A **204**, 406 (1950)
2.199 H.Gummel, M.Lax: Phys. Rev. **97**, 1469 (1955)
2.200 R.Kubo, I.Toyazawa: Prog. Theor. Phys. **13**, 160 (1955)
2.201 G.Richayzen: Proc. R. Soc. A **241**, 480 (1957)
2.202 G.Helmis: Ann. Phys. Leipzig **19**, 141 (1956)
2.203 U.A.Kovarskiy, F.P.Sinyavskiy: Fiz. Tverd. Tela **4**, 3202 (1962)
2.204 E.F.El-Waheidy: Indian J. Phys. **51** A, 278 (1977)
2.205 P.Bräunlich: J. Appl. Phys. **38**, 2516 (1967)
2.206 K.W.Böer, S.Oberländer, J.Voigt: Z. Naturforsch. **13**a, 544 (1958)
2.207 M.Böhm, A.Scharmann: Phys. Status Solidi (a) **4**, 99 (1971)
2.208 P.J.Kemmey, P.D.Townsend, P.W.Levy: Phys. Rev. **155**, 917 (1967)
2.209 P.L.Land: J. Phys. Chem. Sol. **30**, 1693 (1969)
2.210 A.Halperin, M.Leibowitz, M.Schlesinger: Rev. Sci. Instr. **38**, 1168 (1962)
2.211 W.Arnold, H.Sherwood: J. Chem. Phys. **62**, 2 (1959)
2.212 P.J.Kelly, M.J.Laubitz: Can. J. Phys. **45**, 311 (1967)
2.213 H.J.L.Hagebeuk, P.Kivits: Physica **83**B, 289 (1976)
2.214 S.Haridoss: J. Comput. Phys. **26**, 232 (1978)
2.215 C.W.Gear: Comm. A. C. M. **14**, 176, 185 (1971)
2.216 V.V.Antonov-Romanovskii: Izv. Akad. Nauk SSR, Fiz. Ser. **15**, 637 (1951)
2.217 V.Maxia: Lett. Nuovo Cimento **20**, 443 (1977); Phys. Rev. B**17**, 3262 (1978)
2.218 P.Prigogine: *Introduction to Thermodynamics of Irreversible Processes* (Interscience, New York 1961) Chap. 6
2.219 R.H.Bube: Phys. Rev. **106**, 703 (1957)
2.220 A.P.Kulshreshtha, V.A.Goryunov: Sov. Phys.-Sol. State **8**, 1540 (1966)
2.221 R.R.Haering, E.N.Adams: Phys. Rev. **117**, 451 (1960)
2.222 A.Bosacchi, S.Franchi, B.Bosacchi: Phys. Rev. B**10**, 5235 (1974)
2.223 P.Bräunlich: J. Appl. Phys. **39**, 2953 (1968)
2.224 V.P.Zayachkivskii, P.P.Bleisyuk, A.V.Savitskii: Fiz. Tekh. Poluprovodn. **12**, 970 (1978)
2.225 F.E.Williams: J. Phys. Chem. Sol. **12**, 265 (1960)
2.226 A.Halperin, A.A.Braner: Phys. Rev. **117**, 408 (1960)
2.227 R.Chen: J. Electrochem. Soc. (Solid State Science) **116**, 1254 (1969)
2.228 Z. Chvoj: Czech. J. Phys. B**27**, 957 (1977)
2.229 G.Birkle, F.Gavrilov, G.Kitaev: Sov. Phys. J. **20**, 773 (1977)

2.230 J.P.Fillard, J.Gasiot, M.De Murcia: J. Electrostat. **3**, 99 (1977)
       J.Gasiot, M.de Murcia, J.P.Fillard, R.Chen: J. Appl. Phys. **50**, 4345 (1979)
       J.Gasiot, J.P.Fillard: J. Appl. Phys. **48**, 3171 (1977)
2.231 J.Gasiot: Ph. D. Thesis, Univ. of Montpellier, France (1976)
2.232 D.E.Fields, P.R.Moran: Phys. Rev. B**9**, 1836 (1974)
2.233 P.T.Landsberg: J. Phys. D**10**, 2467 (1977)
2.234 D.E.Fields: Thesis, University of Wisconsin (1973)
2.235 G.D.Fullerton: Thesis, University of Wisconsin (1974)
2.236 S.W.S.McKeever, D.M.Hughes: J. Phys. D**8**, 1521 (1975)
2.237 H.Gobrecht, D.Hoffmann: J. Phys. Chem. Sol. **27**, 509 (1966)
2.238 G.Bonfiglioli: "Thermoluminescence: what it can and cannot show" in *Thermoluminescence of Geological Materials*, ed. by D.J.McDougall (Academic Press, London, New York 1968) p. 15
2.239 M.J.Aitken: *Physics and Archaeology* (Clarendon Press, 1974)
2.240 W.Hoogenstraaten: Philips. Res. Rep. **13**, 515 (1958)
2.241 K.Becker, A.Scharmann: *Einführung in die Festkörperdosimetrie* (Thiemig, München 1975)
2.242 G.C.Smith: Phys. **148**, 816 (1966)
2.243 M.Kavansihi, H.Onishi: Proc. 5th Intern. Symp. on Exoelectron Emission and Dosimetry, Zvivkov, CSSR, ed. by A.Bohun, A.Scharmann (Universität Gießen, Germany 1976)
2.244 D.Grogan, J.P.Ashmore, G.Bradley, J.G.Scott: Proc. 5th Intern. Symp. on Exoelectron Emission and Dosimetry, Zvivkov, CSSR, ed. by A.Bohun, A.Scharmann (Universität Gießen, Germany 1976)
2.245 P.Devaux: Thesis (Paris, 1968)
2.246 A.A.Braner, A.Halperin: Phys. Rev. **108**, 932 (1957)
2.247 E.B.Podgorsak: Thesis (Univ. of Wisconsin, 1973)
2.248 P.Fischer, P.Röhl: J. Polym. Sci. **14**, 531 (1976)
2.249 C.Bucci, R.Fieschi, G.Guidi: Phys. Rev. **148**, 816 (1966)
2.250 A.Taylor: Phys. Status Solidi **37**, 401 (1970)
2.251 W.M.Ziniker, J.K.Merrow, J.I.Meuller: J. Phys. Chem. Sol. **33**, 1619 (1972)
2.252 A.Scharmann (ed.): Proc. 5th Conf. on Luminescence and Dosimetry (Universität Gießen, Germany 1977)
2.253 S.A.Yalof, P.Hedvig: Therm. Chem. Acta **17**, 301 (1976)
2.254 J.Kiessling, A.Scharmann: Phys. Status Solidi **32**, 459 (1975)
2.255 G.K.Born, R.J.Grasser, A.Scharmann: Phys. Status Solidi **28**, 583 (1968)
2.256 G.Bonfiglioli: "Criteria for the Design of Thermoluminescence Apparatus" in *Thermoluminescence of Geological Materials*, ed. by D.J.McDougall (Academic Press, New York 1968) p. 169
2.257 M.Castagne, J.Gasiot, J.P.Fillard: J. Phys. E**11**, 345 (1978)
2.258 K.Inabe: J. Phys. E**9**, 931 (1977)
2.259 Z.Chvoj, P.Pokorny: Czech. J. Phys. B**28**, 446 (1978)
2.260 V.Kopane, V.E.Shubin: Instrum. Exp. Tech. **19**, 1228 (1976)
       R.K.Chaudhary, L.Kishore: Cryogenics (GB) **17**, 419 (1977)
2.261 Sovan Han Chor: Thesis, Univ. of Montpellier, France (1975)
2.262 E.B.Podgorsak, P.R.Moran: Phys. Rev. B**8**, 3405 (1973)
2.263 P.R.Moran, D.E.Fields: J. Appl. Phys. **45**, 3266 (1974)
2.264 L.Langouet: C. R. A. S. B**268**, 418 (1969)
2.265 M.de Murcia: Thesis, Univ. of Montpellier, France (1976)
2.266 I.Bonstead, A.Charlesby: Proc. R. Soc. (London) A**316**, 291 (1970)
2.267 I.M.Blair, J.A.Edington: Proc. Apollo XI, Science Conf. **3**, 2001 (1975)
2.268 H.Kluge, V.Siegel: Proc. 5th Intern. Symp. on Exoemission and Dosimetry, Zvivkov, CSSR, ed. by A.Bohun, A.Scharmann (Universität Gießen, 1976)
2.269 P.Bräunlich, A.Scharmann: Nukleonik **4**, 65 (1962)
2.270 F.H.Attix (ed.): Proc. Intern. Conf. on Luminescence Dosimetry, Stanford Univ., 1965 (CONF-650637, NBS, Springfield, Virginia 1967)
2.271 R.Chen: Private communication

2.272 M. Böhm, A. Scharmann: Phys. Status Solidi (a) **5**, 563 (1971)
2.273 M. M. Perlman, S. Unger: J. Phys. D**5**, 2115 (1972)
2.274 C. Popescu, H. K. Henisch: Phys. Rev. B**11**, 1563 (1975)
2.275 H. K. Henisch, C. Popescu: Nature (London) **257**, 363 (1975)
2.276 C. Popescu, H. K. Henisch: J. Phys. Chem. Sol. **37**, 47 (1976)
       H. K. Henisch: J. Electrost. **3**, 233 (1977)
2.277 A. Samoc, M. Samoc, J. Sworakowski: Phys. Status Solidi (a) **39**, 337 (1977)
2.278 H. Arbell, A. Halperin: Phys. Rev. **117**, 45 (1960)
       V. Hoffman: Optical Spectra **12**, 60 (1978)
2.279 M. Böhm, A. Scharmann: Z. Phys. **22**, 313 (1975); Phys. Status Solidi **22**, 143 (1974)
2.280 G. Sawa, K. Kawabe, D. C. Lee, M. Ieda: Jpn. J. Appl. Phys. **13**, 1547 (1974)
2.281 A. Charlesby, R. H. Partridge: Proc. R. Soc. (London) **275**, 312 (1964)
2.282 M. A. Carter, J. Woods: J. Phys. D**6**, 337 (1973)
2.283 R. H. Bube, H. E. McDonald: Phys. Rev. **128**, 2071 (1962)
2.284 I. K. Kaul, A. K. Bhattacharya, D. K. Ganguli, B. F. H. Hess: Mod. Geol. (GB) **6**, 87 (1977)
2.285 E. P. Manche: Rev. Sci. Instr. **49**, 715 (1978)
2.286 M. Castagne: Thesis, Univ. of Montpellier, France (1977)
2.287 P. L. Mattern, K. Lengweiler, P. W. Levy, P. D. Esser: Phys. Rev. Lett. **24**, 1287 (1970)
2.288 W. L. Medlin: "The Nature of Traps and Emission Centers in Thermoluminescence Rock
       Materials" in *Thermoluminescence of Geological Materials*, ed. by D. J. McDougall
       (Academic Press, New York 1968) p. 193
2.289 I. K. Bailiff, D. A. Morris, M. J. Aitken: J. Phys. E**10**, 1156 (1977)
2.290 S. Sanzelle, J. Fain, J. C. Vennat: Rev. Phys. Appl. **12**, 1747 (1977)
2.291 A. C. Carter: Phys. Med. Biol. **22**, 1022 (1977)
2.292 J. P. Fillard, M. de Murcia, J. Gasiot, Sovan Han Chor: J. Phys. E**8**, 993 (1975)
2.293 D. Grogan, J. P. Ashmore, R. Bradley, J. G. Scott: Proc. 5th Intern. Symposium on
       Luminescence and Dosimetry, ed. by A. Scharmann (Universität Gießen, Germany)
2.294 C. Dupuis: Thesis, University of Lille, France (1976)
2.295 P. Scaramelli: Nuovo Cimento **XLV** B, 119 (1966)
2.296 P. D. Townsend, C. D. Clark, P. W. Levy: Phys. Rev. **155**, 908 (1967)
2.297 E. Sonder, W. A. Sibley, W. C. Mallard: Phys. Rev. **159**, 755 (1967)
2.298 B. Bosacchi, R. Fieschi, P. Scaramelli: Phys. Rev. **138**, A 1760 (1965)
2.299 J. Tournon, P. Berge: Phys. Status Solidi **5**, 117 (1964)
2.300 A. A. Braner, M. Israeli: Phys. Rev. **132**, 2501 (1963)
2.301 R. Biederbick, G. Born, A. Hofstaetter, A. Scharmann: Phys. Status Solidi **28**, 583 (1968)
2.302 A. Hofstaetter, J. Planz, A. Scharmann: Z. Naturforsch. **32**a, 957 (1977)

Note Added in Proof

Recently evidence has been accumulating that demonstrates the utility of TSL/TSC techniques in trap level spectroscopy beyond the possibilities described in this chapter. Not only properties of trap levels (situated above the demarcation levels in the case of electron traps) but also those of recombination centers can be studied by these methods. *Fillard* and co-workers have shown that careful measurement of the ratio $\mu_n(T)/\eta\gamma(T)$ in (2.25) by the method of "fractional emptying" together with an independent determination of the temperature behavior of μ_n yields the temperature dependence of $\gamma(T)$ and of the capture cross section for radiative capture of free electrons in recombination centers [J. P. Fillard, J. Gasiot, J. C. Manifacier: Phys. Rev. B **18**, 4497 (1978)].

In an experiment with $CdF_2:Sm^{3+}$ the same technique was employed to determine the activation energy for non-radiative capture of electrons by neutral fluorine interstitials acting as recombination centers (M. deMurcia: USTL, Montpellier, France, private communication, July 1979).

These two examples are of considerable interest for they illustrate the importance of simultaneously measuring a number of different output functions during non-isothermal temperature scans. We are convinced that the future development of TSL/TSC methods as investigative tools of electronic properties of semiconductors and insulators will benefit greatly from similar skillful studies of correlated phenomena during thermally stimulated relaxation in these materials.

3. Space-Charge Spectroscopy in Semiconductors

By D. V. Lang

With 24 Figures

In this chapter we will discuss the special case of thermally stimulated trap-related phenomena in semiconductors. In particular, we will focus on the effects of trapping and emission of carriers at deep energy levels which are located in the space-charge layer of a pn junction or Schottky barrier. A space-charge layer is essentially depleted of mobile carriers and hence is very much like the bulk insulators discussed in Chap. 2. In addition, however, such layers have certain unique features which both simplify the analysis and make possible a totally different class of experimental techniques. We will discuss the various electrical measurements – capacitance, current, and voltage – which have been used to study deep levels in the space-charge layers associated with semiconductor junctions.

Much of the general groundwork for our discussion has been laid in Chap. 2, especially the rate equations which describe the dynamics of carrier capture and emission at deep energy levels. Therefore, we will begin by pointing out those aspects of thermally stimulated processes which are unique to semiconductor space-charge layers. We will then discuss the transient electrical response of semiconductor junctions containing traps, including such topics as the spatial and temporal variations of sensitivity, trap concentration profiles, and the special problems associated with large trap concentrations. Our treatment of the specific experimental techniques will be divided into two parts. First, we will discuss the so-called single-shot techniques which are the simplest and were historically the first developed. Secondly, we will examine in some detail the more recently developed synchronous-detection methods, which are generally referred to as Deep-Level Transient Spectroscopy (DLTS). The discussion of the implementation and interpretation of the various experimental techniques will be followed by a section on carrier capture measurements. This will include a discussion of the corrections to the measured activation energy which are necessary if the capture cross section or energy level are temperature dependent. Finally, we will conclude the chapter with a brief discussion of the instrumentation useful in the practical application of space-charge spectroscopy.

3.1 Unique Aspects of Thermally Stimulated Processes in Semiconductor Junction Space-Charge Layers

3.1.1 Ideal Space-Charge-Layer Properties

The existence of a space-charge layer at a pn junction or Schottky barrier is a general characteristic of semiconductors. Such a layer is necessary to create the electrostatic potential variation needed to counteract the diffusion potential of the carriers across the junction and hence equalize the Fermi level through the material, i.e., maintain thermal equilibrium. The *ideal* space-charge layer (i.e., without traps) for an asymmetric p^+n step junction is shown in Fig. 3.1. For simplicity we will confine most of our examples in this chapter to the case of p^+n junctions or Schottky barriers on n-type material (for which Fig. 3.1 is also applicable). The extension to n^+p junctions is trivial; the situation for arbitrary junction profiles is usually straightforward but often tedious and beyond the scope of this chapter.

We may think of a semiconductor space-charge layer as somewhat like a variable-width insulator, typical thicknesses are in the range of about 0.1 to $10\,\mu m$. Therefore, with reverse bias voltages of 1–100 V, the maximum electric field F_m in the layer is very large, typically of order 10^4–$10^5 V\,cm^{-1}$. From a

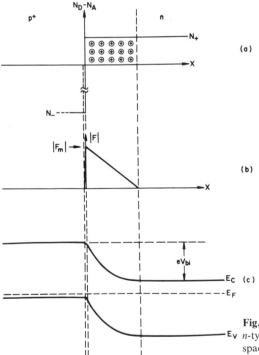

Fig. 3.1a–c. Ideal p^+n step junction (or n-type Schottky barrier) at zero bias (a) space-charge distribution, (b) electric field distribution, (c) band-bending diagram

straightforward solution of Poisson's equation for the asymmetric junction in Fig. 3.1 it can be shown [3.1] that

$$|F_m| = \frac{eN_+ W}{\varepsilon}$$
(3.1)

where

$$W = \left| \sqrt{\frac{2\varepsilon(V_{bi} + V)}{eN_+}} \right.$$
(3.2)

is the width of the space-charge layer, V_{bi} is the built-in potential, V is the applied reverse bias voltage, ε is the dielectric constant of the semiconductor, and e is the electronic unit charge. The capacitance is related to W in exactly the same manner as a parallel plate capacitor, namely

$$C = \frac{\varepsilon A}{W},$$
(3.3)

where A is the area of the junction.

Thus we see that the two principal differences between space-charge layers and bulk insulators are: 1) the width of the layer can be readily adjusted by varying the bias voltage according to (3.2), and 2) a large, spatially varying electric field exists in the space-charge layer even at zero applied bias. The large electric field means that carriers thermally emitted from traps in the space-charge layer are swept out of the layer in a very short time, typically 10^{-10} to 10^{-12} s. Therefore, retrapping effects can be neglected and the analysis of thermal emission transients is considerably simplified as compared to bulk phenomena. The large electric field might also affect the physical properties of the localized defects giving rise to the trap levels. This is both good and bad, for it means on the one hand that it cannot necessarily be assumed that the properties observed are characteristic of the deep level in bulk material at zero field. On the other hand, any electric field effects which exist may be turned to advantage and used to probe the internal symmetry of the localized defect wave function. This latter possibility has not been fully explored but would seem to be a fruitful approach.

The voltage-variable-width feature inherent in semiconductor space-charge layers is basic to nearly all forms of space-charge spectroscopy. Its implications will be discussed in the remainder of this section.

3.1.2 Capacitance and Current Detection of Trap Levels

Trap levels in space-charge layers may be detected either by their effect on the junction *current* or *capacitance*, as illustrated in Fig. 3.2. The current-detection case is almost exactly analogous to the thermally stimulated trap-emission

RF CAPACITANCE MEASUREMENTS
(WIDTH OF SPACE–CHARGE REGION)

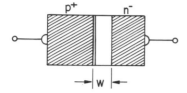

CAPACITOR DISCHARGE (CURRENT) MEASUREMENTS
(dQ/dt OF SPACE–CHARGE REGION)

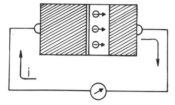

Fig. 3.2. Space-charge spectroscopy of deep levels in semiconductor junctions may be based on either capacitance (top) or current (bottom) measurements

currents which would be seen in a very thin bulk insulator in the absence of retrapping. The capacitance-detection case, which yields the same basic physical information about traps as does current detection, is unique to semiconductor space-charge layers and has no analogy in the bulk insulator case. It is clear from (3.2) that a change in the charge density in the space-charge layer will induce a corresponding change in the width, and hence in the capacitance, of the layer. Thus the space-charge-layer capacitance is a direct measure of the total charge in this layer. Consequently, if the concentration of electrons trapped at deep levels is changed, either by thermal or optical capture or emission, this variation in trapped charge can be readily monitored by observing the corresponding change in the junction capacitance at constant bias voltage. All forms of capacitance spectroscopy are based on this fact. We will discuss the quantitative relationships between the space-charge-layer capacitance and the trapped charge in Sect. 3.2, the purpose of this section is to place semiconductor capacitance measurements into proper perspective relative to the bulk measurements discussed in the other chapters of this volume.

We might view capacitance measurements as utilizing a sort of "charge transducer" effect wherein the semiconductor junction is a combined deep-level sample holder *and* transducer which directly measures the occupation state of the deep levels. Thus the focus in capacitance measurements is *opposite* to that of thermally stimulated conductivity or photoconductivity. In these latter bulk phenomena one measures the *free* carrier concentration and infers from this the properties of traps and recombination centers. In junction capacitance measurements, on the other hand, one directly measures the *trapped* carrier concentration. Thus the trap signals are more-or-less independent from each other and the analysis is considerably simplified. In the case of small trap concentrations where the trap-induced capacitance change ΔC is much less than the overall junction capacitance C, i.e., when $\Delta C \ll C$, the thermal emission capacitance transients are simple exponential decays which are directly proportional to the rate equations for n, the electron occupation of the trap (see Chap. 2). This is the case of the ideal transducer, where the presence of deep levels does not measurably affect the Δn vs $-\Delta C$ relationship. For *large* trap concentrations, however, the situation is much more complex and the capacitance and current transients are nonexponential (see Sect. 3.2.5). In this chapter, unless otherwise specified, we will be considering the $\Delta C \ll C$ case.

The capacitance measurements which we will be discussing are in the high-frequency limit. In this case the measurement frequency is high enough (typically 1 MHz) so that the oscillating voltage applied to the junction by the capacitance meter or bridge is unable to induce changes in the occupancy of the deep levels. In Sect. 3.4.3 we will discuss the opposite case, where via capture and reemission of carriers the deep-level occupation *can* follow the measurement frequency.

3.1.3 Bias Voltage Pulses

The voltage-variable space-charge-layer width also makes possible another unique feature of space-charge spectroscopy, namely, the so-called majority- and minority-carrier pulses. These voltage pulses superimposed on the steady-state reverse bias make it possible to almost totally decouple the measurements of capture and emission processes at deep levels. This is shown most readily in Fig. 3.3 for a majority-carrier pulse in a p^+n junction. Under steady-state conditions the traps in the upper half of the gap (majority-carrier traps in n-type material) are thermally empty. If the bias voltage is momentarily reduced, as in the middle of Fig. 3.3, part of the region which was formerly within the space-charge layer is now in neutral material so that the traps are below the Fermi level. During the time that the bias is at this lower value the deep levels may capture majority carriers and tend to become filled. Immediately after the pulse, the deep levels are again within the space-charge layer where the capture rate is essentially zero. The capacitance will have changed due to the captured carriers, however, and as these carriers are thermally emitted a capacitance or current transient will be produced. Thus a majority-carrier pulse is essentially a means whereby the majority-carrier concentration can be turned *on* and *off* in a small volume of the sample. Majority-carrier *capture* dominates when the pulse is *on*, whereas majority-carrier *emission* dominates when the pulse is *off*.

A minority-carrier pulse (or injection pulse) corresponds to the diode being forward biased during the pulse so that both majority *and* minority carriers are introduced. This makes possible minority-carrier capture, but one should bear in mind that the steady-state occupation of the level corresponds to a balance between majority- and minority-carrier capture. We will discuss these points in more detail in Section 3.5. An injection pulse could also be generated by using

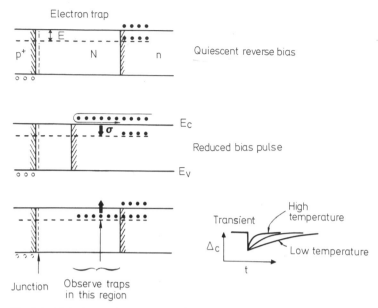

Fig. 3.3. Majority-carrier pulse for p^+n junction with band bending omitted for clarity. Trap occupation and space-charge-layer width are indicated corresponding to the conditions before, during, and after the pulse (from top to bottom of the figure, respectively)

an optical pulse or a pulse of high-energy electrons in an electron microscope. Such techniques are useful in Schottky barriers where forward bias injection is not possible.

Finally, let us point out an additional feature of bias voltage variations, namely, the possibility of obtaining a spatial profile of the trap concentration. This can be done by either varying the steady-state bias voltage or by varying the amplitude of the majority-carrier pulse. In either case the spatial region in which the deep levels are observed is varied and the spatial profile $N(x)$ may be obtained as discussed in Sect. 3.2.4.

3.2 Transient Response of a Semiconductor Junction with Traps

3.2.1 Spatial Variations of Sensitivity

The magnitude of a capacitance or current transient corresponding to thermal emission of carriers from a trap depends on the location of the trap within the space-charge layer. Let us calculate the relative capacitance change $[\Delta C/C]_x$ for $n(x)$ trapped electrons in the interval Δx at x, where $0 < x < W$ corresponds to positions within the n-side of the space-charge layer of a p^+n junction or n-type Schottky barrier such as shown in Fig. 3.1. From Poisson's equation it is straightforward to show that the voltage change induced by trapping $n(x)$ electrons at x is

$$\Delta V = \frac{e}{\varepsilon}[N_+ W \Delta W - n(x) x \Delta x] \tag{3.4}$$

where N_+ is the positive space-charge concentration at W [note that $N_+(x)$ need not be uniform here, as it was in the example in Fig. 3.1]. Since we are interested in capacitance changes at constant bias, we set $\Delta V = 0$ in (3.4) and have

$$\left[\frac{\Delta C}{C}\right]_x = -\frac{n(x)}{N_+ W^2} x \Delta x. \tag{3.5}$$

This result is the capacitance charge-transducer sensitivity calibration for the $\Delta C \ll C$ case. Note that the sensitivity of the junction capacitance to trapped charge varies linearly from *zero* at the junction to a maximum value at the outer edge of the space-charge layer. This means that for inhomogeneous trap distributions capacitance measurements are very insensitive if the traps are located near the junction (near the *surface* for a Schottky barrier). The relationship in (3.5) also means that *motion* of defects within the space-charge layer will create a capacitance change if the motion produces a net change in the first moment of the defect distribution.

It is also of interest to consider the spatial variations of sensitivity for current transients. We will show that for the case of majority-carrier emission the sensitivity is exactly the opposite of (3.5), i.e., maximum at the junction and zero at the outer edge. Let ΔQ be the charge due to the emission of $\Delta n(x)$ electrons from traps in the interval Δx at x, thus

$$\Delta Q = e \Delta n(x) \Delta x. \tag{3.6}$$

For the case of electron emission in n-type material, some fraction of this charge ΔQ_+ never leaves the space-charge layer since it is needed at the outer edge to produce the accompanying capacitance change in (3.5), i.e.,

$$\Delta Q_+ = eAN_+ \Delta W. \tag{3.7}$$

The net charge which is detected in the external circuit ΔQ_{ext} is the difference between ΔQ and ΔQ_+. With the aid of (3.5), this may be written

$$\Delta Q_{ext} = eA \Delta n(x) \Delta x \left(1 - \frac{x}{W}\right). \tag{3.8}$$

For *hole* emission, on the other hand, the net charge reaching the external circuit is ΔQ_+. Thus for a thermal emission transient the measured current density due to deep levels in the interval Δx at x is

$$J(x) = e\Delta x \left\{ \alpha n(x) \left(1 - \frac{x}{W} \right) + \alpha^* [N - n(x)] \frac{x}{W} \right\}, \tag{3.9}$$

where α and α^* are the thermal emission rates for electrons and holes, respectively. Note that in steady state $\alpha n = \alpha^* (1 - n)$ so that the spatially varying terms in (3.9) cancel each other.

The spatial variation of sensitivity for current transients in a p^+n junction in (3.9) is the opposite of the capacitance case for *electron* emission but has the same spatial shape for *hole* emission. The spatially varying terms in (3.9) are actually the displacement current associated with the capacitance change in (3.5). It is therefore not surprising that the sensitivity variations are linear in both cases.

3.2.2 Temporal Variations of Sensitivity

Both capacitance and current transients in the $\Delta C \ll C$ limit are proportional to the trapped electron concentration $n(t)$ for electron emission, or trapped hole concentration $[1 - n(t)]$ for hole emission. The time dependence of n is given by

$$\frac{n(t)}{N} = \begin{cases} 1 & t \leq 0 \\ \dfrac{\alpha^*}{\alpha + \alpha^*} \left\{ 1 + \dfrac{\alpha}{\alpha^*} \exp[-(\alpha + \alpha^*)t] \right\} & t > 0 \end{cases} \tag{3.10}$$

for the case where a majority-carrier pulse at $t = 0$ completely fills the trap (for the n-type example), and by

$$\frac{n(t)}{N} = \begin{cases} 0 & t \leq 0 \\ \dfrac{\alpha^*}{\alpha + \alpha^*} \{ 1 - \exp[-(\alpha + \alpha^*)t] \} & t > 0 \end{cases} \tag{3.11}$$

for the case where an injection pulse at $t = 0$ empties the trap (fills it with holes).

In the context of space-charge spectroscopy, a trap for which $\alpha \gg \alpha^*$ is called a "majority-carrier trap" in n-type material and a "minority-carrier trap" in p-type. The opposite is true for a trap which has $\alpha \ll \alpha^*$. This notation is useful because for capacitance measurements the *sign* of the transient is always *negative* for majority-carrier emission and always *positive* for minority emission, independent of the conductivity type (see also Sect. 3.3.1). Current transients, however, are always of the same sign.

According to (3.5) we see that for $\Delta C \ll C$ the capacitance signal is a direct measure of $n(t)$ and therefore is described by (3.10) or (3.11). The current signal, on the other hand, is proportional to $\alpha n(t)$, for the example of electron emission in n-type material in (3.9). Therefore, the capacitance sensitivity is *independent* of the time constant of the transient whereas the current sensitivity is *directly proportional* to the transient rate. This means that current measurements are most sensitive for fast transients while capacitance is most sensitive for slow transients. We have found that the crossover point between capacitance and current sensitivity for the same trap in the same sample with identical instrumental bandwidths is for time constants roughly of order 1–10 ms.

3.2.3 Transient Magnitudes for Uniformly Distributed Trap Concentrations

In order to evaluate the magnitude of the total signal in a particular sample we must integrate (3.5) or (3.9) over the region of the space-charge layer where emission is actually taking place. To do this we must discuss in more detail the band-diagram of the asymmetric junction case which we are considering. This is shown for zero bias and reverse bias in Fig. 3.4 for the case of uniformly doped *n*-type material, i.e., for a p^+n junction or a Schottky barrier on n-type material.

We must consider two basic regions of the space-charge layer. At the outer edge of the layer there is a transition between the conducting bulk material and the central part of the space-charge layer which is essentially depleted of free carriers. For a particular deep level this gives rise to the so-called *edge region* of thickness λ, defined by the point where the trap level E_T crosses the Fermi level E_F. From Poisson's equation it can readily be shown that

$$\lambda = \sqrt{\frac{2\varepsilon(E_F - E_T)}{e^2 N_+}} \tag{3.12}$$

if N_+ is uniform. Note that in this case λ is independent of bias voltage.

In the edge region the deep level is below the Fermi level and therefore filled with electrons in thermal equilibrium, i.e., there are sufficient free carriers present so that the electron capture rate c_n is much larger than the electron thermal emission rate α. There is also a small additional voltage-dependent edge region λ_v^* near the junction where a sufficient number of free holes have spilled over to make $c_p \gg \alpha^*$ so that the trap is always empty. In the remainder of the space-charge layer the trap level is above the Fermi level and its steady-state occupation is defined by the ratio of thermal emission rates according to (3.10, 11). This inner carrier depletion region is where the capacitance and current transients are generated.

Note that it may seem inconsistent for us to consider here the presence of free carriers within the edge regions of the space-charge layer, but to neglect such carriers in the calculation of W in (3.2). This approach is justified, however,

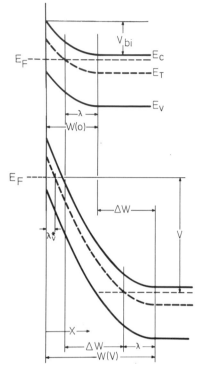

Fig. 3.4. Band-bending diagram for p^+n step junction (or n-type Schottky barrier) at zero bias (top) and at a reverse bias of $V = 3.5\,V_{bi}$ (bottom)

since it has been shown [3.2, 3] that neglect of free carriers in the region $0 < x < W$ introduces negligible errors in calculating the reverse bias capacitance. Thus even though there are sufficient carriers within the edge regions of the space-charge layer to affect the deep-level occupation, there are not enough to significantly change the shape of the junction band bending.

The total capacitance signal produced by a majority-carrier pulse which brings the diode bias to zero for a time long enough to completely fill a trap of concentration N is obtained by integrating (3.5) from $W(0) - \lambda$ to $W(V) - \lambda$, thus

$$\frac{\Delta C}{C} = -\frac{N}{2N_+}\left[1 - 2\frac{\lambda}{W(V)}\left(1 - \frac{C(V)}{C(0)}\right) - \left(\frac{C(V)}{C(0)}\right)^2\right]. \tag{3.13}$$

If one neglected the edge region, the bracketed term in (3.13) would be equal to 1 and the $\Delta C/C$ relationship would reduce to the simple form often given [3.4]. For the example of Fig. 3.4, however, the bracketed term is equal to 0.46, i.e., zero biasing the junction gives slightly less than *half* of the capacitance change that would have been expected if all of the deep levels within the space-charge layer could have been observed. Thus a neglect of the edge region may introduce substantial error in trap concentration measurements, especially at low bias voltages.

For current transients the integral relationship derived from (3.9) for uniformly distributed traps is most conveniently expressed in terms of the *D parameter* introduced by *Grimmeiss* [3.5], namely

$$J(t) = e(x_2 - x_1)\left\{\frac{1}{D}\alpha n(t) + \left(1 - \frac{1}{D}\right)\alpha^*[N - n(t)]\right\}$$ (3.14)

where

$$\frac{1}{D} = \frac{1}{x_2 - x_1}\int_{x_1}^{x_2}\left(1 - \frac{x}{W}\right)dx = 1 - \frac{x_1 + x_2}{2W}.$$ (3.15)

The positions x_1 and x_2 are the points where the spatially varying hole and electron capture rates, respectively, equal the sum of all relevant emission rates. For thermally stimulated processes $x_1 = \lambda_V^*$ and $x_2 = W - \lambda$. In most cases D is near to and slightly less than 2, but in general D depends on temperature, trap depth and spatial distribution, and the optical intensity for photocurrent transients.

3.2.4 Measuring Trap Concentration Profiles

There are many ways of using the bias-dependent width of the space-charge layer to obtain both deep- and shallow-level profiles. Obviously, one cannot deduce the spatial profile of deep levels without first knowing the shallow-level profile. Many schemes for obtaining $N_+(x)$ from the $C - V$ behavior of the junction have been discussed [3.6–9].

One of the most convenient ways of measuring the spatial profiles of deep levels, which is especially useful for low concentrations, is to record the magnitudes of a series of capacitance transients produced with majority-carrier pulses of increasing amplitude. This method has the effect of probing the deep-level concentration inward from W as the majority-carrier pulse amplitude is increased from zero. If we denote by $\delta(\Delta C/C)$ the incremental change in the relative capacitance signal due to the traps filled by the small change δV in the majority-carrier pulse of amplitude V, then a convenient relationship is given [3.10] by

$$\delta\left(\frac{\Delta C}{C}\right) = \left(\frac{\varepsilon}{eW^2 N_+}\right)\frac{N(x)}{N_+(x)}\delta V$$ (3.16)

where x is the width of the space-charge layer during the pulse and C, W, and N_+ are the values corresponding to the steady-state bias between pulses. This relationship is valid for asymmetric junctions or Schottky barriers if $\Delta C \ll C$.

According to (3.16), a linear ΔC vs V plot implies that the trap profile $N(x)$ has the same shape as the shallow-level profile $N_+(x)$. We will show some examples of the use of this relationship in determining deep-level profiles in Sect. 3.4.5.

3.2.5 Capacitance Relationships for Large Trap Concentrations

Capacitance changes in the concentrated-deep-level limit, i.e., $N \gtrsim N_+/3$, are very complex to analyze. A major difficulty is that the capacitance and current transients recorded at constant bias are nonexponential. Since this limit gives rise to the largest signals, it was historically the first to be discovered. Consequently, many treatments of the various possible capacitance relationships have been given in the literature [3.11–21]. Perhaps the most concise and useful is that which relates the concentration N of a uniformly distributed deep-level of energy E_T to the capacitance C_0 immediately following a zero-biasing majority-carrier pulse and the steady-state capacitance C_∞ [3.17, 19, 20], namely,

$$\frac{N}{N_+} = \frac{\left(\dfrac{C_\infty}{C_0}\right)^2 - 1}{\left[1 - \dfrac{C_\infty}{C_0}\sqrt{\dfrac{E_F - E_T}{e(V_{bi} + V)}}\right]^2}. \tag{3.17}$$

It is possible to overcome the problem of nonexponential transients in this regime by utilizing a clever feedback scheme proposed by *Goto* et al. [3.21]. These authors showed that the bias voltage necessary to maintain the diode capacitance at a constant value in a feedback loop is a direct measure of the trapped charge concentration. Thus a measurement of this feedback voltage will produce an exponential transient even for very large trap concentrations. This effect is dramatically demonstrated by Fig. 3.5 which shows the constant-bias and constant-capacitance transients associated with the emission of electrons from the so-called DX center in $Al_xGa_{1-x}As$; in this case $N \cong 8N_+$ [3.22].

3.3 Single-Shot Techniques

3.3.1 Isothermal Transients

The essential physics of all forms of thermally stimulated space-charge spectroscopy – indeed, all of our previous discussion – is based on the thermal-emission transient at some fixed temperature. The more involved techniques which we will introduce later have the benefits of speed and convenience, useful for the study of complex trap spectra, but they cannot produce raw data of any

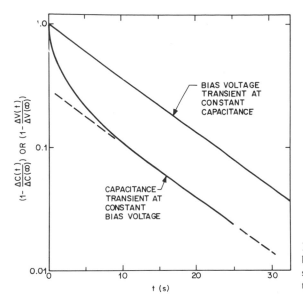

Fig. 3.5. Effect of a large deep-level concentration on the constant-bias and constant-capacitance carrier emission transient

higher quality than can be done with a large number of isothermal transients recorded at many different temperatures.

Isothermal capacitance transients were first proposed as a technique to study traps in semiconductors by *Williams* [3.23], and by *Furukawa* and *Ishibashi* [3.12]. The quantitative expression for a capacitance transient follows from (3.10–12). A capacitance transient due to majority-carrier emission is always *negative*. Figure 3.6 shows such a transient with insets describing the condition of the space-charge layer during the various phases of the transient [3.24]. Similarly, Fig. 3.7 shows a *positive* transient due to minority-carrier emission following an injection pulse [3.24].

Current transients are described by (3.10, 11, 14, 15). Note that the *sign* of current transients is always the same (independent of the nature of the emission process). The current flows in the opposite sense to that in forward bias, thus the voltage drop across the load tends to make the diode bias more positive.

The activation energy $E \equiv E_c - E_T$ for electron emission or $E_A^* \equiv E_T - E_V$ for hole emission can be obtained in the standard manner from the slope of an Arrhenius plot, i.e., a plot of the log of the transient decay rate $\alpha + \alpha^*$ as a function of $1/T$, where T is the absolute temperature at which the transient is recorded. This follows from the relationships

$$\alpha = v \exp(-E/kT) \tag{3.18}$$

and

$$\alpha^* = v^* \exp(-E_A^*/kT) \tag{3.19}$$

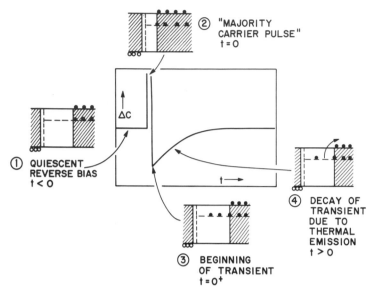

Fig. 3.6. Isothermal capacitance transient for thermal emission from a majority-carrier trap. The insets 1–4 show the conditions of the trap occupation, space-charge-layer width (unshaded), and free carrier concentrations during the various phases of the transient in a p^+n junction

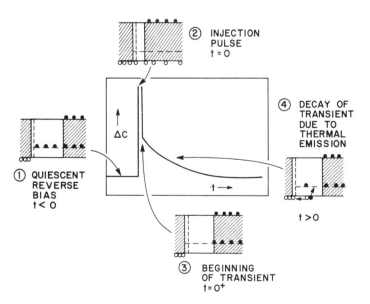

Fig. 3.7. Isothermal capacitance transient for thermal emission from a minority-carrier trap. The insets 1–4 show the conditions of the trap occupation, space-charge-layer width (unshaded), and free carrier concentrations during the various phases of the transient in a p^+n junction

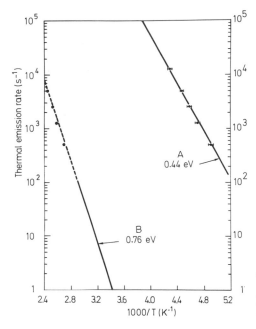

Fig. 3.8. Thermal emission rate α^* versus $1000/T$ for holes trapped on two commonly occurring deep levels of unknown origin in LPE n-GaAs. The data points are DLTS peak positions from Fig. 3.14, the solid lines are isothermal transient data

and the fact that either α or α^* usually dominates the transient rate. An example of such a plot for the hole-emission transients of two common hole traps in GaAs is shown in Fig. 3.8 [3.25]. The activation energy measured in this way is a valid parameter of the trap, but it may not be the true depth of the trap. We will discuss the problems of interpreting activation energies in Sect. 3.5.4 after we have discussed capture processes.

3.3.2 Irreversible Thermal Scans

This class of techniques bears a strong resemblance to many of the thermally stimulated phenomena observed in bulk material. In general, one first prepares the traps in an appropriate initial charge state at some low temperature where thermal emission is negligible. This may be done, for example, by cooling the sample with zero bias (so that all traps are filled with majority carriers) and then applying a reverse bias at low temperature. It is also possible to load the traps with minority carriers by optical excitation or by cooling the sample with forward bias (for the case of a pn junction). Following this initial preparation, the sample is heated at some constant rate q and one observes either current peaks or capacitance steps when the various levels emit their trapped carriers. This is schematically illustrated in Fig. 3.9.

The thermally stimulated current (TSC) method for semiconductor junctions was first proposed by *Driver* and *Wright* [3.26], while the thermally stimulated capacitance (TSCAP) method was first discussed by *Carballes* and

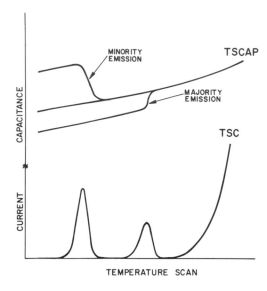

Fig. 3.9. Schematic representation of typical TSCAP and TSC data for a sample containing a majority-carrier trap of concentration N and a shallower minority-carrier trap of concentration $2N$. The increase in current at higher temperatures in the TSC case is the dark current due to steady-state generation and leakage

Lebailly [3.27]. A number of different single-shot thermal techniques have been employed with great success by *Sah* and co-workers to study deep levels in silicon [3.14, 28]. *Buehler* and *Phillips* [3.29] have recently analyzed TSC and TSCAP experiments and have shown that the temperature T_m of the TSC maximum or midpoint of the TSCAP step is related to the activation energy by

$$E = kT_m \ln \left[\frac{vkT_m^2}{q(E + 2kT_m)} \right] \tag{3.20}$$

for the example of electron emission. Thus if $q = 1 \, \text{K s}^{-1}$ and $v = 10^{12} \, \text{s}^{-1}$ we have $E = 28.9 \, kT_m$ for $T_m = 100$ K and $E = 29.5 \, kT_m$ for $T_m = 200$ K.

The thermally scanned techniques are less accurate for determining activation energies than is the measurement of the isothermal transient at several temperatures. The scanning techniques do, however, have an important advantage for the case of samples with several different deep levels. Namely, they make it possible to quickly survey nearly the entire range of trap energies and produce a spectrum of the traps in a particular sample. The TSC and TSCAP methods work reasonably well as survey techniques when the trap concentrations are fairly large, e.g., $N \gtrsim 0.1 |N_D - N_A|$, and the trap energies are fairly deep, e.g., $E \gtrsim 0.3$ eV. In general, one would expect TSCAP measurements to have a better signal-to-noise ratio than TSC because these scanning techniques are in the low-emission-rate limit of temporal sensitivity variations (see Sect. 3.2.2). Because of this temporal variation, the magnitudes of TSC peaks depend on the heating rate while TSCAP steps do not. In addition, as shown by the current increase at higher temperatures in Fig. 3.9, the TSC technique is subject to considerable problems due to junction leakage or steady-state generation currents as the temperature is increased; TSCAP, on the other hand, is less sensitive to diode quality. Another advantage of TSCAP over TSC is a characteristic of capacitance vs current measurements in general, namely, the ability to distinguish between majority- and minority-carrier emission by the *sign* of the capacitance change (see Fig. 3.9). However, TSCAP measurements are difficult if the change in the steady-state capacitance with temperature is much greater than the capacitance steps due to thermal emission of carriers; the TSCAP steps are then superimposed on a steeply sloping baseline and are very difficult to resolve. There may be situations, however, in which TSC is preferable over TSCAP, e.g., majority-carrier traps very near the junction (see Sect. 3.2.1) or traps located in the *i* layer of a p-i-n junction (which induce very small changes in C).

Perhaps the major virtue of the single-shot scanning techniques is their relative simplicity. The synchronous detection DLTS techniques to be discussed in the next section are better survey techniques, but they are somewhat more complex to set up. However, in most cases the added complexity is well worth the effort.

3.4 Deep-Level Transient Spectroscopy (DLTS)

3.4.1 The Rate-Window Concept

The basic idea of the DLTS technique is the rate-window concept. This is illustrated in Fig. 3.10. If we consider a train of repetitive bias pulses applied to the sample, we then have a signal which consists of a series of transients with a constant repetition rate. As the temperature is varied, the time constant of the transients varies exponentially with $1/T$, as discussed in Sect. 3.3.1 and shown in Fig. 3.10. Let us assume that there exists a "black box" with the characteristic that it gives no response to this sequence of transients unless the time constant of the transient decay is near some preselected value – the inverse of the "rate window". The output of this black box as the temperature of the sample is scanned is then a series of peaks as shown in Fig. 3.10. Such a plot is called a DLTS *spectrum*.

A DLTS spectrum looks superficially like the signals produced by the single-shot scanning techniques in the last section. However, there are several important differences. First, the DLTS scan is *reversible* and does not depend on the magnitude of the heating rate. Thus one can scan *up* or *down* in temperature at any rate and can even *stop* the scan on a peak to study the capture properties or spatial profile of that particular trap. This latter possibility is a tremendous advantage relative to the irreversible thermal scan methods. Another advantage is that the baseline is always flat, i.e., there is good common-mode rejection of the steady-state capacitance variations. Therefore it is possible to obtain extremely high sensitivity, e.g., $\Delta C/C \sim 10^{-5}$ for $C \sim 100\,\text{pF}$. This sensitivity may be readily increased by a slower scan with a

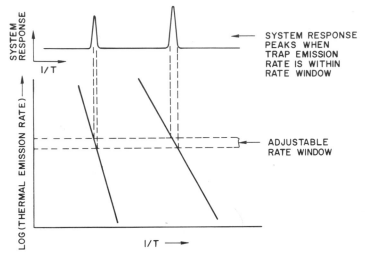

Fig. 3.10. The rate-window concept, which is the basic idea of the DLTS method. This illustrates the peak in the system output response which occurs whenever a component of the decay rate of the input transient signal corresponds to the rate selected by the window

longer final time constant, since the position and shape of the DLTS peaks are independent of the scan rate.

The DLTS method applies equally well to either capacitance or current transients – indeed, to *any* thermally stimulated process which can be repetitively generated by a pulsed excitation. However, there are some differences. First, since the DLTS rate window is usually in the ms to µs regime, current transients are likely to be more sensitive than capacitance transients (see Sect. 3.2.2). This fact is essential to the success of high-resolution DLTS measurements made in a scanning electron microscope, where there are extreme signal-to-noise limitations (see Sect. 3.4.6). However, the price paid for this added sensitivity is loss of the possibility to differentiate between majority- or minority-carrier emission. A second feature of *current* DLTS is the fact that the peak is shifted by somewhat less than half of a linewidth to higher temperature relative to a *capacitance* DLTS peak for the same rate window, i.e., the rate window must be defined slightly differently for current transients. This shift is due to the fact that the magnitude of the current transient is proportional to the thermal emission rate and thus increases very rapidly with temperature, effectively skewing the line-shape toward higher temperature.

3.4.2 Rate-Window Instrumentation

There are a number of ways in which the DLTS rate-window concept may be achieved in practice. The first method proposed in the original DLTS experiments involved the use of a dual-gated integrator (double boxcar) [3.25].

In this method the transient amplitude is sampled at two times t_1 and t_2, after the pulse, as shown in Fig. 3.11. The DLTS signal is the *difference* between the transient amplitude at these two times. Such a difference signal is a standard output feature of a double boxcar. As we can see in Fig. 3.11, there is zero difference between the signal at these two gates for either very slow or very fast transients, corresponding to low or high temperature, respectively. However, when the transient-time constant τ is on the order of the gate separation, a difference signal is generated and the boxcar output passes through a maximum as a function of temperature. This is a DLTS peak. For capacitance measurements the rate window can be expressed in terms of the transient-time constant giving rise to the maximum double-boxcar output [3.25], namely,

$$\tau_{max} = \frac{t_1 - t_2}{\ln(t_1/t_2)}. \tag{3.21}$$

Equation (3.21) is a very good approximation for the case of relatively wide boxcar gates as well, provided t_1 and t_2 are taken as the *midpoint* of each sampling gate. In fact, the best way to use a boxcar is with relatively wide gates. We have found experimentally that the signal-to-noise ratio is proportional to the square root of the gate width, all else being unchanged. Thus the gate configuration shown in Fig. 3.12a might be a typical operating condition.

An alternate means of implementing the rate window is a lock-in amplifier [3.4, 30, 31, 56]. This is equivalent to having the sinusoidally weighted gates shown in Fig. 3.12b. In order to achieve this condition, however, it is necessary to properly adjust the phase of the lock-in relative to the transient signal. An operational definition of the correct phase relationship is to adjust the lock-in so as to be in quadrature with the bias pulse with the pulse generator triggered by the lock-in reference signal. This corresponds to obtaining a null of the lock-in output for the pulse overload recovery transient under conditions where the trap emission transient is zero.

However, a possible problem with the lock-in DLTS method is that some small contribution from the very large pulse recovery transient cannot be totally avoided. This results in excess noise and in possible baseline shifts during the temperature scan. Note that the duration of the pulse overload recovery depends on the output bandwidth of the capacitance meter and may be considerably longer than the actual pulse. Such a problem does not exist for the boxcar or correlator method, however, since the signal is not sampled during the pulse-overload portion of the waveform. The problem may be reduced, but not totally eliminated, in the lock-in case by utilizing a transmission gate before the input of the lock-in to block the signal during the overload transient [3.31]. Such a gate will cut out the jitter of the overload transient but will not stop the DLTS baseline from shifting as the steady-state capacitance of the sample changes with temperature. To solve this latter problem one needs an active dc baseline restorer such as used in the exponential correlator developed by *Miller* et al. [3.30].

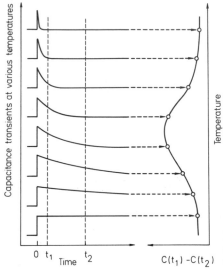

Fig. 3.11. Implementation of a DLTS rate window by means of a double-boxcar integrator with gates set at times t_1 and t_2

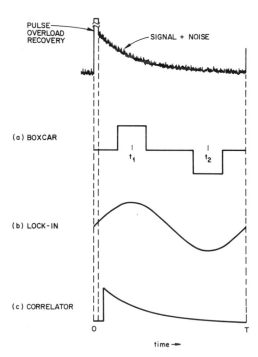

Fig. 3.12a–c. Diagram comparing a typical capacitance transient input signal with three different rate-window operations

The DLTS line shape for the lock-in method has the form [3.4, 30]

$$A(y) = \frac{2\pi(1 - e^{+y})}{4\pi^2 + y^2},$$ (3.22)

where $y = T/\tau$ with T being the reciprocal of the lock-in frequency and τ the transient time constant. The maximum of this line-shape function occurs for $\tau_{max} = 0.424\,T$. This is the lock-in DLTS rate-window relationship for capacitance transients. Note that the maximum of (3.22) differs from the rate window defined by (3.21) for the boxcar gate settings in Fig. 3.12a ($t_1 = T/4, t_2 = 3\,T/4$) by only 7%. This corresponds to a temperature error of $\lesssim 1$ K, which is negligible in most cases. The proper rate window for current transients under these conditions is 2.2 times faster than the capacitance rate window. This is a shift of less than half of a linewidth in (3.22).

In an effort to obtain the highest possible sensitivity in a DLTS measurement, *Miller* et al. [3.30] have analyzed the various techniques using time-domain filter theory. In this analysis the various rate-window techniques are viewed as bipolar weighting functions with which the trap emission signal is multiplied and then integrated to obtain the DLTS signal. Three possible weighting functions are shown in Fig. 3.12. The results show that the exponential weighting function in Fig. 3.12c gives rise to the theoretically *maximum* signal-to-noise ratio for a white noise spectrum. The maximum of the DLTS peak occurs when the time constant of the exponential weighting function is equal to that of the signal.

An instrument which accomplishes this exponential weighting operation is called a *correlator* and has been described by *Miller* et al. [3.4, 30]. An essential feature of this instrument is the active dc baseline restoration circuit. This baseline restorer is essentially a second gate which measures the signal near the end of the transient. Thus the correlator actually has a bipolar weighting function, as it must to achieve the differential measurement mode necessary to produce a DLTS spectrum.

According to the analysis of *Miller* et al. [3.30] the exponential correlator achieves a signal-to-noise ratio which is about two times higher than the lock-in amplifier. The signal-to-noise ratio of the double boxcar is roughly comparable to the lock-in for the gates shown in Fig. 3.12a, but decreases as the square root of the gate width for narrower gates. Thus for very narrow gates the boxcar is quite a bit noisier than the other two techniques.

A *single*-gate-boxcar DLTS scheme has been proposed by *Wessels* [3.32] for the case of current transients. The previously discussed rate-window techniques are, of course, also applicable to current transients, if one takes into account that the peaks are slightly shifted to higher temperature because of the rapid increase in the current transient magnitude with temperature. This rapid increase also means, however, that a *single* gate will generate a DLTS peak. For capacitance transients a single gate will produce a *step* for dc coupling [3.33] and a *peak* for ac coupling between the capacitance meter and boxcar. The single-gate schemes are not terribly satisfactory, however. The ac-coupled method has the same problem with fluctuations and drifts due to the pulse-overload-recovery transient as does the lock-in method, while the single-gate *current* transient is sensitive to the steady-state leakage current of the sample, which may become significant at high temperatures and cause a baseline shift.

As we mentioned earlier, there is no superior way to gather raw data than to measure the entire transient at many different temperatures. Such a procedure would be prohibitively time consuming if done manually in a sample with many traps. Thus the various survey techniques, such as DLTS, have a distinct advantage in speed at the expense of not recording all of the possible data, such as the actual transient time dependences. However, the use of automated digital data acquisition by a computer makes the analog rate-window techniques unnecessary. With a computer it is possible to rapidly record the actual transients at various temperatures during a thermal scan and to process this data in a variety of different ways. For example, an Arrhenius plot such as in Fig. 3.8 could be produced simultaneously for all of the traps in the sample with just a single scan of temperature. The *same* stored data could also be analyzed using the digital equivalent of one of the weighting functions in Fig. 3.12 to produce a DLTS spectrum.

It is also possible to use a computer to simulate a double boxcar directly, by digitizing the signal sampled during the two gates. This was used by *Lefevre* and *Schulz* [3.34] to obtain the DLTS spectrum corresponding to traps located

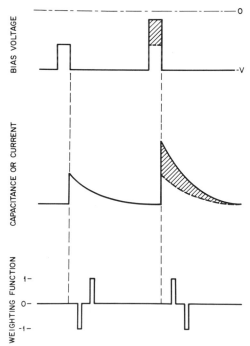

Fig. 3.13. Diagram showing how alternating bias pulses of two different amplitudes produce a DLTS signal corresponding to an internal region of the space-charge layer when analyzed with the alternating weighting function shown. This is called double DLTS

in a narrow region in the *interior* of the space-charge layer. The method, termed *double* DLTS (DDLTS), consists of applying alternating bias pulses of two different amplitudes, as shown in Fig. 3.13. If the digitized gate information is processed as $C(t_1) - C(t_2)$ for one of the pulse amplitudes and as $C(t_2) - C(t_1)$ for the other pulse, then the resultant DLTS signal corresponds to the difference between the transients (shown shaded). The spatial location of the traps giving rise to this "difference transient" corresponds to the range of space-charge-layer widths produced by the shaded region of bias voltages which is the difference between the two bias pulses. The DDLTS method has some advantages if the thermal emission rate of the trapped carriers is electric-field dependent so that the DLTS peak position depends on the location of the trap within the space-charge layer. This method also discriminates against low-frequency noise, i.e., noise of a much lower frequency than the pulse repetition rate. Apparently, some Schottky barrier samples are subject to such noise [3.34].

The shaded difference transient in Fig. 3.13 can also be obtained manually by subtracting one isothermal transient from the other. This was first proposed by *Sah* and *Fu* [3.35] and has been used to study the electric-field dependence of thermal emission rates for deep levels in Si [3.36–38] and GaP [3.10].

3.4.3 Admittance Spectroscopy

If one measures the capacitance of a pn junction or Schottky barrier as a function of temperature at rather low frequencies, e.g., 100 Hz, one finds a trap-related increase in the capacitance at a temperature where the emission and capture rates at the trap are as fast as the measurement frequency [3.4, 39]. In all of our previous discussions we had assumed that the measurement frequency was sufficiently high so that this never occurred. The resulting series of capacitance steps in the low-frequency case look very much like a TSCAP curve (see Fig. 3.9), except that the frequency-response effect does not depend on the direction or rate of the thermal scan. For any such step in the capacitance response corresponding to a frequency-dependent element being able to follow the driving voltage as the temperature is increased, there is necessarily also a *peak* in the ac conductance response of the sample. Measurement of the conductance peaks during a thermal scan was proposed by *Losee* [3.39], and is called admittance spectroscopy.

We have included this technique in the section on DLTS measurements because the conductance peak can actually be viewed as a type of rate-window method for current transients. In this case, however, the bias voltage is not pulsed but is sinusoidally varying. The sinusoidal rate-window weighting function (Fig. 3.12b) is provided by the phase-sensitive detector of the conductance bridge or meter. Indeed, the ac conductance is often measured using a lock-in amplifier with the circuit and phase setting almost identical to that which would be used for the lock-in DLTS method in the case of current transients. The single exception is the triggered pulse generator which converts the lock-in reference signal into a repetitive series of bias pulses in the DLTS case. For the conductance case the lock-in reference signal is the sinusoidal voltage drive which is applied directly to the sample.

When admittance spectroscopy is viewed in this manner it is easy to see that it has an important limitation relative to other DLTS measurements. The timing of the carrier-capture process cannot be varied independently from the carrier-emission process – both are fixed by sinusoidal waveforms. The more conventional DLTS measurements, on the other hand, separate the carrier-capture and rate-window functions by the use of bias pulses. Thus it is possible to measure the carrier-capture cross sections of the trap, as we will discuss in Sect. 3.5. It is also possible with normal DLTS to study minority-carrier capture and emission; this is not possible with admittance spectroscopy. There are situations, however, for which admittance spectroscopy is better suited than DLTS, for example, in p-i-n junctions where bias pulses do not penetrate into the i layer.

3.4.4 Determination of Activation Energies

The best way to determine the activation energy of a deep level is to construct an Arrhenius plot such as in Fig. 3.8. This can be done simultaneously for nearly all of the traps in a sample by recording several DLTS spectra using different rate windows. An example of such data is shown in Fig. 3.14 for the common, but as yet unidentified, hole traps denoted A and B which are nearly always present in GaAs grown by liquid-phase epitaxy (LPE) [3.25, 40]. An Arrhenius plot is constructed from these data by plotting the log of the rate window from (3.21) vs the inverse temperature of the DLTS peak T_m^{-1}. The data points in Fig. 3.8 correspond to the rate-window vs peak-temperature data in Fig. 3.14. The relationship of the measured activation energy to the true depth of the deep level is complex, however, and will be discussed in Sect. 3.5.4.

It is also possible to obtain somewhat more limited emission rate data, spanning about one decade, by using the appropriate analytical line-shape function, such as (3.22) for lock-in DLTS. If the FWHM of the lock-in DLTS peak is $T_2 - T_1$, where T_1 and T_2 are the temperatures corresponding to the half-height on the low- and high-temperature side of the peak, respectively, then it can be shown from (3.22) that the activation energy is [3.4]

$$E = \frac{k \ln(15)}{1/T_1 - 1/T_2} - 2kT_m . \tag{3.23}$$

This is a reasonably useful approximate method if the peak is well resolved and the transient is exponential. However, since one does not know a priori from only DLTS data whether or not a particular peak is due to an exponential transient, one must use some care in applying (3.23). Nonexponential transients

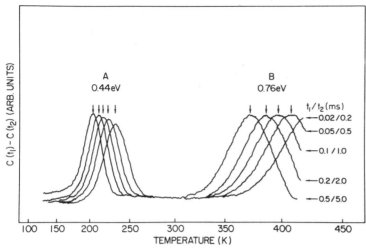

Fig. 3.14. DLTS hole-emission spectra of two hole traps in LPE n-GaAs. The values of t_1 and t_2 corresponding to the various scans are indicated. The arrows indicate the values of T_m used in Fig. 8

typically give rise to DLTS peaks which are broader than the ideal line shape given by (3.22, 23). Even though the ideal line-shape functions are not valid in such cases, the Arrhenius plots constructed from DLTS rate-window vs T_m data for nonexponential transients are still well defined, since the DLTS peak is always composed of a linear superposition of exponential-transient peaks in the $\Delta C \ll C$ case.

A very rough measure of the activation energy of a trap may be estimated directly from the temperature T_m of the maximum of the DLTS peak. This is analogous to the determination of energies from TSC peak positions according to (3.20) and is just about as accurate. If we assume that $v = 10^{12}\ s^{-1}$, then with a rate window of $\tau_{max}^{-1} = 50\ s^{-1}$ we find $E \approx 23.7\ kT_m$. Since E depends only logarithmically on v, this temperature–energy conversion factor is accurate to approximately $\pm 10\%$.

Perhaps one of the most fruitful ways to employ DLTS measurements is to use the spectroscopic nature of the entire DLTS signal vs temperature scan and compare the trap spectra of various samples by always using the same standardized rate window. We have found that $50\ s^{-1}$ is a convenient choice for this routine-survey mode. As shown in Fig. 3.15 for the example of n-GaAs [3.40] it is possible to distinguish characteristic peaks due to various deep levels even for rather complicated spectra. A unique feature of DLTS relative to *optical* spectroscopy is that the various DLTS peak heights are *directly proportional* to the respective deep-level concentrations. Thus *quantitative* analysis of electrically active defects is readily possible with a sensitivity of at least $10^{12}\ cm^{-3}$ in samples with a shallow-level doping of $10^{16}\ cm^{-3}$. With more lightly doped samples the sensitivity is even more impressive, e.g., $<10^{10}\ cm^{-3}$ traps if $|N_D - N_A| = 10^{14}\ cm^{-3}$.

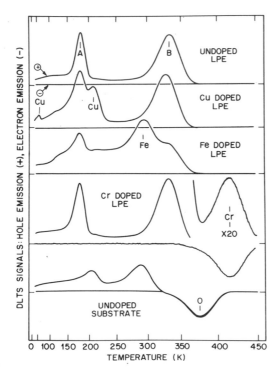

Fig. 3.15. DLTS spectra of five GaAs p^+n junctions with the deep-level doping indicated. The rate window was $50 \, s^{-1}$

3.4.5 Trap Concentration Profiles

In Sect. 3.2.4 we discussed a basic relationship useful in determining the spatial profile of trap concentrations, see (3.16). When this is coupled with the DLTS method, a very powerful profiling technique emerges which is capable of simultaneously obtaining the profiles of many different traps of very low concentration in the same sample. Figure 3.16 shows a series of four DLTS spectra of four common hole traps in p-GaAs recorded for different values of reverse bias so that the observed traps are located at various distances x from the n^+p junction [3.41]. By using majority-carrier pulses with an amplitude small compared to the steady-state bias voltage, it was possible to obtain good spatial resolution in these spectra. A series of such spectra taken with various steady-state voltages and pulse amplitudes were analyzed using (3.16) with the shallow-level profile known from *Miller* feedback profiler [3.9] measurements. The trap profiles obtained in this way are shown in Fig. 3.17. Note the good spatial resolution ($<0.1 \, \mu m$) and sensitivity (low 10^{13} range). This example was a p-type LPE GaAs layer grown on an n^+-substrate. The data clearly show minute levels of Cu and Fe contamination (0.01–0.03 ppm) which have apparently diffused into the epitaxial layer from the substrate; levels A and B have apparently been gettered by the n^+-layer.

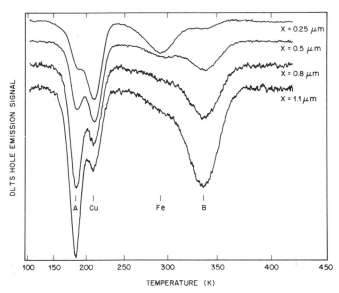

Fig. 3.16. DLTS spectra of an n^+p GaAs junction with spatially varying concentrations of four deep levels. The bias and majority-carrier pulse amplitude were chosen to correspond to the values of x shown. The gain was chosen for each scan to make the concentration scale the same for all four spectra. The rate window was $50 \, s^{-1}$

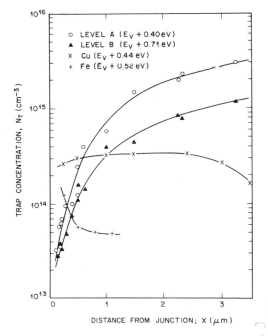

Fig. 3.17. Concentration profiles of the four deep levels shown in Fig. 3.16

3.4.6 Scanning DLTS

The spatial information gained by the profiling method just discussed is only in one dimension, namely, that perpendicular to the junction plane. It is possible to obtain spatial information *within* the junction plane by combining the DLTS technique with the $x - y$ scanning capabilities of a scanning electron microscope (SEM) [3.42]. The idea of such a measurement is quite simple in principle but rather difficult to successfully achieve in practice. In the scanning DLTS (SDLTS) case, the traps are filled by the e–h pairs which are generated by the high-energy electron beam of the microscope. If this beam is pulsed on and off in the same manner as the bias pulses used in normal DLTS measurements, the resulting emission transients can be detected when the beam is off, making it possible to produce a DLTS spectrum. If the temperature of the sample is adjusted to correspond to a DLTS peak, an image of the spatial variation of the trap responsible for that peak can be formed by using the DLTS signal magnitude to modulate the intensity of a CRT display which is rastered in synchronism with the pulsed electron beam.

The major problem with SDLTS measurements is the extremely small thermal-emission signal generated within the small beam-excited region. For example, if the beam were to fill defects in a 5 μm diameter area in a sample 0.5 mm in diameter, then the SDLTS signal would be only 10^{-4} of the normal DLTS signal for a particular deep level in this sample. The SDLTS sensitivity is inversely proportional to the *square* of the spatial resolution. One way to overcome this extreme signal-to-noise problem is to use current transients in the μs region. For reasons discussed in Sect. 3.2.2, such transients are considerably more sensitive than capacitance transients. With this method it has been possible to achieve sensitivities of $\lesssim 10^{15} \, \text{cm}^{-3}$ with fairly low resolution ($\sim 20 \, \mu\text{m}$) and resolutions as high as $\sim 2 \, \mu\text{m}$ with lower sensitivity [3.42]. The usefulness of this variation of the DLTS method has only begun to be explored, but it is clear that such spatial resolution capabilities will greatly aid in the study of certain classes of inhomogeneously distributed deep levels in semiconductors.

3.5 Carrier Capture Measurements

3.5.1 Majority-Carrier Pulse

It is clear from Fig. 3.2 that if the duration of the majority-carrier pulse is less than the time necessary to fill the deep levels, the resulting emission transient will be less than its maximum possible value. Thus, for example, DLTS spectra recorded with bias pulses of various lengths may have differing peak heights as shown in Fig. 3.18. It is often more convenient to stop the DLTS scan on a particular peak and record the variation of peak height as a function of pulse width. Such data are shown for the Cr level in LPE n-GaAs in Fig. 3.19, as an

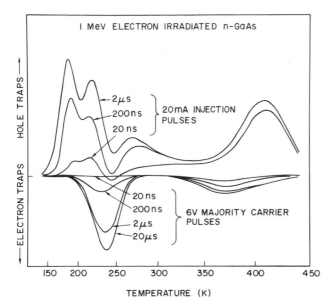

Fig. 3.18. DLTS spectra of deep levels produced by 1 MeV electron irradiation of a p^+n GaAs junction. The effect of incomplete filling of the levels for short pulses is shown. The rate window was 5.1×10^3 s^{-1}

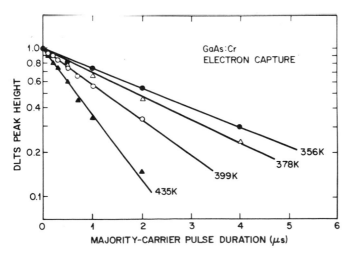

Fig. 3.19. DLTS signal versus majority-carrier pulse durtation for electron capture at Cr levels in LPE n-GaAs at four different temperatures

example. In this case, since we are interested in the majority-carrier capture cross section of a minority-carrier trap, a *double*-pulse sequence has been used. The first pulse is an injection pulse which totally fills the Cr hole trap with holes. The second pulse is a majority-carrier pulse which is applied before the trapped holes are thermally emitted. Thus the electron capture during this second pulse *reduces* the observed DLTS hole-emission signal. For a majority-carrier trap, on the other hand, the signal would *increase* with majority-carrier pulse duration.

Such carrier-capture measurements are typically done at low enough temperatures so that the majority-carrier capture rate (c_n for electrons in n-type material or c_p for holes in p-type material) is much faster than the thermal emission rates α or α^*. Thus, for example, the time dependence of the trapped electron concentration (in n-type material) during a majority-carrier pulse is given by

$$n(t) = N[1 - \exp(-c_n t)] . \tag{3.24}$$

Hole capture is given by an analogous expression. The capture rates are defined in terms of capture cross sections S_n and S_p, average thermal velocities $\langle v_n \rangle$ and $\langle v_p \rangle$, and free carrier concentration n_c and p, as

$$c_n \equiv S_n \langle v_n \rangle n_c$$
$$c_p \equiv S_p \langle v_p \rangle p . \tag{3.25}$$

The experimental data, such as in Fig. 3.19, correspond to the values of $n(t)$ in (3.24) at the *end* of the majority-carrier pulse. Thus a plot of the DLTS or isothermal-transient signal (for the case of minority-carrier traps with the double-pulse sequence) or N *minus* this signal (for the case of majority-carrier traps with a single pulse) versus the pulse duration gives the majority-carrier capture rate directly. From this it is a straightforward matter to obtain the capture cross section from (3.25). Note, however, that $n_c < (N_D - N_A)$ and $p < (N_A - N_D)$ at low temperatures. Capacitance profiling methods [3.6-9] measure $|N_D - N_A|$, not the free-carrier concentration, thus Hall data on the temperature dependence of the free-carrier concentration in similar samples must be known in order to obtain the capture cross sections.

In the edge region of the space-charge layer during the pulse the carrier concentration is much smaller than in neutral material and, in addition, is spatially varying (see Sect. 3.2.3). Therefore, there will be a distribution of majority-carrier capture rates on the edge region which will be much slower than in the remainder of the volume covered by the pulse. This will tend to produce a long, nonexponential tail on the initially-exponential capture data. The presence of such a tail can be seen in Fig. 3.18, where the signal increases rapidly for very short pulses, followed by a small, slower increase for much longer pulses. This effect can be minimized and almost totally eliminated, however, by using capacitance transients with rather large reverse bias voltages (> 10 V); because then the pulse-edge region is a smaller fraction of the total space-charge-layer width and, furthermore, is located in the low-sensitivity region near $x = 0$ (see Sect. 3.2.1).

3.5.2 Minority-Carrier (Injection) Pulse

The measurement of minority-carrier capture is operationally very similar to the majority-carrier case just discussed. However, there are several important differences which complicate matters. First, during the injection pulse there are *both* minority and majority carriers present. Thus, for example, the time dependence of the captured electron concentration at an initially empty electron trap in p-type material is given by

$$n(t) = \frac{c_n N}{c_n + c_p} \{1 - \exp[-(c_n + c_p)t]\} . \tag{3.26}$$

An exactly analogous expression for trapped holes may be written for hole traps in n-type material. Note that the observed capture rate in (3.26) is the *sum* of the electron and hole capture rates and that the steady-state value of n is not necessarily equal to N, but depends on the relative values of c_n and c_p. If the minority-carrier capture rate is much larger than the majority-carrier rate, then we say that the trap is "saturable", i.e., a sufficiently long and intense injection pulse will fill the trap with minority carriers. If, on the other hand, the opposite is true and the majority-carrier capture rate is largest, the trap is "unsaturable" and the minority-carrier occupation will always be near zero. Therefore, an unsaturable minority-carrier trap will not give rise to a transient following an injection pulse and hence will be undetected. It may be possible, however, to detect such deep levels by filling the state optically.

A second difference between minority- and majority-carrier capture is that the minority-carrier concentration needed to obtain the capture cross section from (3.25) depends on the magnitude of the injection-pulse current density J. For the example of electron injection into p-type material, the injected electron concentration n_c is given by [3.43]

$$n_c = \left(\frac{uL_n}{eD_n}\right) J \tag{3.27}$$

where L_n and D_n are the electron diffusion length and diffusion constant, respectively, and where u is the electron injection efficiency of the pn junction, i.e., the fraction of the total forward current corresponding to electron injection

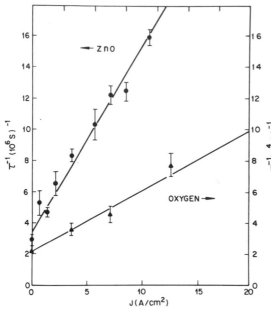

Fig. 3.20. Rate of increase of trapped minority carriers as a function of injection-pulse-current density for the ZnO and oxygen centers in p-type GaP

into the p-side, given by [3.1]

$$u = \left(1 + \frac{D_p L_n p_p}{D_n L_p n_n}\right)^{-1}, \tag{3.28}$$

where n_n and p_p are the free electron and hole concentrations on the n- and p-side of the junction, respectively. Analogous expressions follow readily for hole injection into the n-side of a pn junction.

Thus, according to (3.25–27), the time constant of the increase in the minority-carrier emission signal as a function of the injection-pulse width is a function of the injected current density J during the pulse. An example of the capture-rate variation versus injection-pulse current is shown in Fig. 3.20. This corresponds to injected electron capture at the oxygen and ZnO centers in p-type GaP at 190 K [3.10]. The intercept at zero current is the majority-carrier capture rate while the component of the measured rate which is linear with injected current is the minority-carrier capture rate. This method is considerably more time consuming and somewhat less accurate than the majority-carrier capture measurements of the last section. The inaccuracy stems from the difficulty in knowing the properties of the junction well enough to calculate the minority-carrier concentration from (3.27, 28) with great precision. Therefore it is always advisable to measure capture cross sections by the majority-carrier-pulse method, if this is at all possible.

3.5.3 Optically Generated Carriers

In some cases the carrier capture time is shorter than the duration of the shortest pulse (typically ~ 10 ns). It is then possible to measure the capture cross section by observing the rate of capture of optically generated carriers in the space-charge layer [3.14, 44]. The technique consists of illuminating a semitransparent Schottky barrier with above-band-gap light and measuring the time constant of the capacitance transient due to the capture of photogenerated carriers. If the average drift velocity μF in the space-charge layer is smaller than the thermal velocity $\langle v \rangle$, then the electron density, for example, in the space-charge layer is related to the measured photocurrent J by

$$n_c = J/\mu_n eF \tag{3.29}$$

where the average value of the electric field F must be calculated from the doping profile (see Fig. 3.1) and represents the largest source of error in this measurement. In general, electrons and holes are both present in such an experiment (unless the light is totally absorbed near the surface) so that the capture transient is described by an equation like (3.26).

3.5.4 Temperature-Dependent Capture

In most considerations of thermally stimulated processes it is assumed that the capture cross sections S_n and S_p are independent of temperature. While this may be the case for some deep levels, it is not in general true. In fact, many deep levels in III–V semiconductors, for example, have capture cross sections which are *thermally activated* [3.44, 45], as shown in Fig. 3.21. On the basis of data such as in Fig. 3.21, as well as optical Stokes shifts of various deep levels, it was shown [3.44] that an important, previously neglected, nonradiative capture mechanism for many deep levels is multiphonon emission via lattice relaxation.

It is also possible to have capture cross sections which *decrease* with increasing temperature [3.46]. This is a well-known phenomenon for shallow donors and acceptors at low temperatures (<77 K) and can be explained by capture into excited Coulomb states followed by a cascade of single-phonon emissions into the ground state [3.47]. Similarly, thermal reemission from the excited states of a deep level will also give rise to an effective capture cross section which decreases with temperature [3.44, 48].

The occurrence of temperature-dependent capture cross sections is important for space-charge spectroscopy because it affects the measured activation energy for thermal emission and hence must be taken into account in order to obtain the *true* depth of the level. Thus, for example, if

$$S_n = S_\infty \exp(-E_s/kT), \tag{3.30}$$

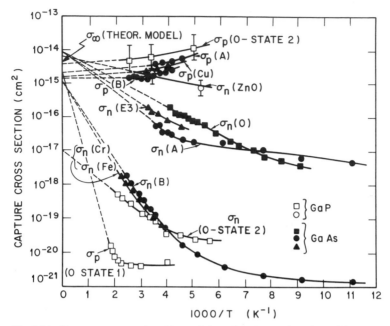

Fig. 3.21. Capture cross section (denoted here by σ) as a function of inverse temperature for seven deep levels in GaAs and two in GaP

such as at higher temperatures in Fig. 3.21, then (3.18) must be modified to become

$$\alpha = \frac{S_\infty \langle v_n \rangle N_c}{g} \exp[-(E+E_s)/kT], \tag{3.31}$$

where g is the degeneracy factor for the deep level [3.46]. Equation (3.31) is a straightforward consequence of the steady-state detailed balance between emission and capture [3.41]. The measured activation energy E_{meas} is related to the true energies of the system by

$$E_{meas} = E + E_s + 2kT, \tag{3.32}$$

where the $2kT$ term arises from the T^2 dependence of $\langle v_n \rangle N_c$. This term is not needed if the Arrhenius plot is constructed as $\log \alpha T^{-2}$ vs $1/T$.

For cases where E_s is negative, i.e., the cross section *decreases* with temperature due to reemission from an excited state of depth E_s, (3.32) is still valid. In this case the measured activation energy E_{meas} is the energy difference between the ground state and excited state [3.48]. This behavior occurs when the thermal emission rate from the excited state is faster than the transition rate from the excited state to the ground state.

In addition to the energy corrections due to the temperature dependence of capture cross sections, there are also corrections to the measured activation energy due to the temperature dependence of the deep-level energy itself. It is helpful to describe the temperature dependence of the deep-level ionization energy in thermodynamic terms. This does not necessarily provide any new physical insight but does make available the powerful machinery of thermodynamic relationships. We first note that the energy E in (3.18) [E_A^* in (3.19)] is the standard Gibbs free energy of the ionization reaction, usually denoted by G. This is the proper free energy to use in a solid, since the pressure P is constant. The slope of an Arrhenius plot is given by

$$\frac{d}{d\frac{1}{T}}\left(\frac{G}{T}\right)_P = G + \frac{1}{T}\left(\frac{dG}{d\frac{1}{T}}\right)_P. \qquad (3.33)$$

By making use of the thermodynamic identity

$$\left(\frac{dG}{dT}\right)_P = -S, \qquad (3.34)$$

where S is the entropy, we see that the right-hand side of (3.33) is equal to $G + TS$, which is the definition of enthalpy, H. Note that at $T=0$ the enthalpy H and the free energy G are equal. Indeed, it is straightforward to show from the identity in (3.34) that at a given temperature T the value of G obtained by extrapolating the tangent of the true $G(T)$ curve to $T=0$ is identically equal to $H(T)$. Thus H is the proper thermodynamic term for what is often referred to as the "extrapolated $T=0$ energy" of a deep level [3.5, 49, 50]. Since $H = G + TS$, we can see that the parameter which expresses the difference between H and G is the entropy S. Thus an ionization energy which is independent of temperature may also be described by saying that the ionization reaction produces no change in entropy.

By using these proper thermodynamic quantities for the example of electron emission, we may rewrite (3.18) as

$$\alpha = v\exp(-G/kT) = v\exp(S/k)\exp(-H/kT). \qquad (3.35)$$

Thus, since the activation energy as determined from an Arrhenius plot is actually H [after the corrections of (3.32) have been applied], we see that an added entropy term is present in the exponential prefactor. Typical values of the entropy change for ionization reactions in semiconductors are $0 \lesssim S \lesssim 6k$ [3.51, 52]. Therefore, the exponential entropy factor in (3.35) may be as high as 400. Neglect of this factor could thus lead to large errors in determining the capture cross section from (3.31) by using the activation energy as determined from the slope of an Arrhenius plot. Indeed, such discrepancies have been noted [53, 54] and can be explained by the additional factor in (3.35) [3.50].

Care must also be taken in relating energies determined from Arrhenius plots to energies measured optically, since in general the former is H while the latter is G [3.49]. Thus in addition to the problems associated with Stokes shifts and temperature-dependent capture, optical and thermal measurements can only be compared with high accuracy when the temperature dependence of the energy level, i.e., the entropy of ionization, is known.

3.6 Instrumentation

3.6.1 Standard Capacitance Meter with Pulse Transformer

One of the most straightforward ways of making capacitance measurements is to use a standard commercially available 1 − MHz capacitance meter, such as a Boonton model 72B or equivalent. Such a meter works very well for measuring the steady-state capacitance, but has certain shortcomings for measuring capacitance transients. First, its rather slow output time constant (typically ~ 1 ms) gives rise to a rather long pulse-overload-recovery transient, which is a problem for lock-in DLTS measurements and also makes it impossible to measure transients faster than 1 ms. Second, there is no provision made for the introduction of a fast bias pulse. The normal bias input reaches the sample through a heavily filtered dc network which has such a long time constant that it is impossible to use µs pulses.

An alternate scheme, which is adaptable to any type of meter, is the pulse-transformer method proposed by *Henry* et al. [3.43]. The schematic diagram of this pulse-transformer circuit is shown in Fig. 3.22. The transformer can be externally added to the capacitance meter by connecting the sample in series with the secondary winding of the transformer. The voltage of the pulse is measured with a low-capacitance oscilloscope probe connected to the sample lead just as it enters the temperature-control dewar. The current of an injection pulse can be measured with a commerically available current probe.

In practice, the limited bandwidth of most pulse transformers [3.55] makes it impossible to cover with a single transformer the entire ns to ms range often needed for majority-carrier and injection pulses. Thus to cover this range in making carrier capture measurements one typically must use several transformers best suited to different parts of this range. A good choice for routine measurements with a 10 V reverse bias is a 10 V − 10 µs mayority-carrier pulse and a 10 A cm^{-2} − 10 µs injection pulse. A problem that is often encountered with injection pulses is that the pulsed current is driven into the internal circuitry of the capacitance meter and causes instabilities and additional noise at the output. Such effects do not seem to occur with majority-carrier pulses.

A final note for anyone contemplating the use of a capacitance meter in this application is that the signal-to-noise ratio of the meter is proportional to the amplitude of the rf drive level used in measuring the capacitance. This is usually limited to 15 mV$_{rms}$ in order to be well within the small-signal limit for the

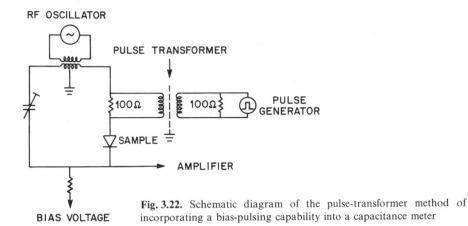

Fig. 3.22. Schematic diagram of the pulse-transformer method of incorporating a bias-pulsing capability into a capacitance meter

measurement of the capacitance of transistors. A much larger drive amplitude can be tolerated – indeed, is *beneficial* – in capacitance spectroscopy measurements. Some manufacturers, e.g., Boonton Electronics, offer a higher drive-voltage option (e.g., $100\,\text{mV}_{\text{rms}}$) which is very desirable. The resulting noise level on a DLTS scan at $\tau_{\text{max}}^{-1} = 50\,\text{s}^{-1}$ is less than $10^{-3}\,\text{pF}$.

3.6.2 Fast Capacitance Bridge

The capacitance-meter method of measuring thermal-emission transients has the advantages of ease and simplicity. However, to overcome some of its shortcomings it is necessary to build a faster capacitance bridge from individual components [3.10]. Such a bridge is quite analogous to that which might be used to detect magnetic resonance signals. An updated version of the original bridge [3.10] is shown in Fig. 3.23. This bridge is capable of sub-μs pulse-overload recovery, limited by the recovery time of the amplifiers from overload. In addition, since the bias pulses are directly coupled to the sample, it is possible to span the entire ns–*to*–ms output range of the pulse generator without changing pulse transformers. This is ideal for capture-cross-section measurements.

The frequency at which the bridge is operated is not critical. Obviously, one wants a rather high frequency to obtain the fastest possible overload recovery. However, the frequency must not exceed the limit imposed by the RC time constant of the sample itself. This may be as low as a few MHz for Schottky barriers on thin epitaxial layers grown on insulating substrates. The range of 10–20 MHz seems like a good compromise for most situations.

The purpose of the phase shifter and attenuator which are in parallel with the sample is to provide an impedance of roughly equal magnitude to that of the sample but 180° out of phase. By adjusting this network ("balancing the bridge"), the input to the rf amplifier is kept small enough to be within its linear

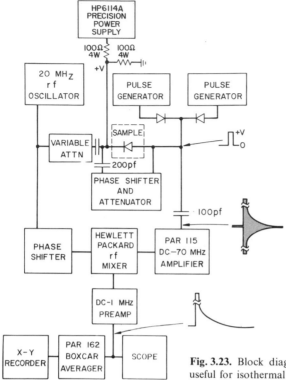

Fig. 3.23. Block diagram of a fast capacitance bridge useful for isothermal transient or DLTS measurements

range. We have found that a passive $50-\Omega$ delay line in series with a tapped variable low-inductance resistor is best. One may also use various transformer-capacitor schemes to accomplish the balancing of the bridge. However, we have found that any circuit containing inductance gives rise to unsatisfactory bias-pulse response, especially for injection pulses.

One of the principal shortcomings of the bridge in Fig. 3.23 is that it is not calibrated, and furthermore its amplitude response depends on the capacitance of the sample. This problem arises because the demands for fast recovery and direct-coupled 50Ω pulse generators dictate attempts at proper 50Ω terminations as much as possible. This, however, gives rise to bridge impedances comparable to that of the sample so that the amplitude response depends on the sample impedance. Thus we have devised a relatively simple means for quickly calibrating the bridge response for each sample *in situ*.

The key to the calibration procedure is a power supply with digital control via thumb-wheel switches so that a small, reproducible voltage step ΔV can be applied to the sample. This voltage step produces a small capacitance change ΔC which can be measured with a calibrated capacitance meter or bridge. By choosing ΔV small enough so that the resulting signal does not overload the bridge in Fig. 3.23, it is possible to measure the bridge output change corresponding to ΔV and thus calibrate the bridge. Such a calibration is, of course, necessary to obtain deep-level concentrations, but is not needed for energy or capture-rate measurements.

3.6.3 DLTS Spectrometer

This subject has already been partially discussed in Sect. 3.4.2 in connection with the various means of implementing the DLTS rate-window concept. We wish to briefly discuss here the overall experimental system for DLTS. The main elements of a DLTS spectrometer are shown in Fig. 3.24. This system is a combination of the elements needed for isothermal transient measurements (i.e., pulsed bias supply and capacitance meter) with those needed for irreversible thermal scanning techniques such as TSCAP (i.e., temperature scan control and X–Y recorder). In addition, of course, one needs some sort of DLTS rate-window instrumentation (i.e., double boxcar, lock-in, or correlator).

The elements shown in Fig. 3.24 may be implemented in a number of different ways. For example, the bias supply/pulse generator/capacitance meter combination could be either that shown in Fig. 3.22 or the fast bridge shown in Fig. 3.23. The dewar may be of the cold-finger type [3.29] or a heated-gas-flow design. In general the major design rule for the temperature scanning system is to use thermal masses which are as small as possible and to measure the temperature with either a thermocouple or diode placed very close to the

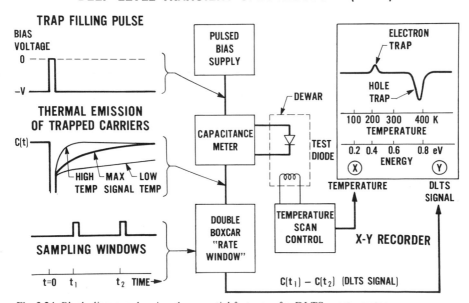

Fig. 3.24. Block diagram showing the essential features of a DLTS spectrometer

sample and in good thermal contact with it. Accurate temperature measurement is perhaps the most difficult experimental problem of thermal scanning techniques. One may test for good thermal coupling between the thermal sensor and the sample by verifying that DLTS peaks occur at the same temperature for both increasing and decreasing thermal scans. Indeed, the thermal emission rate of a particular trap makes a very good thermometer. The electronics for the temperature scan control may be as simple as a heater driven by a constant current or a more complex servo-type temperature controller driven by a voltage ramp. This latter type is advantageous for measuring capture cross sections or concentration profiles since it is possible to stop the scan at a DLTS peak and regulate the temperature at that point.

The DLTS spectrum shown on the $X - Y$ recorder in Fig. 3.24 is meant to be only illustrative but shows a typical temperature scan range (50–450 K) with a rough energy scale corresponding to $v = 10^{12} \, \text{s}^{-1}$ and $\tau_{max}^{-1} = 50 \, \text{s}^{-1}$ ($E = 23.7 \, kT$). Such a scan can be achieved in about 5–10 min with a temperature accuracy of 1–2 K; this corresponds to $q \sim 1 \, \text{K} \, \text{s}^{-1}$.

A DLTS spectrometer using current transients would be essentially the same as Fig. 3.24 except that the capacitance meter would be replaced by a current amplifier. For SDLTS measurements in an SEM the DLTS signal at the boxcar output would be applied to a CRT intensity control for image formation and to the $X - Y$ recorder, as shown, for examining spectra from localized regions with the microscope beam held stationary.

We have seen in this chapter that the general DLTS spectrometer in Fig. 3.24 makes possible a wide variety of measurements on deep levels in semiconductor space-charge layers. One can measure the thermal ionization energy, electron and hole capture cross sections, concentration, and spatial distribution of both radiative and nonradiative defects with extremely high sensitivity. This opens a whole new horizon for deep-level studies in semiconductors.

References

3.1   S.M.Sze: *Physics of Semiconductor Devices* (Wiley, New York 1969)
3.2   C.G.B.Garrett, W.H.Brattain: Phys. Rev. **99**, 376–387 (1955)
3.3   G.Lubberts, B.C.Burkey: Solid-State Electron. **18**, 805–809 (1975)
3.4   G.L.Miller, D.V.Lang, L.C.Kimerling: Ann. Rev. Mater. Sci. **7**, 377–448 (1977)
3.5   H.G.Grimmeiss: Ann. Rev. Mater. Sci. **7**, 341–376 (1977)
3.6   J.Hildebrand, R.D.Gold: RCA Rev. **21**, 245 (1960)
3.7   P.J.Baxandall, D.J.Colliver, A.F.Fray: J. Sci. Instrum. **4**, 213 (1971)
3.8   J.A.Copeland: IEEE Trans. ED-**16**, 445 (1969)
3.9   G.L.Miller: IEEE Trans. ED-**19**, 1103 (1972)
3.10 D.V.Lang: J. Appl. Phys. **45**, 3014–3022 (1974)
3.11 A.M.Goodman: J. Appl. Phys. **34**, 329–338 (1963)
3.12 Y.Furukawa, Y.Ishibashi: Jpn. J. Appl. Phys. **5**, 837 (1966); **6**, 503 (1967)
3.13 R.R.Senechal, J.Basinski: J. Appl. Phys. **39**, 3723–3731, 4581–4589 (1968)
3.14 C.T.Sah, L.Forbes, L.L.Rosier, A.F.Tasch,Jr.: Solid-State Electron. **13**, 759–788 (1970)

3.15 Y. Zohta: Appl. Phys. Lett. **17**, 284–286 (1970)
3.16 K. Hesse, H. Strack: Solid-State Electron. **15**, 767–774 (1972)
3.17 G. H. Glover: IEEE Trans. ED-**19**, 138–143 (1972)
3.18 Y. Zohta, Y. Okmura: Appl. Phys. Lett. **21**, 117–119 (1972)
3.19 M. Bleicher, E. Lange: Solid-State Electron. **16**, 375–380 (1973)
3.20 T. Ikoma, B. Jeppson: Jpn. J. Appl. Phys. **12**, 1011–1019 (1973)
3.21 G. Goto, S. Yanagisawa, O. Wada, H. Takanashi: Jpn. J. Appl. Phys. **13**, 1127–1133 (1974)
3.22 D. V. Lang, R. A. Logan, M. Jaros: Phys. Rev. B (to be published)
3.23 R. Williams: J. Appl. Phys. **37**, 3411 (1966)
3.24 D. V. Lang, L. C. Kimerling: Proc. Intern. Conf. Lattice Defects in Semiconductors, Freiberg 1974 (Institute of Physics Conf. Series No. 23, London 1975) p. 581
3.25 D. V. Lang: J. Appl. Phys. **45**, 3022 (1974)   ·
3.26 M. C. Driver, G. T. Wright: Proc. Phys. Soc. (London) **81**, 141 (1963)
3.27 J. C. Carballes, J. Lebailly: Solid State Commun. **6**, 167 (1968)
3.28 C. T. Sah: Solid-State Electron. **19**, 975–990 (1977)
3.29 M. G. Buehler, W. E. Phillips: Solid-State Electron. **19**, 777–788 (1976)
3.30 G. L. Miller, J. V. Ramirez, D. A. H. Robinson: J. Appl. Phys. **46**, 2638–2644 (1975)
3.31 M. D. Miller, D. R. Patterson: Rev. Sci. Instrum. **48**, 237–239 (1977)
3.32 B. W. Wessels: J. Appl. Phys. **47**, 1131–1133 (1976)
3.33 O. Wada, S. Yanagisawa, H. Takanashi: Appl. Phys. **13**, 5–13 (1977)
3.34 H. Lefevre, M. Schulz: Appl. Phys. **7**, 45–53 (1977)
3.35 C. T. Sah, H. S. Fu: Phys. Status Solidi (a) **14**, 59–70 (1972)
3.36 J. M. Herman III, C. T. Sah: Phys. Status Solidi (a) **14**, 405–415 (1972)
3.37 L. D. Yau, W. W. Chan, C. T. Sah: Phys. Status Solidi (a) **14**, 655–662 (1972)
3.38 L. D. Yau, C. T. Sah: Solid-State Electron. **17**, 193–201 (1974)
3.39 D. L. Losee: Appl. Phys. Lett. **21**, 54 (1972); J. Appl. Phys. **46**, 2204–2214 (1975)
3.40 D. V. Lang, R. A. Logan: J. Electron. Mat. **5**, 1053 (1975)
3.41 D. V. Lang, R. A. Logan: J. Appl. Phys. **47**, 1533–1537 (1976)
3.42 P. M. Petroff, D. V. Lang: Appl. Phys. Lett. **31**, 60–62 (1977)
3.43 C. H. Henry, H. Kukimoto, G. L. Miller, F. R. Merritt: Phys. Rev. B**7**, 2499–2507 (1973)
3.44 C. H. Henry, D. V. Lang: Phys. Rev. B**15**, 989–1016 (1977)
3.45 D. V. Lang, C. H. Henry: Phys. Rev. Lett. **35**, 1525 (1975)
3.46 A. G. Milnes: *Deep Impurities in Semiconductors* (Wiley, New York 1973)
3.47 M. Lax: Phys. Rev. **119**, 1502 (1960)
3.48 R. M. Gibb, G. J. Rees, B. W. Thomas, B. L. H. Wilson, B. Hamilton, D. R. Wight, N. F. Mott: Philos. Mag. **36**, 1021–1034 (1977)
3.49 C. M. Penchina, J. S. Moore: Phys. Rev. B**9**, 5217–5221 (1974)
3.50 A. Mircea, A. Mitonneau, J. Vannimenus: J. Phys. (Paris) Lett. **38**, L41–L43 (1977)
3.51 C. D. Thurmond: J. Electrochem. Soc. **122**, 1133 (1975)
3.52 J. A. Van Vechten, C. D. Thurmond: Phys. Rev. B**14**, 3539–3550 (1976)
3.53 A. Mircea, A. Mitonneau: Appl. Phys. **8**, 15 (1975)
3.54 D. V. Lang, A. Y. Cho, A. C. Gossard, M. Ilegems, W. Wiegmann: J. Appl. Phys. **47**, 2558 (1976)
3.55 Pulse Engineering Inc., San Diego, California 92112 USA
3.56 L. C. Kimerling: IEEE Trans. NS-**23**, 1497–1505 (1976)

4. Field-Induced Thermally Stimulated Currents

J. Vanderschueren and J. Gasiot

With 56 Figures

The method of field-induced thermally stimulated currents (FITSC) consists of measuring, with a definite heating scheme, the currents generated by the buildup and/or the release of a polarized state in a solid dielectric sandwiched between two electrodes. The general experimental procedure usually involves four steps [4.1]: 1) the application of a dc bias V_p at a starting temperature T_p; 2) cooling under this bias to some lower temperature T_0; 3) change of the bias at T_0 to another value V_d; and 4) heating at a constant rate while maintaining the new bias and recording the current as a function of temperature. If the bias V_d is zero, current peaks are observed during the thermally activated transition from the polarized state to the equilibrium state; this technique, generally known as the ionic thermocurrent technique (ITC) or thermally stimulated depolarization current technique (TSDC), is the most widely applied and emphasis will thus be given to it in this paper. But, if the bias V_p is zero, current peaks superimposed onto the normal dc conduction current are obtained as a result of the opposite process, i.e., thermally activated transition from the equilibrium state to the polarized state; they are generally designated by the term thermally stimulated polarization currents (TSPC). In the general case where both V_p and V_d are unequal to zero, the currents observed are obviously governed by the combined effects of polarization and depolarization mechanisms.

The field-induced thermally stimulated current method is thus a general method of investigating the electrical properties of high-resistivity solids via the study of thermal relaxation effects and, as such, offers an attractive alternative to the conventional bridge methods or current-voltage-temperature measurements. As far as electronic carriers are concerned, it is closely related to the thermally stimulated conductivity method (TSC) (see Chap. 2). The mathematics is similar to that found in other nonisothermal techniques such as thermoluminescence, thermally stimulated electronic emission, differential scanning calorimetry or thermogravimetric analysis.

Our purpose is, on one hand, to outline the more general theories related to the main polarization processes (dipolar, space-charge and interfacial polarizations) and to compare their predictions with typical experimental data, and, on the other, to discuss the various means of identifying the microscopic origin of a given current spectrum. Primary emphasis will be given to the most commonly used TSDC technique.

Section 4.2 deals with the basic principles of the method. It includes a review of the microscopic mechanisms responsible for the polarization and of the various charging processes used to induce a stored charge in dielectrics. In Sect. 4.3 a brief description of the experimental technique is given. Sections 4.4–4.7 are devoted to TSDC resulting from dipolar relaxation. The theoretical treatments, involving systems characterized by a single relaxation time and a distribution of relaxation times, are presented in Sects. 4.4, 5. The various methods derived for evaluating the dipolar relaxation parameters from experimental data are discussed in detail in Sect. 4.6 and the possible correlations of these data with the more conventional dielectric measurements are treated in Sect. 4.7. The TSDC theories of space-charge relaxation involving carrier drift and trapping processes are summarized in Sects. 4.8, 9; also discussed in these sections are the various means of determination of the conductivity or trap parameters. The theories of TSDC resulting from an interfacial polarization are briefly touched upon in Sect. 4.10. Section 4.11 is concerned with the TSPC technique; dipolar and space-charge theories are reviewed. In Sect. 4.12 an attempt is made to collect the various experimental procedures which allow us to differentiate the polarization mechanisms in a given material. Finally, Sect. 4.13 summarizes the main advantages and limitations of the FITSC method, outlines some of the areas where our understanding of the experimentally observed phenomena is still incomplete, and suggests some problems where additional experimental and theoretical studies are needed.

4.1 Background

Until recently TSDC measurements alone have been used and they have been essentially applied to the study of electrets, i.e., permanently charged dielectrics [4.2]. Their use dates back to *Frei* and *Groetzinger* who in 1936 proposed to enhance the mobility of the frozen-in charges by slowly heating the electret between two electrodes connected to an ammeter for measuring the discharging current [4.3]. The technique was later extensively used by other authors working in the field of electrets such as *Gross* [4.4], *Wikstroem* [4.5], *Gubkin* and *Matsonashvili* [4.6], *Murphy* [4.7] and others. All these authors made use of an arbitrarily programmed temperature rise and most of their studies were carried out with complex materials such as waxes, resins, ceramics or plastics, which prevented the proper identification of the fundamental mechanisms involved. It was only in 1964 that a first theoretical basis of the TSDC phenomenon could be given by *Bucci* and *Fieschi* based on their work on point defect dipoles in ionic crystals [4.8] and, it still took several years before TSDC was recognized as a method for studying all the fundamental mechanisms of charge storage and release in nonmetallic solids. ·

Denoted as ionic thermocurrents or ionic thermoconductivity method (ITC) by *Bucci* and *Fieschi*, the TSDC procedure thereafter received a number of confusing terms, due to the fact that it was reinitiated and developed by several investigators using quite different starting points and who, most of the time, were not aware of preceding work in the field: electret thermal analysis [4.9], thermally stimulated discharge [4.10], thermal current spectra [4.11], thermally stimulated depolarization [4.12, 13], thermally activated depolarization [4.14]. Furthermore, the short-circuit currents observed during heating of samples, previously polarized by some other means than the classical field procedure, sometimes received special names; such is the case, for example, with the currents obtained after simultaneous application of field and irradiation at low temperature: thermally stimulated currents under the condition of persistent internal polarization [4.15], radiation-induced thermally activated depolarization [4.14], etc. Up to now, none of these terms has been universally accepted. In the following, we will always make use of the term thermally stimulated depolarization currents (TSDC) which seems to us the most appropriate and descriptive of the actual phenomena observed.

The TSPC method had also been used for a long time to investigate special phenomena such as pyrolysis and transition processes [4.16–18] before its ability to elucidate the general relaxation behavior of dielectrics was emphasized [4.19, 20]. First used under the name of dynamic electrothermal analysis [4.21], it was later considered by *Devaux* and *Schott* as a special case of TSC for studying the trapping properties of materials under blocking conditions [4.22] and was then called thermally activated polarization by *Moran* and *Fields* in their study of dipolar complexes in ionic crystals [4.14], and thermally stimulated polarization by *Müller* [4.19], and *Mc Keever* and *Hughes* [4.20] in their investigations of space charge and dipolar polarization phenomena. This last term is now the most widely accepted.

The general FITSC method, which involves the application of different nonzero voltages in the cooling and heating steps, has only been used by a few authors [4.1, 23]. It was called stimulated dielectric relaxation currents method (SDRC) by *Simmons* and *Taylor* in their study of metal–insulator–metal systems in which the electrodes provided blocking contacts with the insulator [4.23].

During the past few years, the experimental and theoretical development of FITSC related to various types of charge storage mechanisms has reached a high, but often confusing, level which makes it desirable to clarify the whole situation and to outline their differences and their common features. As a matter of fact, a great number of theoretical models have been formulated describing the temperature dependence of the current (mainly TSDC). However, the proposed mathematical expressions, generally based on too simplified and unrealistic assumptions, predict similar functional relationships for most of the polarization processes. Since, in addition, the microscopic parameters involved cannot be measured, the comparison between theory and experiment is rarely definitive. In fact, it is one of the major problems of these measurements to unequivocally determine the physical origin of the observed current peaks which must obviously be known prior to any mathematical analysis. This is not an easy task and a great deal of controversy still surrounds the interpretation of experimental data for most of the materials tested.

4.2 Polarization Mechanisms and Basic Principles of the FITSC Method

The polarization of a solid dielectric submitted to an external electric field may occur by a number of mechanisms involving either microscopic or macroscopic charge displacement.

I) *The electronic polarization* is the fastest process, requiring about 10^{-15} s. It results from the deformation of the electronic shell.

II) *The atomic polarization*, requiring about 10^{-14} to 10^{-12} s, results from the atomic displacement in molecules with heteropolar bonds. This is also a deformation effect.

III) *The orientational or dipolar polarization* occurs in materials containing permanent molecular or ionic dipoles. Depending upon the frictional resistance of the medium, the time required for this process can be as low as 10^{-12} s or so long that no relaxation is observed under the conditions of observation.

IV) *The translational or space-charge polarization*, which is observed in materials containing intrinsic free charges (ions or electrons or both), is due to a macroscopic charge transferred towards the electrodes acting as total or partial barriers. The time required can vary from milliseconds to years.

V) *The interfacial polarization*, sometimes referred to as Maxwell–Wagner–Sillars polarization (MWS polarization), is characteristic of systems with a heterogeneous structure. It results from the formation of charged layers at the interfaces, due to unequal conduction currents within the various phases. The time scale is also from milliseconds to years. Nearly equivalent to an interfacial polarization is the migration of free charges over microscopic distances with subsequent trapping (*Gerson–Rohrbaugh*'s model [4.24]). This bulk process can also be considered as somewhat similar to an ionic dipolar polarization.

All the above-mentioned effects are relevant to internal polarization phenomena, i.e., they are produced by the rotation or migration of charges originating from and remaining within the dielectric. They lead therefore to surface charges which have the opposite polarity to those of the polarizing electrode (heterocharges).

VI) A space-charge polarization of extrinsic origin (homocharge) may be additionally created in dielectrics submitted to high fields when excess electronic or ionic charge carriers are generated either by injection mechanisms from the electrodes (Schottky emission) or from Townsend breakdowns in the surrounding atmosphere (for imperfect electrode–dielectric contacts). A similar type of polarization is obtained by submitting dielectrics to electronic bombardment or corona discharges.

Owing to the possibility of deep trapping phenomena, a field-induced homocharge injection usually leads, in high-resistivity materials, to a polarized state characterized by a long lifetime (electret effect) without requiring further experimental procedures. However, this is generally not so for a polarization of the heterocharge type which rapidly disappears when the field is removed because thermal motions tend to reestablish the disorder in the elementary dipole directions and internal fields constrain the free carriers to return to their normal distribution. To obtain a permanent heterocharge in shorted samples then necessitates the combined use of electric field and heating–cooling sequences allowing the polarization to be "frozen-in".

4.2.1 Freezing-in the Heterocharge Polarization. The Thermoelectret State

Since internal friction and ionic mobility depend exponentially on temperature, heating a dielectric to a high temperature T_p markedly enhances the response time of permanent dipoles and internal free charges to an electric field and usually allows the equilibrium polarization to be reached in a reasonably short time. If the field is then maintained while cooling the sample to a temperature T_0 sufficiently low to increase the relaxation times of dipoles and ions to values of hours or more, these are practically "frozen" in the electrical configuration reached at T_p and, consequently, do not respond when the field is switched off; only the electronic and atomic components of the polarization instantaneously adjust to the new conditions since they are intramolecular, and thus nearly temperature-independent effects. The dielectric, maintained at T_0, is now bearing a persistent charge and is called a thermoelectret (electret obtained via application of an electric field during a thermal cycle).

It should be noted that a polarization of the heterocharge type can also be induced in a dielectric as a result of electron or ion trapping if the electric field is applied while irradiating the sample with light (photoelectret effect) or ionizing radiation (radioelectret effect).

4.2.2 Charging and Discharging Currents

The Isothermal Case

When a polarized sample is placed between the electrodes of a capacitor, every increase or decrease of the polarization frees image charges which flow back through the external circuit where a current may then be recorded. At low temperature, where any orientational or translational motion is hindered, only the charging and discharging currents of the ordinary dielectric capacitor are observed and the minimum rise time of the current pulses is then a few tenths of a second. At higher temperature, on the other hand, orientational and translational motion can occur and the corresponding polarization and depolarization currents dominate. These are the well-known transient currents which generally decay, at least over several decades of time, according to the Curie–Von Schweidler law. These transient currents have been widely used for investigating buildup and dissipation of internal polarization but the observed time dependence does not usually permit any discrimination to be made between the various polarization models [4.25, 26].

The Nonisothermal Case: TSDC and TSPC

Thermally Stimulated Depolarization Currents. By heating a polarized sample up to or above the polarization temperature, the release of charges is gradually sped up and, when the half-life of this process becomes comparable with the time scale of the experiment (determined by the heating rate), discharge becomes measurable and gives rise in the external circuit to a current which first increases with increasing temperature, and then decays when the supply of charges is depleted (Fig. 4.1). A current peak thus will be observed at a temperature where dipolar disorientation, ionic migration or carrier release from traps is activated and, as the total polarization usually arises from a combination of several individual effects with various relaxation times, a complete picture of the temperature-dependent relaxations will in principle be obtained (Fig. 4.2). As far as dipolar phenomena are concerned, such a TSDC relaxation spectrum is somewhat similar to the dielectric loss curves plotted as a function of temperature (ε''-T measurements). The TSDC method is, however, inherently more sensitive and owing to its low equivalent frequency (Sect. 4.7) the resolution is usually much better.

On the other hand, depending on the decay process and its efficiency in releasing charge, the current passing through the ammeter will be equal to or less than the net polarization of the polarized sample. For a dipolar relaxation, the efficiency will obviously be 100 %, but this will not usually be true for space-charge or interfacial relaxation (Sects. 4.8–10). An additional difficulty could also occur when homo- and heterocharges coexist. If the homocharge decays by external conduction, the homo- and heterocurrents will have the same direction, but if the decay proceeds by internal conduction, the direction of the corresponding external current will be opposite to the normal heterocurrent, resulting in possible current reversals if the relaxation times of the two types of charges differ widely (Fig. 4.1).

Thermally Stimulated Polarization Currents. If an unpolarized state is first fixed at a low temperature by cooling the sample under short-circuit conditions and an electric field is then applied during a subsequent definite heating, the thermally stimulated transition from neutrality to a polarized state can be followed by registering the charging current as a function of temperature. If dipolar or ionic processes are involved, these thermally stimulated polarization currents will be characterized by peaks roughly similar to those occurring in

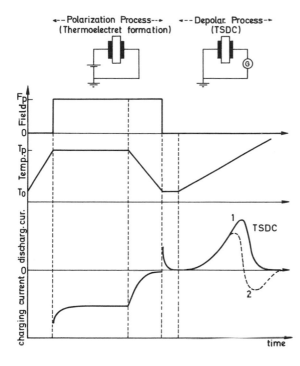

Fig. 4.1. Thermoelectret formation and principle of the TSDC method: (1) heterocharge only present; (2) coexistence of hetero- and homocharge

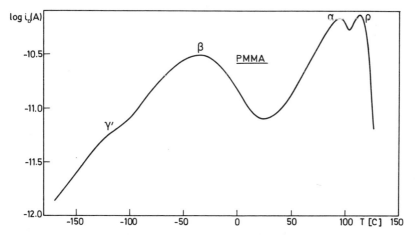

Fig. 4.2. Example of an experimental TSDC spectrum. The sample is poly(methyl methacrylate), polarized at 393 K with $F_p = 2 \cdot 10^4$ V/cm. The peaks are attributed to ionic migration (ϱ), cooperative motions of large parts of the polar polymer chain (α), local dipolar motions (β), and absorbed water (γ') [4.27]

TSDC experiments but they will obviously appear in a reverse direction, at least at the beginning of the spectrum (Sect. 4.11). This method can thus provide complementary information on the relaxational behavior of dielectrics but is subject to additional difficulties resulting from the possible superposition of internal conduction or injection currents.

4.3 Experimental Considerations

The experimental techniques of investigating the thermally stimulated process and the general principles of complete data acquisition with simple treatment have been reviewed in detail elsewhere in this volume and, therefore, will not be discussed further. Let us simply recall that the essential parts of the equipment needed are I) a thermostated sample holder which can be evacuated or filled with an inert gas to ensure a good thermal equilibrium between the sample and heater (this gas, usually nitrogen or helium, should in most instances be dry or of known water vapor content since the electrical properties of most materials are notoriously sensitive to traces of moisture), II) a heating unit which should be designed to give a variable but linear heating rate over a wide range of temperatures (in FITSC measurements, where the more commonly used sample thicknesses range from a few microns to a few millimeters and the insulators for the electrodes are usually characterized by small heat capacity, the heating rate is often chosen relatively low – 0.01 to $0.2\,\mathrm{K\,s^{-1}}$ – in order to reduce possible temperature gradients across the material investigated), III) a stabilized dc generator, IV) a sensitive current detector which typically should be able to measure currents in the range 10^{-4} to $10^{-14}\,\mathrm{A}$ and V) a current–temperature recording unit which may be an $X - Y$ recorder or a two-channel stripchart recorder.

The specimens are generally used in form of films, plates or single crystals but polycrystalline compact or lightly compressed powder have also been tested [4.28]. Like in conventional conductivity measurements [4.29], they must ideally be provided with guarded electrodes in order to avoid field inhomogeneities and minimize surface conduction. To reach the highest possible sensitivity when investigating bulk processes, a direct contact between sample and electrodes must be achieved and maintained during the whole experiment; while many different electrode systems have been studied to this end, the most convenient ones seem to be evaporated films of silver, aluminum or gold. When measuring free-charge displacement currents, on the other hand, it is often advisable to use blocking electrodes or electrodes insulated from the samples by air gaps or thin foils of teflon, mylar or mica [4.19, 30, 31].

Further details related to the possible spurious currents arising from various experimental arrangements in FITSC experiments are given in Sect. 4.12.1.

4.4 Theory of Dipolar TSDC in Dielectrics
with a Single Relaxation Time

The structural interpretation of dielectric relaxation processes occurring in many polar materials is usually approached by assuming impaired motions or limited jumps of permanent electric dipoles. In molecular compounds, for example, relaxation can be considered as arising from the hindered rotation of the molecule as a whole, of small units of the molecule or of some flexible group around its bond to the main chain, while in ionic crystals, it can be mainly associated with ionic jumps between neighboring sites (ion–vacancy pairs). Whatever the case, dipolar mechanisms such as these are spatially uniform, which means that in absence of interacting processes, they can be treated theoretically with a minimal number of simplifying assumptions and can be relatively easily distinguished from space-charge formation (Sect. 4.12).

It is known from conventional dielectric measurements that materials obeying the classical Debye treatment with a single relaxation time are rather rare [4.38]. They are mainly found among certain dilute solutions of polar molecules in nonpolar solvents and in some solids such as divalent ion-doped alkali halides where each cation impurity–vacancy dipole may be hoped to relax independently when the concentration of impurities is low. A number of measurements on ionic crystals of the NaCl type using the TSDC technique have been reported in the literature [4.8, 39–42]. They have generally confirmed Debye behavior and, on the basis of these results, the first consistent theory of TSDC applied to dipolar processes was put forward in 1964 by *Bucci* and *Fieschi* [4.8]. A similar theory was independently proposed by *Nedetzka* et al. in 1969 to explain the nonisothermal polarization properties of molecular organic materials [4.12]. Both were in fact dipolar forms of the theories used for describing the thermally stimulated conductivity and thermoluminescence induced by thermal detrapping of charge carriers in dielectrics and semiconductors [4.2] and let to similar results as far as the shape of the peaks and the determination of the characteristic parameters were concerned. Since then the theory has been generalized and considerably extended, in particular to include data obtained in materials with distributed relaxation times, by *Gross* [4.32], *Vassilev* et al. [4.33, 34], *Solunov* and *Ponevsky* [4.35, 36], and especially *Van Turnhout* [4.10, 30] and *Ong* and *Van Turnhout* [4.37] (Sect. 4.5).

In most of these theoretical treatments, the polarized material is assumed to be free of charge carriers, so that the internal field and the dipolar polarization can be considered as space independent. In practice, however, dipolar and space-charge polarizations often coexist, particularly in thermoelectrets formed at high temperature (Fig. 4.2), and the electric field and polarization must then be considered as averaged over the thickness of the sample [4.30]. Furthermore, the simultaneous displacement of free charge carriers and dipoles during the polarization process may lead to a particular situation where the internal electric field is nearly zero, so that no preferred orientation of dipoles occurs. On the other hand, this implies that, after removing the voltage at low temperature, an internal electric field, due to the fixed space charges, exists inside the sample and that, during the subsequent heating, this field will be responsible for a gradual orientation of the dipoles, leading to a current peak quite similar to a true disorientational peak [4.19]. The possibility of observing such a process, which in fact is an "internal thermally stimulated polarization process", has been demonstrated by *Van Turnhout* [4.30] in irradiated monoelectrets (i.e., unipolarly charged electrets), but this point is still open to discussion for classical thermoelectrets.

In the following, we will only consider the theories of TSDC resulting from a homogeneous polarization occurring in dielectrics free of charges.

4.4.1 The Bucci–Fieschi Theory

The time and temperature dependence of the dipolar polarization is determined by the competition between the orienting action of the field and the randomizing action of thermal motions. Assuming an ideal rotational friction model (Debye) or a symmetrical two-site barrier model (Fröhlich) for establishing the polarization, the buildup of polarization P in a unit volume of the material during time t after the application of an electric field F_p at a temperature T_p is described, in the elementary theory of dielectrics, by an exponential function of time

$$P(t) = P_e \left[1 - \exp\left(-\frac{t}{\tau} \right) \right], \tag{4.1}$$

where τ is the dipolar relaxation time and P_e is the equilibrium or steady-state polarization which, for all but the lowest temperatures and very high fields (i.e., below electrical saturation) has been shown by *Langevin* to be

$$P_e = \frac{s N_d p_\mu^2 \kappa F_p}{k T_p}. \tag{4.2}$$

In this expression s is a geometrical factor depending on the possible dipolar orientation (for free rotating dipoles $s = 1/3$ while for nearest-neighbor face-centered vacancy positions in ionic crystals, $s = 2/3$), N_d the concentration of dipoles, p_μ their electrical moment, k Boltzmann's constant and κF_p the local directing electric field operating on the dipoles [4.43].

Provided that the relaxation times for polarizing and depolarizing the dielectric can be considered identical, the decay of polarization after removal of the field at $t = \infty$ is given by

$$P(t) = P_e \exp\left(-\frac{t}{\tau} \right) \tag{4.3}$$

and the corresponding depolarization current density can be written

$$J(t) = -\frac{dP(t)}{dt} = \frac{P(t)}{\tau}. \tag{4.4}$$

In order to obtain the current density produced by the progressive decrease in polarization in the course of a TSDC experiment, where time and temperature are simultaneously varied, the differentiation must be performed in terms of the new variable T. This parameter can be introduced by assuming a simple temperature program, most generally a linearly increasing temperature from a temperature T_0, so that

$$T = T_0 + qt, \tag{4.5}$$

where $q = dT/dt$ is the heating rate.
  Rewriting (4.3) to be

$$P(t) = P_e \left[\exp\left(-\int_0^t \frac{dt}{\tau} \right) \right] \tag{4.6}$$

and postulating that I) this relation also holds for varying temperature, II) the initially frozen-in polarization $P(T_0)$ is equal to the equilibrium polarization reached at the polarizing temperature $P_e(T_p)$ and III) the temperature variation of τ is given by an Arrhenius-type equation

$$\tau(T) = \tau_0 \exp\left(\frac{E}{kT} \right), \tag{4.7}$$

where τ_0 is the relaxation time at infinite temperature (τ_0^{-1} is the characteristic frequency factor and is usually directly related to the vibrational frequency of the material) and E is the activation energy of dipolar orientation or disorientation, the current density J_D during a TSDC experiment is

$$J_D(T) = \frac{P_e(T_p)}{\tau_0} \exp\left(-\frac{E}{kT} \right) \exp\left[-\frac{1}{q\tau_0} \int_{T_0}^T \exp\left(-\frac{E}{kT'} \right) dT' \right]. \tag{4.8}$$

This expression, similar to those describing thermoluminescence or thermally stimulated conductivity processes obeying first-order kinetics [4.2], represents an asymmetrical "glow curve" (Fig. 4.1) the amplitude of which is a linear function of the previously applied field. The first exponential, which dominates in the low-temperature range, is responsible for the initial increase of the current with temperature (increase of mobility of the rotating dipoles), while the second exponential, which dominates at high temperature, gradually slows down the current rise and then depresses it very rapidly, especially for high activation energies (progressive exhaustion of the induced polarization). It is obvious from (4.8) that the only parameters affecting the shape of the curve will be the heating rate q in combination with the characteristic frequency factor τ_0^{-1} and the activation energy E. Their mutual influence can be best appreciated in the following transcendental equation obtained by differentiating (4.8) to arrive at the maximum temperature T_m of the peak

$$T_m = \left[\frac{E}{k} q\tau_0 \exp\left(\frac{E}{kT_m}\right)\right]^{1/2}. \tag{4.9}$$

Using typical values in this expression, it is readily seen that small variations in q or τ_0 do not markedly affect the quantity $E/kT_m \equiv y_m$ (changing q by one order of magnitude, for example, and starting from values of y_m lying in the usual range of 15–40, y_m varies no more than 1–3 units) and, consequently, involve only small changes in curve shape.

On the other hand, (4.9) shows that the position of a dipolar TSDC peak will be an increasing function of the parameters q, τ_0 and E and that, for a given q, it will have a fixed value characteristic of the dielectric since then it only depends on material constants.[1]

The important role played by the heating rate in the amplitude (4.8) as well as in the position of the peak (4.9) is obvious: When the heating rate increases, the initial polarization has to be released in a shorter time while the dielectric responds less quickly. Thus the peak increases and shifts to a higher temperature. As seen in Fig. 4.3 however, a significant displacement can only be observed for large variations in the heating rate, which is also apparent from (4.9). If one chooses typical values $E = 0.6\,\text{eV}$, $\tau_0 = 10^{-10}\,\text{s}$, $q = 10^{-2}\,\text{K s}^{-1}$, for example, a ten percent increase of q will only change the peak position from 242.2 to 242.9 K. This implies that small departures from linearity during temperature programming will not be too critical.

In principle, another important consequence of a variation in heating rate will be to change the resolution of the TSDC spectrum [4.30]. When several relaxation peaks overlap each other, a decrease in heating rate should increase the resolving power in the same way as in dielectric measurements the resolution increases by lowering the measuring frequency [4.46]. In practice, however, the use of large differences in heating rate is usually not possible, owing to probable temperature lags, and the relative shifts and shape variations of neighboring peaks will thus be small, unless these are governed by quite different τ_0 and E values. Furthermore, a decrease in heating rate will simultaneously involve a decrease in current intensity and thus in the signal-to-

---

1 As derived by *Vassilev* et al. [4.44], the dependence of the maximum temperature on the heating rate may be limited under certain conditions, in particular for the relaxation processes occurring in the vicinity of the glass transition temperature (T_g) in polymers and glasses (T_g can be defined as the temperature at which the bend in the measured volume–temperature curves occurs at the lowest practical rate of cooling). In poly (methyl methacrylate), for example, ($T_g = 278\,\text{K}$), a saturated value $T_m = 397\,\text{K}$ is to be expected for heating rates $q > 8\,10^{-2}\,\text{K s}^{-1}$.

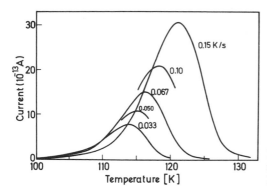

Fig. 4.3. Effect of variation in heating rate on TSDC observed in AgCl: 700 ppm Ni [4.45]

noise ratio. The choice of a heating rate for a given experiment will therefore rarely be determined by resolution considerations alone but rather by a compromise taking into account the signal intensity and the possibility of temperature gradients in the sample (Sect. 4.3).

4.4.2 Equivalent-Model Theory

It is well known that a number of dielectric relaxation effects can be satisfactorily described by means of linear systems and circuit theory [4.38]. As pointed out by *Gross*, however, the use of such equivalent models to calculate thermally stimulated depolarization currents is subject to severe restrictions [4.47]. As a matter of fact, under nonisothermal conditions, the values of model elements do not remain constant but are functions of the independent variable (time) via temperature-induced changes in resistance and capacitance and, therefore, we cannot presume *a priori* that models which are equivalent at constant temperature remain so during temperature variations. The simple Maxwell and Wagner models of a dielectric (Fig. 4.4), for example, although strictly equivalent under isothermal conditions, lead to externally released charges differing from each other during a temperature increase. As a consequence, the corresponding TSDC expressions are also different. An equation similar to (4.8) can only be derived from the Maxwell model, which, unlike the Wagner model, does not exhibit internal leakage in short-circuit and is thus charge-invariant [4.32].

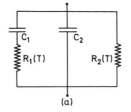
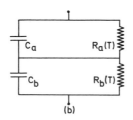

Fig. 4.4a and b. The Maxwell (**a**) and Wagner (**b**) models of a dielectric

4.4.3 Temperature Dependence of the Relaxation Time

The Arrhenius expression used in the *Bucci–Fieschi* theory is usually appropriate only in the description of the temperature-dependent properties of a Debye or Fröhlich model. As such, it has been found to fit satisfactorily experimental data in many cases, particularly for ionic solids [4.38]. On the other hand, considerable deviations from this law have been reported in a number of dielectrics, especially organics, where other relaxation models are involved. Even in certain ionic crystals the validity of the Arrhenius formula may be questionable. This is the case, for example, in ion-doped lithium hydrides where its use in analyzing the relaxation processes at temperatures below the Debye temperature often leads to unrealistic τ_0 values [4.48]. Other $\tau(T)$ functions are then required. Except when quantum mechanical models are involved they still contain, however, an exponential term (the so-called shift factor) and differ only in detail depending on the physical model or the material envisaged.

Where a rate process is assumed, the preexponential factor is no longer temperature independent. The following expression, derived from the chemical rate theory of *Eyring* [4.49] has been used by some authors, in particular for organic materials [4.50, 51]

$$\tau(T) = \tau_0 T^{-1} \exp\left(\frac{E}{kT}\right). \tag{4.10}$$

It is apparent, however, that under normal conditions ($E \gg kT$) this additional small temperature dependence remains negligible with regard to the exponential factor and that the corresponding TSDC signal is not substantially modified in shape or in position. This obviously implies that relaxation times of the Eyring and Arrhenius type can hardly be differentiated solely on the basis of TSDC data.

In polymers or, generally speaking, in glass-forming materials, the relaxation processes due to local motions occurring at low temperature can usually be described using $\tau(T)$ functions of the Arrhenius or Eyring type but this is no longer true at temperatures higher than the glass transition temperature T_g. In this temperature range, as shown by *Williams*, *Landel* and *Ferry* [4.52], the relaxation time often obeys the empirical equation

$$\tau(T) = \tau_0 \exp\left[-\frac{U_1(T-T_g)}{U_2+T-T_g}\right], \tag{4.11}$$

where U_1 and U_2 are universal constants. This so-called WLF equation, which has lately received a plausible theoretical basis in terms of concepts of free volume [4.53, 54] or statistical thermodynamics [4.55], can easily be converted into a form similar to the Arrhenius formula [4.30]

$$\tau(T) = \tau_0' \exp\left[\frac{E_w}{k(T-T_\infty)}\right], \tag{4.12}$$

where $\tau_0' = \tau_0/\exp U_1$, $E_w = kU_1U_2$ and $T_\infty = T_g - U_2$. Consequently, this equation leads to TSDC peaks with shape properties nearly similar to the classical case [4.30], except that their initial slope can no longer be proportional to E/kT (Sect. 4.6.2). Another particularity of this equation is that it suggests that the ordering in the material increases with falling temperature, until relaxation becomes infinitely slow at T_∞ and, therefore, it is no longer valid for $T < T_\infty$. In fact, below T_g, the measured temperature dependence of the relaxation time closely resembles an Arrhenius shift and the data are thus often represented by a hybrid of (4.7) and (4.11) [4.56]. Another means of extending the applicability of the WLF expression is to introduce in (4.11) an effective temperature T_e instead of T [4.57]. Dipolar relaxation can also proceed via tunneling. However, since in this case τ is only weakly T-dependent, this process cannot be monitored by TSDC [4.28].

4.4.4 Analytical Expressions

The integral in (4.8) is similar to those appearing in the usual TSC or TL expressions, the approximative resolution of which has been discussed by many authors [4.58–60]. It can be expressed in terms of the exponential integral, the asymptotic expansion of which is, in the simplest case of an Arrhenius shift,

$$\int_0^T \exp(-y)dy = \frac{\exp(-y)}{y^2}\left(1 - \frac{2!}{y} + \frac{3!}{y^2}\cdots\right), \tag{4.13}$$

where $y = E/kT$. Furthermore, since y is large in practice, it is legitimate to drop all but the first term in the series:

$$J_D(T) \simeq \frac{P_e(T_p)}{\tau_0}\exp\left(-\frac{E}{kT}\right)\exp\left[-\frac{1}{q\tau_0}\frac{kT^2}{E}\exp\left(-\frac{E}{kT}\right)\right]. \tag{4.14}$$

Similar expressions are easily obtained when other $\tau(T)$ functions are involved except that, with a relaxation time obeying the Eyring theory, T^2 must be replaced by T^3 [4.12] while, with a relaxation time of the WLF type, T must be replaced by $(T - T_\infty)$ or $(T_e - T_\infty)$ [4.30].

4.4.5 Steady-State and Frozen-In Polarization

In deriving (4.8), it was implicitly assumed that the frozen-in polarization of the sample $P(T_0)$ before a TSDC experiment is started, can be considered as equal to the equilibrium polarization at T_p, i.e.,

$$P(T_0) \equiv P_e(T_p). \tag{4.15}$$

In fact, this identity can only be postulated if two conditions are initially fulfilled:

1) The first one is obvious and easily realized in practice: the equilibrium polarization must be reached at T_p $[t_p \gg \tau(T_p)]$ and no significant discharge must occur at $T_0(T_0 \ll T_p)$. Otherwise, $P(T_0)$ will be a function of the prior temperature programs and must then be related to the formation conditions following [4.30]

$$P(T_0) = P_e(T_p) H(T,t) \tag{4.16}$$

where $H(T,t)$ is a function characterizing the filling state of the polarized sample

$$H(T,t) = \left\{ 1 - \exp\left[-\frac{t_p}{\tau(T_p)} - \frac{1}{q} \int_{T_p}^{T_0} \frac{dT}{\tau(T)} \right] \right\} \exp\left[-\frac{t_0}{\tau(T_0)} \right]. \tag{4.17}$$

Here t_0 is the short-circuit storage time at T_0. In this expression, the first term determines the degree of filling produced by the polarizing field during the isothermal formation at T_p and the cooling period from T_p to T_0 whereas the exponential term accounts for the decay during the storage period.

2) The second condition is that no change in the initially induced equilibrium polarization occurs during cooling. Unfortunately, except for a temperature-independent equilibrium polarization, this is an unrealistic condition since it corresponds to an instantaneous cooling rate $(-dT/dt \to \infty)$. In fact, during cooling the polarization is necessarily increasing with decreasing temperature following the Langevin function or any other temperature-dependent relation until the relaxation time is not essentially larger compared to the time interval during which the sample is being cooled. As shown by Harasta [4.61], this increase in polarization can be quite large, depending on the cooling rate used (Fig. 4.5). It obviously follows that the final frozen-in polarization will always be larger than the equilibrium polarization reached at T_p and will depend on the cooling rate. This also implies that the magnitude and area of a TSDC peak, which are proportional to $P(T_0)$, will be functions of the cooling rate. This phenomenon, experimentally observed in various materials such as ionic solids [4.61] or polymers [4.62], has been overlooked by most authors working on thermally stimulated depolarization currents. As we shall see later, however, neglecting it can lead to important errors when the relaxation parameters included in the Langevin function (N_d and p_μ) are determined directly from the TSDC curve (Sect. 4.6.3). Furthermore, an additional difficulty for the evaluation of these TSDC data can also arise when the Langevin function does not hold in the temperature range investigated. This is the case, for example, in the vicinity of the glass transition temperature of polymers, where an abrupt variation of the P vs T relationship is often observed [4.46].

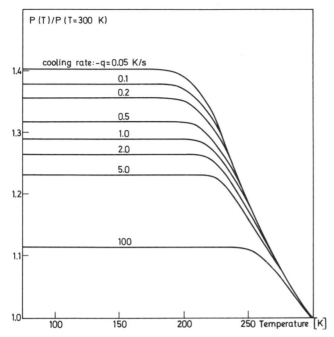

Fig. 4.5. Increase of stored polarization during cooling under constant field in the hypothesis of equilibrium polarization obeying the Langevin function. $\tau_0 = 10^{-13}$ s, $E = 0.67$ eV, $N_d = 10^{18}/\text{cm}^3$, $F_p = 10^3$ V/cm, $p_\mu = 14.4$ Debye [4.61]

4.5 Theory of Dipolar TSDC in Dielectrics with a Distribution in Relaxation Times

A great number of conventional dielectric measurements have shown that most of the relaxation processes occurring in solids deviate considerably from the simple Debye behavior exemplified by (4.1) [4.38, 46]. The method of fitting the experimental data consists then of expressing the dielectric constant and loss by means of equations involving empirical parameters characteristic of some kind of distribution function of relaxation times. Physically, the existence of a spectrum of relaxation times can be explained by several mechanisms, including dipole–dipole interactions, variation in size and shape of the rotating dipolar entities, anisotropy of the internal field in which the dipoles reorient, etc., which is especially plausible not only in complicated molecules such as polymers, where there are many possibilities of internal rotation, bending and twisting, each with a corresponding characteristic relaxation time, but also in any material with a high dipole concentration.

Considering an Arrhenius equation for describing the temperature dependence of relaxation times, either the pre-exponential factor or the activation energy, or both, may be distributed. There is an important difference between

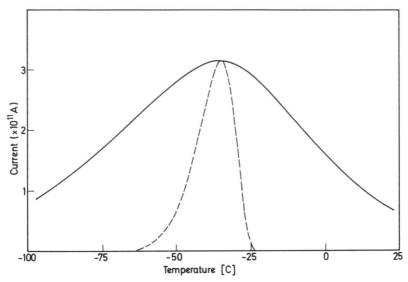

Fig. 4.6. Comparison of the TSDC β peak of Poly(methyl methacrylate) (———) with a theoretical peak computed according to (4.8) using the value of activation energy deduced from independent ac measurements (0.9 eV) (– – –). The two curves are normalized at the maximum

these two types of distribution in that, to account for a given distribution in relaxation times, a much larger range in τ_0 than in E is needed. A choice between these various possibilities is, however, often difficult and will depend on the material and range of temperature involved. In glassy polymeric systems for example, the assumption of a reasonably narrow range of energy barriers to dipolar reorientation is sufficient to lead to a large spectrum of relaxation times, allowing a realistic description of experimental results (Sect. 4.6.5).

The need for taking into consideration such distributions to interpret TSDC data has been emphasized by many authors. It is a very general fact in polymers [4.30, 33–37] for example, that the peaks extend over a wide temperature range and are flatter and more symmetrical than expected from the simple expression (4.8) (Fig. 4.6). It also has often been observed in a number of materials, organic [4.63] as well as inorganic [4.64–67], that the maximum temperature does not behave as (4.9), i.e., is not independent of, the polarization conditions, and that the amount of broadening of a peak is a direct function of dipole concentration [4.64]. These facts are often strong arguments in favor of distributed processes (Sect. 4.5.4).

4.5.1 General TSDC Formulations

As shown by *Gross* [4.32] and *Van Turnhout* [4.30], the relaxation processes involving a spectrum of relaxation times can still be described by the theory outlined in the preceding section, provided that an appropriate distribution function is introduced. For a continuous distribution in τ_0, for example [4.37, 65], and assuming that the dipole–dipole interactions are negligible and that the relaxation times τ_i of all the individual relaxations experience the same shift factor (i.e., that the time–temperature superposition principle is respected), the total polarization can be obtained by summing the differential polarization contributions over the whole relaxation time range

$$P(t) = \int_0^\infty P_i(t, \tau_0) d\tau_0 . \tag{4.18}$$

Taking into account the formation and storage conditions and neglecting the temperature dependence of the equilibrium polarization (Sect. 4.4.5), we can then write at each temperature during a TSDC experiment [4.37]

$$P(T) = P_e(T_p) \int_0^\infty H(\tau_0) f(\tau_0) \exp\left[-\frac{1}{q\tau_0} \int_{T_0}^T \exp\left(-\frac{E}{kT'}\right) dT'\right] d\tau_0 , \tag{4.19}$$

where $H(\tau_0)$ is the filling state function of each individual polarization and $f(\tau_0)$ is the normalized distribution function

$$\int_0^\infty f(\tau_0) d\tau_0 = 1 . \tag{4.20}$$

By differentiating (4.19), the current density then takes the form

$$J_D(T) = P_e(T_p) \exp\left(-\frac{E}{kT}\right) \int_0^\infty \tau_0^{-1} f(\tau_0) H(\tau_0) \exp\left[-\frac{1}{q\tau_0} \int_{T_0}^T \exp\left(-\frac{E}{kT'}\right) dT'\right] d\tau_0 . \tag{4.21}$$

A similar expression is obtained for a temperature-independent distribution function of activation energy, $g(E)$, provided that the pre-exponential factor τ_0 is then supposed to be common to all relaxation times [4.37, 68, 69]

$$J_D(T) = P_e(t_p) \tau_0^{-1} \int_0^\infty g(E) H(E) \exp\left[-\frac{E}{kT} - \frac{1}{q\tau_0} \int_{T_0}^T \exp\left(-\frac{E}{kT'}\right) dT'\right] dE . \tag{4.22}$$

Whatever the type of distribution involved, the current density remains proportional to $P_e(T_p)$ and thus to the polarizing field F_p. On the other hand, the peaks will obviously be broadened and, owing to the appearance of the filling state functions $H(\tau_0)$ or $H(E)$ within the integral, not only their amplitude but also their shape and position will be strongly affected by the formation and storage conditions. This feature, already predicted by *Bucci* and *Riva* [4.70], is intuitively obvious since each individual polarization will be specifically affected and thus will contribute more or less to the final TSDC spectrum.

Table 4.1. Most common distribution functions applicable to dipolar relaxation processes

References	$f'(u)$	Limits of τ	Limits of the distribution parameters
Symmetrical distributions			
Fröhlich-Gevers [4.74,75]	$(u_2 - u_1)^{-1}$	$\tau_1 < \tau < \tau_2$	
Wagner [4.71]	$\dfrac{\bar{\alpha}}{\sqrt{\pi}}\exp\left(-\bar{\alpha}u^2\right)$	$-\infty < \tau < +\infty$	$0 < \bar{\alpha} < \infty$
Cole-Cole [4.76]	$\dfrac{\sin(\bar{\beta}\pi)}{2\pi[\cosh(\bar{\beta}u) + \cos(\bar{\beta}\pi)]}$	$-\infty < \tau < +\infty$	$0 < \bar{\beta} \leq 1$
Fuoss-Kirkwood [4.77]	$\dfrac{\bar{\gamma}\cos(\bar{\gamma}\pi/2)\cosh(\bar{\gamma}u)}{\pi[\cos^2(\bar{\gamma}\pi/2) + \sinh^2(\bar{\gamma}u)]}$	$-\infty < \tau < +\infty$	$0 < \bar{\gamma} \leq 1$
Asymmetrical distributions			
Davidson-Cole [4.78]	$\begin{cases} \dfrac{\sin(\bar{\delta}\pi)}{\pi}\left[\dfrac{1}{\exp(-u)-1}\right]^{\bar{\delta}} \\ 0 \end{cases}$	$\begin{array}{l} -\infty < \tau < \tau_m \\ \tau > \tau_m \end{array}$	$0 < \bar{\delta} \leq 1$
Havriliak-Negami [4.79]	$(1/\pi)\exp[u\bar{\varepsilon}(1-\bar{\eta})]$ $\cdot\sin\left\{\bar{\varepsilon}\arctan\left[\dfrac{\sin\pi(1-\bar{\eta})}{\exp[u(1-\bar{\eta})]+\cos\pi(1-\bar{\eta})}\right]\right\}$ $\cdot\{\exp[2u(1-\bar{\eta})] + 2\exp[u(1-\bar{\eta})]\cos[\pi(1-\bar{\eta})] + 1\}^{-\bar{\varepsilon}/2}$	$-\infty < \tau < +\infty$	$\begin{array}{l} 0 < \bar{\varepsilon} \leq 1 \\ 0 \leq \bar{\eta} < 1 \end{array}$

4.5.2 The Distribution Functions

The first mathematical distribution function was derived in 1913 by *Wagner* [4.71] by assuming that an infinite number of independent causes disturb an original relaxation time τ_m. This assumption leads to a Gaussian probability function which causes the dielectric constant and loss to deviate from the Debye curve in the same way as found experimentally. Since then, several other forms of distribution functions have been suggested by many authors. Most of these functions have been compiled by *Gross* [4.72] and *Van Roggen* [4.73] and those most frequently used for describing dielectric and TSDC data are assembled in Table 4.1. For convenience, they are expressed in terms of the variable $u = \ln(\tau/\tau_m)$, so that

$$\int_{-\infty}^{+\infty} f'(u)du = 1 \,. \tag{4.23}$$

Following the type of distribution involved (τ_0 or E), u will be equal to $\ln \tau_0/\tau_{0_m}$ or $(E - E_m)/kT$, τ_{0_m} and E_m being the most probable values of τ_0 and E, respectively, see (4.7).

The Fröhlich–Gevers function is a box distribution which has been used essentially only as a teaching model. The Cole–Cole and Fuoss–Kirkwood functions, empirically derived from dielectric data relating to a considerable number of liquids and solids [4.76, 77, 80], are physically more plausible as they do not differ much from a Gaussian distribution. As shown by *Kauzmann* [4.50], these three distributions can be made to coincide fairly closely by a suitable choice of the parameters $\bar{\alpha}$, $\bar{\beta}$ and $\bar{\gamma}$. Both, the Cole–Cole and Fuoss–Kirkwood distributions, fall off more slowly towards extreme values of u than the Gaussian distribution (Fig. 4.7).

The asymmetrical distributions are also empirical functions used to conveniently represent the asymmetrical dielectric loss curves observed in certain materials such as polyhydroxycompounds [4.78], tricresylphosphate [4.81] or

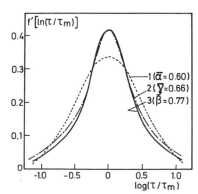

Fig. 4.7. Shape of some common symmetrical distribution functions: (*1*) Wagner; (*2*) Fuoss-Kirkwood; (*3*) Cole-Cole

polar polymers in the glass transition range [4.79, 82–84]. The Davidson–Cole distribution is a special function which increases without limit as τ/τ_m increases from zero to unity and is zero for all values of τ/τ_m greater than one, while the Havriliak–Negami function, where $\bar{\eta}$ and $\bar{\varepsilon}$ respectively represent the breadth and the skewness of the distribution, is a generalization of the Cole–Cole and Davidson–Cole functions.

A number of dielectric measurements show clearly a gradual narrowing of loss peaks with rising temperature, suggesting that the distribution function is temperature dependent [4.46]. For analyzing TSDC data which are, by nature, nonisothermal measurements, the knowledge of this dependence would be particularly important. It is, however, a very complicated problem since both the most probable relaxation time and the width of the distribution change with temperature during a measuring run. Realizing that none of the functions discussed above is really suitable for a detailed investigation of such behavior, *Mrazek* has proposed a special type of temperature-dependent distribution, based on a two-site model characterized by an exponential distribution of the interaction energy [4.85–87]. The corresponding expression for the current density leads to TSDC peaks with the usual shape but its general validity is still to be proved in concrete experimental cases.

4.5.3 TSDC Model Calculations Involving Specific Distribution Functions

An analytical solution of the $J_D(T)$ function is only feasible for a box distribution in τ_0 [4.30, 37]. This distribution is, unfortunately, physically unrealistic and leads to TSDC curves having a special asymmetrical shape, which has rarely been observed. In all the other cases, the Laplace transforms of (4.21, 22) have to be evaluated numerically.

The case of the symmetrical distributions listed in Table 4.1 has been extensively discussed by *Van Turnhout* [4.30, 37]. Figure 4.8 shows some theoretical TSDC curves typical of distribution in τ_0, calculated by this author after truncation of the infinite integration interval at a point for which the integrand has decreased to 10^{-5} times of its maximum value. Owing to the

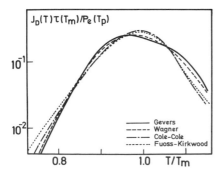

Fig. 4.8. TSDC calculated for Gevers ($\tau_m/\tau_2 = 10$), Wagner ($\bar{a} = 0.5$), Cole-Cole ($\bar{\beta} = 0.8$) and Fuoss-Kirkwood ($\bar{\gamma} = 0.7$) distributions in $\tau_0(E/kT_m = 20)$ [4.30]

similarity between the Wagner, Cole–Cole and Fuoss–Kirkwood distributions, the peaks can be made approximately coincident by properly choosing the distribution parameters, which obviously implies that distinguishing between these various functions in practical cases will be virtually impossible by just considering the shape characteristics of experimental curves. In fact, a precise identification of a real distribution will be a difficult – and often not very significant – task. It requires either – after postulating a priori a given initial distribution – the use of nomograms or fitting procedures, which is very time consuming, or point-by-point calculations of the distribution from the TSDC peak, which can only be done approximately by introducing a number of new simplifying assumptions (Sect. 4.6.5). Only at both temperature extremes, the current markedly depends on the type of distribution involved. This is, in theory, an important feature to be taken into consideration for determining relaxation parameters from the initial rise (Sect. 4.6.4).

No systematic model calculation has been reported for asymmetrical distributions but several distribution functions, deduced from TSDC experimental data, have been shown to correspond approximately either to a Davidson–Cole function [4.88] or a Havriliak–Negami function [4.89–92] (Sect. 4.6.5).

4.5.4 Experimental Distinction Between Distributed and Nondistributed Processes

As noted in Sect. 4.5.1, varying the polarizing and storage conditions during the thermoelectret formation of samples where a distribution of relaxation times is present, will result in more or less pronounced truncations of the original distributions, which will change the shape and position of the corresponding TSDC peaks. In Fig. 4.9, for example, the maximum temperature of a theoretical TSDC peak involving a Cole–Cole distribution in τ_0 is shown to be an increasing function of the polarization temperature T_p until saturation is reached. In the same way, the peak position depends on the polarization time t_p (for low T_p) and the initial storage temperature T_0 (unless $T_0 \ll T_p$) [4.30, 37]. All these conclusions are obviously also valid when distributions in E are involved.

In fact, such behavior is currently observed in a number of materials. This is illustrated in Fig. 4.10 for the β peak of poly(methyl methacrylate), which is seen to noticeably shift and vary by increasing the storage temperature. Similar variations have been found in polymers [4.27, 30, 69], phosphate glasses [4.65, 66], or ion-doped calcium fluoride [4.64].

In addition, it must be emphasized that, if the temperature dependence of the equilibrium polarization is not negligible, the cooling rate will also have a pronounced effect on the properties of a distributed peak. As a matter of fact, the progressive increase in total polarization during the cooling step will be the result of unequal contributions of individual processes and the subsequent TSDC peak will no longer be truly characteristic of the original distribution

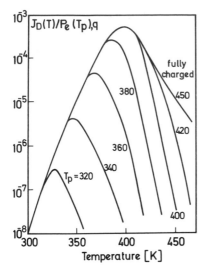

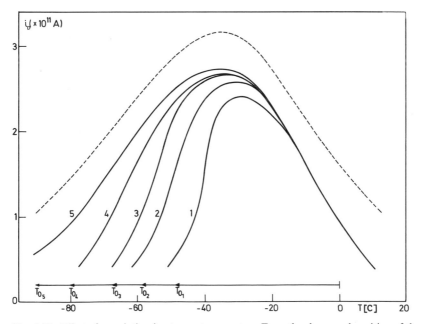

Fig. 4.9. Effect of a variation in polarizing temperature T_p on the shape and position of a TSDC peak calculated with a Cole-Cole distribution in τ_0 and $\bar{\beta} = 0.5$, $E = 60$ kcal/m (2.6 eV), $T_m = 400$ K, $q = 1.2$ K/min and $t_p = 60$ min [4.30]

Fig. 4.10. Effect of a variation in storage temperature T_0 on the shape and position of the TSDC β peak of poly(methyl methacrylate); $T_p = 273$ K, $t_p = 30$ min, $F_p = 2 \cdot 10^4$ V/cm, $q = 5$ K/min

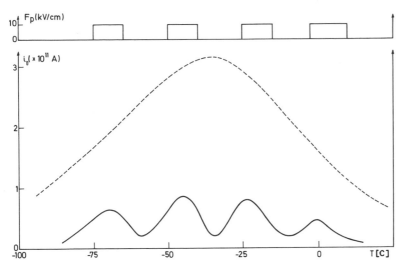

Fig. 4.11. Application of the fractional polarization method to the β peak of poly(methyl metha-crylate). Upper part: polarizing steps applied during linear cooling; lower part: partial TSDC peaks obtained during linear heating at $q=2$ K/min. The dashed curve corresponds to the β saturated peak ($T_p=313$ K, $T_0=77$ K, $F_p=10^4$ V/cm, $t_p=30$ min) [4.96]

function prevailing at T_p but rather of a hybrid one depending on the cooling rate [4.93]. In view of published results on the influence of the cooling rate, this prediction is confirmed by the variations in shape and position of the peaks observed in certain polymers [4.62].

Considering the special case of an asymmetrical two-site model, *Adachi* and *Shibata* [4.94, 95] have been able to show by model calculations that the shape and position of a TSDC-distributed peak can also be changed by the polarizing field. This follows from the fact that since the relaxation process is more affected by the depth of the shallower site, the contribution of two-site pairs characterized by larger differences in the depths of the potential well is more significant at high fields (i.e., the activation energy of the relaxation process becomes smaller at high fields) and the TSDC peak will thus shift towards lower temperatures and expand onto the lower temperature side with increasing fields.

The observation of such significant modifications in peak properties when the polarizing conditions are varied is a first and powerful argument in favor of the existence of a distribution in relaxation times. Under certain conditions, however, similar properties can be theoretically predicted for nondistributed peaks resulting from a space-charge polarization (Sect. 4.8.1). An unambiguous and rapid way to reveal a distribution is then to use a fractional polarization technique [4.96]. This method is illustrated in Fig. 4.11.

During a slow linear cooling, the electric field is not applied continuously but in several steps, separated by short-circuit periods. In this way, if the initial process is distributed, during the subsequent heating a corresponding number of partial peaks must appear, each characteristic of some components of the distribution. Theoretically, the number of peaks observed is not limited.

Practically, more or less severe overlapping occurs when the number of polarization steps increases in a given temperature range and no more than three or four well-isolated partial peaks can be obtained in one experiment related to a given relaxation process.

By performing a series of such experiments where the temperature ranges of the polarizing steps are systematically varied, it is also possible, in principle, to distinguish between continuous and discrete distributions. In a number of polymers, for example, it was shown that the maximum of the partial peaks so obtained could be situated everywhere in the temperature range of the original saturated peaks, which obviously implies the existence of initially continuous distributions [4.96]. Similar results were obtained using the technique of "thermal sampling" which consisted of "sampling" the relaxation process within a narrow temperature range by means of polarization, followed by depolarization a few degrees K lower in order to isolate some of the relaxation components [4.97–100]. *Chatain* et al., on the other hand, have claimed the existence of discrete distributions in several polymers, especially polyamides, each isolated relaxation being characterized by a relaxation time obeying an Arrhenius equation [4.101–103]. In fact, the distinction between continuous and discrete distributions is a somewhat tricky problem since, when there are more than two Debye processes forming a single maximum, it can be shown theoretically that the components cannot be separated and a continuous distribution is detected [4.100].

4.5.5 The Jonscher Model

In contrast to the preceding analyses based on the classical concept of distribution of Debye-like relaxation processes, *Jonscher* has suggested that the flatness and broadness of dielectric loss or TSDC peaks occurring in several dielectrics do not necessarily follow from a distribution of relaxation times but can be interpreted by means of a general physical model based on the concept of many-body interactions between dipoles or hopping charge carriers, electronic as well as ionic, leading to a finite range of mutual screening [4.104–106]. This attractive model shows good agreement with isothermal dielectric data of a number of materials, but uncertainties are encountered when applying it to nonisothermal conditions. As a matter of fact, the model then predicts [4.107] on one hand, that the TSDC peaks should show no evidence of shifting properties in variable polarizing temperature experiments or fractional polarization experiments, which is a behavior rarely reported in literature, and, on the other hand, that the activation energy determined from the current in the rising part of the TSDC peak (Sect. 4.6.4) should be much less than the loss peak activation energy. This latter behavior was clearly found in certain materials such as polymers even if the distributed character of the relaxations involved was taken into consideration by using the fractional polarization technique [4.108]. In the present state of theory and experiment it seems that, although the TSDC measurements performed on dielectrics departing from the Debye behavior usually show clear evidences of continuous or quasi-continuous distributions, the possibility that such distributions involve "non-Debye-like" processes should not be disregarded.

4.6 Evaluation of Relaxation Parameters from TSDC Dipolar Data

The most important parameters which have to be known for fully characterizing a classical dipolar relaxation process are I) the orientational activation energy E (or, more generally speaking, the temperature shift of the relaxation time), which ranges normally between some hundredths of an electron volt and a few electron volts (1–100 kcal/mole)[2], II) the pre-exponential factor τ_0, which is typically of the order $10^{+13}\,s^{-1}$ (but values from 10^{+8} to $10^{+14}\,s^{-1}$ are not uncommon), III) the equilibrium polarization P_e which, via the Langevin function, can lead to the knowledge of the dipole concentration N_d or the dipole moment p_μ if either quantity is known, and IV) (eventually) the distribution function $f(\tau)$.

It is clear from the similarity between the TSDC dipolar equations, on one hand, and the first-order kinetics TSC and TL equations, on the other, that the methods of parameter calculation derived for one of these techniques will be, in theory, easily transposable to one another. Since the early work of *Randall* and *Wilkins* [4.109], a number of approximate methods, the most important of them being concerned with the determination of activation energy, have been worked out for the analysis of TSC and TL glow peaks [4.2, 110, 111]. However, in absence of some outside knowledge of the defect structure of the material investigated, their practical application has been shown to be severely limited because any particular curve can then be analyzed on the basis of a large variety of more or less sophisticated models, each of which yields different values for the pertinent parameters [4.112, 113]. The restrictions are, fortunately, not so drastic for analyzing TSDC dipolar peaks, as far as processes with single relaxation time are involved, since all the usual models predict well-defined, first-order kinetics, which allows us to legitimately use the TSC and TL methods derived in this hypothesis (Sect. 4.6.2). It must be borne in mind, however, that such single processes are rather scarce. Where a distribution of relaxation times is present, the situation becomes by far more complicated and, in fact, very similar to the TSC and TL cases. Unambiguous information on the relaxation parameters usually depends on previous knowledge of the type and width of the distribution (Sect. 4.6.4).

Even for nondistributed processes, most of the usual methods are only applicable to well-isolated peaks. Before briefly reviewing them, we shall

---

2 1 eV = 23.054 kcal/mole.

discuss the properties of TSDC peaks for materials where overlapping relaxations are present.

4.6.1 The Peak Cleaning Technique

It is frequently found in the experimental practice of TSDC as well as conventional dielectric measurements that overlapping discharge processes are present in most regions of the spectra, so that the precise determination of relaxation parameters requires previous separation of the specific processes involved. This is hardly feasible from ac data and, therefore, correct analysis means using a vast range of frequency and temperature. By contrast, and provided that the relaxations involved are nondistributed, the TSDC method allows the easy and quick resolution of the overall spectrum. There are two ways in which this can be done.

The first one, common to all nonisothermal techniques, was developed by a number of authors from TL experiments [4.114–116] and was applied to TSC and TSDC data by *Heijne* [4.117], and *Creswell* and *Perlman* [4.11], respectively. Its principle is illustrated in Fig. 4.12. Suppose, for example, we have two peaks 1 and 2, whose maximum temperatures T_{m_1} and T_{m_2} ($T_{m_1} < T_{m_2}$) are close enough to each other as to overlap. After the whole curve has been obtained, a second thermal cycle is started but, instead of recording the discharge at one time, we first discharge the lower temperature peak 1 by heating until T_a ($T_{m_1} < T_a < T_{m_2}$), then cool the sample again and finally obtain the discharge of peak 2, which is pure or nearly pure.

The second cleaning method, specific to TSDC measurements, is due to *Bucci* et al. [4.39] (Fig. 4.13). It consists of first polarizing the material at a temperature T_b such that $T_{m_1} < T_b < T_{m_2}$ for an interval of time $t_p \simeq \tau_1(T_b) \ll \tau_2(T_b)$, so that the dipoles of type 1 are polarized at saturation while the dipoles of type 2 remain practically distributed at random; the resulting TSDC curve will then show only peak 1. Peak 2 can also be isolated in a similar way by using $T_c > T_{m_2}$ and removing the field at a temperature T_d such that $T_{m_1} < T_d < T_{m_2}$.

Wherever a distribution of relaxation times exists, the foregoing procedures are obviously no longer applicable since, although still producing a separation of peaks, they simultaneously involve a more or less severe truncation of the original distributions, leading to partial TSDC peaks which are no longer characteristic of the material, but rather of the polarization or depolarization conditions used (Sect. 4.6.4).

4.6.2 Determination of the Activation Energy for Processes with a Single Relaxation Time

The activation energy of a nondistributed relaxation process can be, in theory, easily calculated from a single TSDC experiment by means of some characteristic elements of the peak such as its halfwidth, inflection point or initial part of the current rise. More sophisticated methods make use of several different

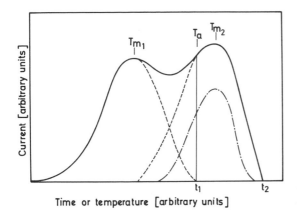

Fig. 4.12. Principle of the peak cleaning technique of Perlman and Unger (schematic) [4.140]

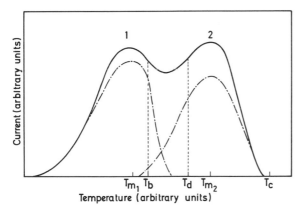

Fig. 4.13. Principle of the peak cleaning technique of *Bucci* et al. (schematic). The dashed curves represent the TSDC peaks isolated during independent experiments by polarizing the sample from T_b to the lowest available temperature (peak 1) and from T_c to T_d (peak 2) [4.39]

heating rates or call upon all the measured values of current and temperature either via iterative or nomogram procedures or via a series of graphical integrations. All but the last one were first derived from TSC or TL data [4.2, 110, 111]. The most useful ones and, in fact, the most used are undoubtedly the initial rise and graphical integration methods, one of them, at least, being always easily applicable to a previously cleaned peak. The other methods, usually less accurate, will only be briefly discussed here.

The *initial rise method* of *Garlick* and *Gibson* [4.118] is based on the fact that, the integral term in the $J_D(T)$ function being negligible at $T \ll T_m$ (Sect. 4.4.1), the first exponential dominates the temperature rise of the initial current, so that

$$\ln J_D(T) \simeq \text{const.} - \frac{E}{kT} \text{ (Arrhenius shift)} \tag{4.24}$$

$$\simeq \text{const.} + \ln T - \frac{E}{kT} \text{ (Eyring shift)} \tag{4.25}$$

$$\simeq \text{const.} - \frac{E_w}{k(T - T_\infty)} \text{ (WLF shift)}. \tag{4.26}$$

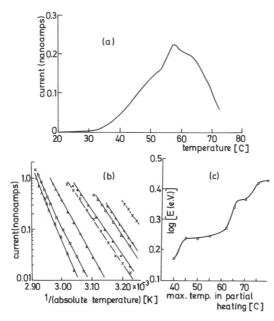

Fig. 4.14a–c. Application of the step-heating technique for resolving a complex TSDC spectrum (carnauba wax): (**a**) TSDC ($T_p = 323$ K, $t_p = 3$ h, $F_p = 8 \cdot 10^3$ V/cm); (**b**) slopes obtained during successive partial heatings up to 313, 318, 323, 328, 333, 338, 343, and 348 K; (**c**) plot of the apparent activation energy calculated from the preceding slopes against the maximum temperature reached in each partial heating [4.121]

The Arrhenius and Eyring shifts are usually indistinguishable (Sect. 4.4.3) and the activation energy can be determined with good accuracy by plotting $\ln J_D(T)$ against $1/T$ or $1/(T - T_\infty)$. In first approximation, a straight line is obtained, the slope of which gives $-E/k$.

This method is generally advocated as very satisfactory and, as such, is widely used. It has several advantages, indeed, as it necessitates neither a linear heating rate nor a precise knowledge of the absolute temperature (it is readily seen, for example, that a thermal gradient of 5 K in a sample leads to a relative error $\Delta E/E$ less than 3 % at 300 K). It is however somewhat limited because use of only the initial part of the curve is permitted, which may force one into the region where the uncertainty in background current is important [4.119, 120]. On the other hand, the method is still applicable for overlapping relaxations even if they cannot be isolated by cleaning, provided that a series of partial heatings, each separated by rapid cooling, are performed up to gradually increasing temperatures that span the whole temperature range of the spectrum [4.116]. As far as TSDC measurements are concerned this is the so-called *step-heating* or *partial heating* [4.11] or *multi-stage* [4.30] *technique*, which is illustrated in Fig. 4.14. In temperature regions where two processes with different energies interfere, the initial slope gradually increases due to the increasing contribution of the process with the larger energy. On the other hand, the further occurrence of several rises having the same slope suggests that they are related to a single process (Fig. 4.14b). This is more clearly seen by plotting, according to *Creswell* and *Perlman* [4.11], the activation energy deduced of each initial rise against the maximum temperature reached in each partial heating: the "flat" spots appearing on the curve then correspond to the activation energies of the individual peaks (Fig. 4.14c).

The *graphical integration method* was proposed independently by *Bucci* et al. [4.39], and *Laj* and *Berge* [4.122]. It directly follows from (4.4), on one hand, and (4.7), (4.10) or (4.12), on the other, that one can write for the entire range of temperatures covered by a TSDC peak:

$$\ln \tau(T) = \ln \left[\int_{t(T)}^{\infty} J(t')dt' \right] - \ln J_D(T)$$

$$= \ln \tau_0 + \frac{E}{kT} \text{ (Arrhenius shift)} \qquad (4.27)$$

$$= \ln \tau_0 - \ln T + \frac{E}{kT} \text{ (Eyring shift)} \qquad (4.28)$$

$$= \ln \tau_0' + \frac{E_W}{k(T - T_\infty)} \text{ (WLF shift).} \qquad (4.29)$$

The integral can be obtained with accuracy by graphical integrations of the peak and it is thus possible to plot $\ln \tau(T)$ as a function of $1/T$ or $1/(T - T_\infty)$. The straight line obtained, often referred as the *Bucci–Fieschi–Guidi* or *BFG plot*, has a slope equal to E/k. Like the initial rise method, this procedure does not necessarily require a linear heating rate. Furthermore, it has the advantage of using data of the whole curve. It has been applied successfully by a number of authors, mainly for determining the activation energy of ion–vacancy pairs in doped ionic crystal (Fig. 4.15).

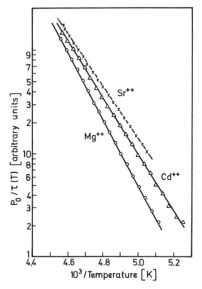

Fig. 4.15. Application of the graphical integration method to TSDC obtained from LiF crystals doped with various divalent impurities [4.42]

From TSC and TL theories a number of approximate methods for calculating the activation energy have been proposed, which are based on the halfwidth of the peak on each side of the maximum, i.e., the temperature intervals in which $J_D(T)$ passes between the half maximum and full maximum value. Those derived for first-order kinetics are, in theory, applicable to dipolar TSDC data obeying an Arrhenius shift. An example is the simple *Grossweiner formula* [4.123]

$$E = 1.51 \frac{kT_1 T_m}{T_m - T_1},$$

(4.30)

where T_1 is the temperature at which peak half-height occurs in the increasing part of the curve.

The so-called *varying heating rate methods* are based on the shift of the TSDC maximum with heating rate. Various ways of plotting the results have been proposed. From (4.9) it is readily seen that the activation energy for an Arrhenius shift can be determined by using two heating rates q_1 and q_2 and measuring the corresponding values T_{m_1} and T_{m_2} of the maximum [4.124, 125]

$$E = \frac{kT_{m_1} T_{m_2}}{T_{m_1} - T_{m_2}} \ln\left(\frac{q_1 T_{m_2}^2}{q_2 T_{m_1}^2}\right).$$

(4.31)

A better procedure is that suggested by *Hoogenstraaten* [4.115]. A number of heating rates are used and T_m is determined as a function of q. A plot of $\ln(T_m^2/q)$ against $1/T_m$ should yield a straight line, from the slope of which the activation energy can be calculated. These methods are similar to those used in isothermal dielectric experiments, in which the activation energy is deduced from the shift of a relaxation peak with frequency but they will usually lead to less precise results. In ac measurements, as a matter of fact, the frequency can readily be changed over several decades with good accuracy.

This is not so here (Sect. 4.4.1) and the success of these methods for the analysis of TSDC curves will thus depend on how much the position of a given peak can be shifted. They will, therefore, be applicable only when the activation energies are small.

Finally, simplified expressions of the current density such as those discussed in Sect. 4.4.4, can be used to calculate a family of theoretical TSDC curves as a nomogram which is then compared to the experimental data [4.126, 127]. Alternatively, an iterative procedure such as that proposed by *Cowell* and *Woods* can be employed [4.60].

4.6.3 Determination of Relaxation Parameters Other than E for Processes with a Single Relaxation Time

Knowing the activation energy by employing one of the previous methods and the type of temperature dependence of the relaxation time, the characteristic frequency factor τ_0^{-1} can be calculated from the relation giving the maximum temperature of the TSDC peak as a function of E and τ_0 (4.9). This parameter can also be directly obtained from a BFG plot.

Provided the TSDC peak corresponds to a saturated process and no change in polarization has occurred during the cooling and storage steps, the equilibrium polarization $P_e(T_p)$ may be obtained by integrating the current density

$$P_e(T_p) = \int_0^\infty J_D(T)\, dT$$

(4.32)

and the dipolar relaxation strength $(\varepsilon_s - \varepsilon_\infty)$ can then be easily calculated by means of the following equation, well known from the elementary theory of dielectrics [4.38]

$$P_e(T_p) = \varepsilon_0 (\varepsilon_s - \varepsilon_\infty) T_p F_p. \tag{4.33}$$

Here ε_0 is the permittivity of free space and ε_s and ε_∞ are the limiting values of the real part of the permittivity of the material at low and high frequency, respectively. In fact, it is obvious that, owing to the unavoidable increase in polarization occurring during the cooling step (Sect. 4.4.5), such calculations will always include a systematic error, the importance of which depends particularly on the cooling rate used. Consequently, the same will be true for the further determination of any parameter directly related to the equilibrium polarization or the relaxation strength, such as the total number of dipoles per unit volume N_d or the dipole moment p_μ (4.2).

It must be emphasized that, even if the equilibrium polarization is accurately known, obtaining precise values of these parameters depends on several additional conditions requiring previous knowledge of the material. The obvious condition is that the temperature dependence of the equilibrium polarization is adequately described by an equation of the Langevin type, then either N_d or p_μ must be previously estimated from independent measurements or physical–chemical considerations and, then the geometrical factor s and the local field correcting factor κ must also be known. Although these conditions are rarely fulfilled, this calculation procedure is still widely used to investigate the variation of the number of dipoles associated with solid-state reactions or defect production in ionic solids. The results should be viewed with caution and great care should be taken in correlating them with dielectric data obtained from isothermal measurements.

For relaxations of the WLF type, it was shown in the preceding section that a plot of $\ln \tau(T)$ versus $1/(T - T_\infty)$ must be linear. This allows the determination, by trial and error, of the critical temperature T_∞, together with the characteristic parameters U_1 and U_2. By using a theoretical WLF expression in terms of free volume [4.53, 54], parameters such as the fractional free volume and the expansion coefficient of the free volume can then be determined [4.128].

4.6.4 Determination of Relaxation Parameters for Processes with a Distribution of Relaxation Times

Most of the methods for calculating E described in Sect. 4.6.2 are based upon theories which are strictly valid only for systems having a single relaxation time and therefore do not hold as such for distributed processes. Except for the *varying heating rate methods*, they invariably yield too low a value of the activation energy, either by overestimating the role of the slowest relaxation times (*initial rise method*) or by taking into account too high a number of components (*halfwidth, BFG and fitting methods*). By slightly modifying the basic equations, however, and provided that the material is previously polarized at saturation and the type and width of the distribution function are known, it remains theoretically possible to calculate the activation energy E (for distributions in τ_0) or the mean activation energy E_m (for distributions in E) by means of these methods. When no information is available on the distribution, and this is nearly always the case, only the *step-heating technique* and, to a lesser degree, the *varying heating rates methods can* be expected to be used with some success.

For symmetrical distributions in τ_0 *Van Turnhout* has shown by means of model calculations that up to the lower halfwidth temperature, the initial current rise takes the form [4.30]

$$\ln J_D(T) \simeq \text{const.} - \frac{wE}{kT}, \qquad (4.34)$$

where $w = 1$ for the Wagner and Fröhlich–Gevers distributions, $w = \bar{\beta}$ for the Cole–Cole distribution and $w = \bar{\gamma}$ for the Fuoss–Kirkwood distribution. On the other hand, no simple relation can be derived for distributions in E.

Even if the distribution function is previously known, the practical application of (4.34) is severely limited for the real initial rise is rarely apparent due to overlapping phenomena and no cleaning technique can be applied. The determination of the activation energy is then best performed by the *step-heating method*, which causes the initial distribution to gradually change by exhausting the fast polarization components. For Wagner and Fröhlich–Gevers distributions in τ_0 the initial rise does not vary, while for Cole–Cole and Fuoss–Kirkwood distributions it becomes rapidly constant with a slope $-E/k$ [4.30, 37]. The variations are more pronounced when distributions in E are involved. Due to the increasing contribution of components characterized by a higher activation energy, the initial slope is a continuously increasing function of the maximum temperature reached in each partial heating so the E_m value can only be approximately determined from the particular slope corresponding to the step carried out close to the maximum temperature of the original saturated peak. The same trend is obviously expected when using thermal sampling or fractional polarization techniques (Sect. 4.5.4). Such behavior is commonly found in polymers, especially for the peaks related to local relaxations [4.30, 69, 96, 108] (Fig. 4.16). It has also been reported in some doped ionic materials [4.64]. It must be emphasized, however, that the occurrence of a systematic variation in apparent activation energy is not always indicative of a distributed process as it can also be found when important overlapping of discrete processes exists (compare Figs. 4.14c and 4.16).

Due to the determining role played by the type and width of the distribution on successive partial integrations of a TSDC peak, the BFG *plots* related to distributed processes will only be linear at low temperature, even for a distribution in τ_0. However, they usually will remain straight up to higher

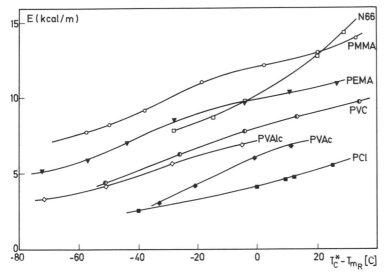

Fig. 4.16. Application of the fractional polarization technique to the β peaks of several polymers. The apparent activation energy calculated from each initial slope is plotted as a function of the difference $(T_C{}^* - T_{mR})$, where $T_C{}^*$ is the temperature of cancelling the field and T_{mR} is the maximum temperature of the original β peak (the mean activation energy is then obtained for $T_C{}^* - T_{mR} \simeq 0$) [4.96]

temperatures compared to the *initial rise plot*, which enables us to find a more accurate value for the apparent activation energy, via (4.34) [4.30]. The practical application of this equation is obviously subject to the same restrictive conditions as those mentioned for the *initial rise method*. For distributions in E, the BFG *plots* can sometimes be approximated by straight lines, especially when a Gaussian distribution is involved [4.69], but usually they will have a complex form, from which no quantitative evaluation can be expected.

The *varying heating rate methods* are the only ones which remain in theory applicable without modification to distributed processes, even for distributions in E, since the maximum temperature of a TSDC peak is expected to be essentially determined by the dipoles having the average value of activation energy. Unfortunately, we have seen that its accuracy is rather low and this is more especially true for distributed peaks, the maxima of which are often "flat" and thus difficult to determine with precision.

The use *of curve fitting methods* is obviously always theoretically possible from the general integral expressions (4.21) or (4.22). They are, however, very time consuming as they require not only having previously made a choice of a specific distribution but also having adjusted at least two unknown parameters, i.e., E or τ_0, on one hand, and the width of the distribution on the other. This procedure has been tested in polymers by *Van Turnhout* for the simplest case of a Fröhlich–Gevers distribution in τ_0 [4.30, 37] and in phosphate glasses by *Thurzo* et al. on the basis of a Gaussian distribution in τ_0 [4.65].

Besides these classical methods, various methods have been proposed to directly determine the shift factor $a^*(T) = \tau(T)/\tau(T+\Delta T)$ in various temperature intervals ΔT, either by differentiating the current with respect to the inverse of the heating rate or by using two or more discharge curves measured by increasing the temperature in an arbitrary way with time [4.99].

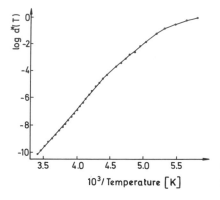

Fig. 4.17. The function $a^*(T)$ determined from TSDC data for the β relaxation in poly(methyl methacrylate) [4.99]

The knowledge of the shift factor in a wide temperature range is of particular interest when distributions in E are involved, for it yields the mean activation energy with accuracy and possible variations of this value as a function of temperature. For the β relaxation of poly (methyl methacrylate) for example, the plot of $\log a^*(T)$ vs $1/T$ is not linear at lower temperatures, indicating a gradual decrease of the mean activation energy (Fig. 4.17).

The determination of the other relaxation parameters, i.e., characteristic frequency, equilibrium polarization and relaxation strength, is little affected by the existence of a distribution of relaxation times. For a pure distribution in E, for example, the characteristic frequency factor can still be calculated from the equation giving the maximum temperature of the peak as a function of E and τ_0, provided that E_m is used instead of E, while the relaxation strength can still be obtained from the area underlying the peak via (4.32, 33). These procedures are subject to the same restrictive conditions as those outlined for non-distributed processes (Sect. 4.6.3).

4.6.5 Experimental Distinction Between a Distribution in τ_0 and a Distribution in E: Evaluation of the Distribution Function

Usually, an overall distributed peak can be formally described by a distribution in E as well as in τ_0 and, before attempting to calculate a specific function from TSDC data, it is thus necessary to have previous arguments in favor of one or the other. These can be found by means of the *step-heating method*, which allows one to follow the gradual variation of apparent E or τ_0 values. Alternatively, this variation can be studied by the *fractional polarization technique* or the *thermal sampling technique* (Sect. 4.5.4). The temperature dependence of τ_0 and E obtained from computer simulations of a series of thermal sampling spectra for Fuoss–Kirkwood distributions are depicted in Fig. 4.18. When a distribution in τ_0 is involved, the calculated values of this parameter gradually decrease with increasing temperature while, for a distribution in E, they remain located close to the initially assumed value. The calculated activation energies obviously vary in an opposite manner.

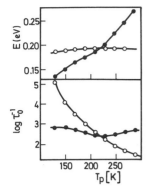

Fig. 4.18. Plots of τ_0 and E against T_p for a Fuoss-Kirkwood distribution in τ_0 (O) and E (●), calculated for simulated thermal sampling spectra with polarizing steps $(T_p - T_0) = 4$ K [4.100]

Another criterion for the existence of a distribution in τ_0 has been proposed by *Solunov* and *Ponevsky* [4.36]. Such a distribution obeys the time–temperature superposition principle and it is easy to show that the relation between currents and temperatures measured at various heating rates at which the same polarization is obtained can be written as

$$\frac{J_D(T_1^*, q_1) T_1^{*2}}{J_D(T_{1,m}^*, q_1) T_{1,m}^{*2}} \simeq \frac{J_D(T_i^*, q_i) T_i^{*2}}{J_D(T_{i,m}^*, q_i) T_{i,m}^{*2}}. \tag{4.35}$$

This equation implies that the plots of $J_D(T)/J_D(T_m)$ vs T^2/T_m^2, obtained for different heating rates, must coincide.

Knowing the type of distributed parameter to be taken into consideration, a distribution function of relaxation times can be computed by fitting theoretically determined current profiles to the experimentally observed TSDC peaks. A spectrum $f(\tau_0)$ or $g(E)$ is assumed, the current density is plotted as a function of temperature and the fit between the calculated and experimental curves is then successively improved. This procedure is obviously very laborious and various workers have therefore derived approximate solutions allowing a relative distribution function to be determined directly from the experimental data. All these calculations are made by assuming there is no interaction between dipoles, i.e., a TSDC distributed peak can be expressed as a summation of many dipolar processes, each being characterized by a single relaxation time. Furthermore, it is always implicitly assumed that the equilibrium polarization and the width of the distribution are temperature independent. For most materials, however, these conditions are not fulfilled (the width of a distribution, for example, is usually a decreasing function of temperature [4.46]) and this then implies that the TSDC peaks, which are by nature registered under nonisothermal conditions in samples previously polarized also in nonisothermal conditions, will be governed by a hybrid distribution depending on the experimental conditions rather than by the original distribution characterizing the material at a given temperature. It thus appears that distributions calculated from such TSDC measurements cannot be expected to be more than rough approximations, particularly for relaxation processes extending over a large temperature range.

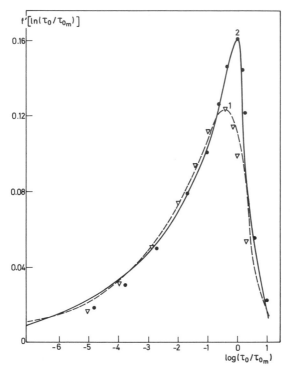

Fig. 4.19. Distribution of relaxation times in poly(bisphenol-A carbonate) (α relaxation) calculated from TSDC data according to (4.36) (dashed curve), and (4.37) (solid line). The experimental curves are compared with theoretical Havriliak-Negami functions: ∇ best fit with the first-order approximation ($1 - \bar{\eta} = 0.75$; $\bar{\varepsilon} = 0.25$); ● best fit with the second-order approximation ($1 - \bar{\eta} = 0.81$, $\bar{\varepsilon} = 0.27$) [4.89]

As shown by *Van Turnhout* [4.30], simplified expressions allowing the evaluation of distribution functions in τ_0 from the TSDC data can be derived by means of an inverse Laplace transform of (4.21). Starting from the approximation suggested by *Schwarzl* and *Staverman* for isothermal measurements [4.129], and assuming relaxation times obeying an Arrhenius equation, this author has proposed, for the logarithmic distribution, the two following first-order and second-order approximations:

$$f'_1\left(\frac{\tau_0}{\tau_{0_m}}\right) \simeq \frac{1}{P_e(T_p)}\left[\frac{kT^2}{qE} J_D(T)\right]_{\xi = \tau_0} \tag{4.36}$$

$$f'_2\left(\frac{\tau_0}{\tau_{0_m}}\right) \simeq \left[f'_1\left(\frac{\tau_0}{\tau_{0_m}}\right) - \frac{1}{qP_e(T_p)}\left(\frac{kT^2}{E}\right)^2 \frac{dJ_D(T)}{dT}\right]_{\xi = 2\tau_0} \tag{4.37}$$

When the type of temperature dependence of the relaxation times is unknown, these relationships are to be expressed in terms of the general shift factor $a^*(T)$, which must then be determined from independent TSDC data or dielectric measurements [4.34, 92, 99]. The distribution function calculated by means of (4.36, 37) from TSDC data related to the glass transition of polycarbonate is represented in Fig. 4.19. As expected from independent dielectric measurements, it is strongly asymmetrical and can be approximated by a Havriliak–Negami distribution [4.89, 90]. Such calculations, however, are somewhat arbitrary not only owing to the above-mentioned limitations of the method but also because they presuppose a constant activation energy, which cannot be rigorously true in view of results obtained from step-heating measurements.

Approximate methods for determining a distribution in activation energy, which is the most frequently encountered case in experimental practice, have been obtained by a number of authors. Up to now, none has been convincingly tested by means of detailed and systematic comparisons with isothermal dielectric data and it is thus difficult to make a rational choise between them.

A first-order approximation has been first derived by *Van Turnhout* from (4.22), which gives [4.30, 37]:

$$g[E(T)] \simeq \frac{TJ_D(T)}{P_e(T_p)qE(T)},\tag{4.38}$$

where $E(T)$ is related to T by the transcendental equation

$$E(T)\exp\left[\frac{E(T)}{kT}\right] = \frac{kT^2}{\tau_0 q}.\tag{4.39}$$

This equation has been used for several low temperature relaxations in polymers.

In the method advocated by *Fischer* and *Rohl* [4.69], the activation energy spectrum is obtained by means of a similar step-by-step procedure involving the further assumption of a linear relationship between E and T while in the method proposed by *Hino* [4.97], the distribution function $N(E)$ defined by $N_d = \int_0^\infty N(E)dE$ is determined by solving a set of n simultaneous equations describing relations between $J_D(T_i)$ and $N_d(E_j)$, obtained by inserting n values of T_i $(i=1,2,...,n)$ in

$$J_D(T_i) = \sum_{j=1}^{n} N_d(E_j)\frac{p_\mu^2 F_p}{3kT_p}\tau_{0j}^{-1}\exp\left[-\frac{1}{q\tau_0}\int_{T_0}^{T_i}\exp\left(-\frac{E_j}{kT'}\right)dT'\right]\exp\left(-\frac{E_j}{kT_i}\right)\Delta E_j.\tag{4.40}$$

These equations are represented in a matrix form whose elements are determined by using E_j values obtained experimentally by the *thermal sampling technique* applied in the whole temperature range of the original saturated peak. The spectrum $N(E_j)$ can then be calculated, provided that the dipole moments are known. The method has been applied by *Hino* et al. for determining the distribution function in poly(ethylene terephtalate) and SiO_2 films [4.97, 130].

A determination of the distribution function based on the time dependence of the polarization at a given T_p has been proposed by *Ikeda* and *Matsuda* [4.131] on the basis of the *Primak* theory for isothermal annealing experiments [4.132]. The distribution so obtained for the γ peak of Nylon 6 is compared, in Fig. 4.20, with the spectrum calculated from the profile of a saturated peak. The two curves are seen to agree both in shape and broadness but the maxima of activation energy are significantly different.

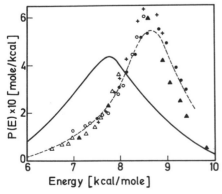

Fig. 4.20. Comparison of the activation energy spectra of nylon 6 (γ relaxation region) as calculated from the time dependence of polarization (dashed curve) and from the corresponding saturated TSDC peak (solid line). The spectra are given in terms of a distribution $P(E)$ defined by $P_0(T_p) = \int_0^\infty P(E)\,dE$ [4.131]

4.7 Correlation Between Dipolar TSDC and Conventional Dielectric Measurements

To facilitate the molecular interpretation of TSDC results and possibly to use this technique as an alternative to the measurement of dielectric losses as a function of frequency and temperature, it is of major importance to establish, both theoretically and experimentally, correlations between TSDC and conventional ac data. Unfortunately, the exact relations between thermally stimulated depolarization currents and isothermal or isochronous dielectric losses are given by Fourier transforms and manageable expressions can therefore be obtained only by means of a time–frequency transformation.

4.7.1 The Equivalent Frequency of TSDC Measurements

The equivalent frequency of TSDC measurements, i.e., the frequency at which ac experiments should be performed in order to obtain a loss peak having the same maximal temperature as the TSDC peak, can be estimated by relating the heating rate q and the ac angular frequency $\omega = 1/\tau$ via the equation that defines the maximal temperature of the TSDC peak [4.88]. For an Arrhenius shift involving a constant activation energy, for example, the equivalent frequency takes the form

$$f_{eq} = \frac{\omega}{2\pi} = \frac{1}{2\pi} \frac{qE}{kT_m^2} \qquad (4.41)$$

while, for relaxations obeying a WLF relation, we obtain

$$f_{eq} = \frac{1}{2\pi} \frac{qU_1 U_2}{(U_2 + T_m - T_g)^2}. \qquad (4.42)$$

Adopting typical values of the TSDC parameters q, E and T_m, on one hand, and q, U_1, U_2, T_m and T_g, on the other hand, these equations lead to $f_{eq} \simeq 10^{-4}$–10^{-1} Hz, i.e., frequencies which are much lower than those which we are able to use under normal conditions in ac measurements. This implies, in particular, that the assessment of TSDC peaks to specific depolarization processes is generally not possible on the basis of a direct comparison between dc and ac experimental data. It can only be decided whether the position of the current peak is compatible with a given dielectric relaxation by extrapolating, to the equivalent frequencies, the dielectric function $\log f = f(1/T_m^*)$, where T_m^* is the maximum temperature of the loss curve at frequency f. The application of this procedure to an Arrhenius-like relaxation and a WLF-like relaxation of poly(vinylidene fluoride) is shown in Fig. 4.21. The data derived from TSDC and ac experiments are seen to join smoothly, demonstrating self-consistency of the analysis.

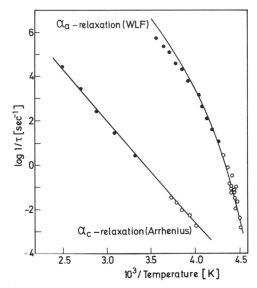

Fig. 4.21. Temperature shifts of the α_a and α_c relaxations of poly(vinylidene fluoride) with frequency (ac measurements, ●) and heating rate (TSDC measurements, ○) [4.88]

As another consequence of this low equivalent frequency of the TSDC method, the location of the peaks measured at low heating rates is often found to closely correspond to dilatometric transition temperatures. In polymers, for example, there is good agreement between the maximum temperatures of the TSDC α peaks and the glass transition temperatures [4.27, 30, 88, 133–137].

4.7.2 Calculation of Dielectric Constant-Temperature Plots from TSDC Data

By extending *Hamon's* [4.138] as well as *Schwarzl* and *Struik's* [4.139] approximation methods for isothermal transient currents to TSDC, *Van Turnhout* has derived the following first-order approximation formula, valid for an Arrhenius shift with a constant activation energy (single relaxation time or distribution in τ_0 [4.30]):

$$\varepsilon_0 F_p \varepsilon''[f'_{eq}(T), T] \simeq 1.47 \frac{kT^2}{qE} J_D(T), \qquad (4.43)$$

where $\varepsilon''[f'_{eq}(T), T]$ is the imaginary part of dielectric constant at an equivalent frequency f'_{eq} related to T by

$$f'_{eq}(T) = 0.113 \frac{qE}{kT^2}. \qquad (4.44)$$

Equation (4.43) yields the $\varepsilon''[f'_{eq}(T), T]$ values point by point from the current density, provided E is known. The plot will be a special $\varepsilon''(T)$ function in

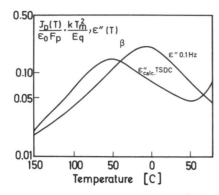

Fig. 4.22. $\varepsilon''(T)$ data at 0.1 Hz of the β peak of poly(methyl methacrylate) compared with $\varepsilon''(T)$ data calculated from TSDC measurements according to (4.45) [4.30]

which each of the ε'' values is related to a variable frequency. The frequency variations, however, can often be neglected allowing us to replace $f'_{eq}(T)$ by $f'_{eq}(T_m)$, which leads to the more manageable expression

$$\varepsilon_0 F_p \varepsilon'' [f'_{eq}(T_m), T] \simeq 1.47 \frac{kT_m^2}{qE} J_D(T), \tag{4.45}$$

where

$$f'_{eq}(T_m) = 0.113 \frac{qE}{kT_m^2}. \tag{4.46}$$

This procedure should allow us to extend sinusoidal measurements to the ultra-low frequency range in a manner similar to the dc *step-response technique*. Practically, however, its use is rather limited because, even for a single Debye relaxation, model calculations show that it yields ε'' values which, for $T > T_m$, rapidly deviates from the theoretical curve. In addition, the conversion definitely becomes incorrect when a distribution in activation energy is involved and it leads, in particular, to loss peaks which are clearly too low (Fig. 4.22).

4.7.3 Calculation of Dielectric Constant-Frequency Plots from TSDC Data

Simplified methods for relating the dielectric constant and dielectric losses at all frequencies to the measured currents and microscopic parameters of the relaxation have been suggested by *Hino* [4.130], and *Perlman* and *Unger* [4.140]. Starting from a single Debye relaxation model, these last authors assumed that the frozen-in polarization $P(T_0)$ in a TSDC experiment is equivalent to the Langevin equilibrium polarization $P_e(T_p)$, which, by using

(4.2, 33), leads to the following expression:

$$(\varepsilon_S - \varepsilon_\infty)_T = \frac{P(T_0)}{\varepsilon_0 F_p} \cdot \frac{T_p}{T^*}, \tag{4.47}$$

where T^* is the temperature at which the comparative dielectric measurements are to be made. Taking into account the familiar Debye equations [4.38]

$$\varepsilon' = \varepsilon_\infty + \frac{\varepsilon_S - \varepsilon_\infty}{1 + \omega^2 \tau^2} \tag{4.48}$$

$$\varepsilon'' = \frac{(\varepsilon_S - \varepsilon_\infty)\omega\tau}{1 + \omega^2 \tau^2}, \tag{4.49}$$

where ω is the angular frequency and assuming a linear superposition to be valid for all the present discrete relaxations, we then obtain

$$\varepsilon''_{T*} = \frac{1}{\varepsilon_0 F_p} \frac{T_p}{T^*} \sum_{j=1}^{n} \frac{\omega\tau_j P_j(T_0)}{1 + \omega^2 \tau_j^2} \tag{4.50}$$

and

$$(\varepsilon' - \varepsilon_\infty)_{T*} = \frac{1}{\varepsilon_0 F_p} \frac{T_p}{T^*} \sum_{j=1}^{n} \frac{P_j(T_0)}{1 + \omega^2 \tau_j^2}, \tag{4.51}$$

where τ_j is the Debye relaxation time for system j.

The practical application of these expressions to calculate dielectric constant and loss from measured TSDC curves is obviously subject to several restrictions resulting from the initial assumptions on the frozen-in polarization (Sect. 4.4.5) and, in addition, it means that all the involved relaxations must be previously isolated in order to determine the P_j and τ_j values (Sect. 4.6.1). They remain valid in theory for relaxation processes characterized by continuous

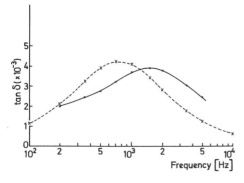

Fig. 4.23. Loss tangent (tg $\delta = \varepsilon''/\varepsilon'$) versus frequency for KCl: Eu^{++} (0.3 mole %) at 351 K from ac measurements (solid line) and calculated from TSDC measurements (dashed curve) [4.140]

distributions provided that the sums are replaced by integrals and the distribution functions involved are previously known.

The method has been tested by *Perlman* and *Unger* in the system $KCl:Eu^{++}$ [4.140]. Good agreement in shape and height with the usual bridge measurements has been obtained but a large displacement of the curves along the frequency axis occurs, which is attributed to the large inherent error in the determination of the relaxation times (Fig. 4.23).

4.8 Theory of TSDC Involving Space-Charge Processes: The Charge Motion Model

The space-charge polarization, i.e., the polarization due to excess charges which cause the material to be spatially not neutral, is a much more complex phenomenon than the dipolar polarization. Even if injection of extrinsic carriers is prevented by the use of insulating electrodes, several processes can be involved simultaneously and their parameters are not only time and space dependent but also depend on many variables atypical of the material. In heterocharged dielectrics, for example, where the presence of space charges arises from field-induced transport of pre-existing and/or field-generated charge carriers, a full description of the polarization process will require taking into account the counteracting action of diffusion, the loss of migrating charges by recombination with opposite carriers, the more or less blocking effect of the electrode–dielectric interfaces and the possible trapping properties of the material. The same is obviously true for the depolarization process, where carrier drift now occurs under the influence of local electric fields.

When depolarization is studied by the TSDC method, an additional difficulty arises from the fact that, contrary to the dipole reorientation process which always has an efficiency of 100%, only part of the space-charge decay may be observed in shorted samples. This lack of efficiency, which affects carrier drift as well as diffusion, is a consequence of three main phenomena [4.30, 141]: I) the partial dissipation of the excess charges by the space-independent intrinsic conductivity passes unnoticed in the external circuit, II) the release of the image charges induced at the electrodes is incomplete as a result of their partial neutralization by the excess charges, and III) the current released by diffusion depends on the blocking character of the electrodes. As a general rule, the detection efficiencies expected in usual TSDC experiments will be no more than 0–15% for the carrier drift and 0–100% for the diffusion process. It is possible, however, to significantly increase the efficiency by using an insulating electrode adjacent to the nonmetallized side of the sample (insulating foil or air gap). Since the insulating electrode blocks any charge exchange, all image charges previously induced at the noncontacting electrode will be released during the TSDC run and it will be possible to observe the decay resulting from intrinsic conduction.

In view of the foregoing considerations, it is clear that the theory of space-charge polarization is very intricate. In particular, the choice of a simplified model for a given material and the distinction between electronic and ionic carriers is not an easy task, for much controversy still surrounds the nature of the conduction mechanisms in dielectrics. Two basic models are usually considered as starting points for a TSDC theory: In the charge motion model, the current is assumed to be essentially governed by the bulk conductivity of the material (electronic or ionic), irrespective of its possible trapping properties, while in the trapping model, the current is assumed to result only from carriers (usually electronic) released into the conduction (or valence) band as the charge distribution returns to equilibrium.

In the following, we will discuss only the main features of the theories based on the charge motion model in heterocharged dielectrics, i.e., materials polarized via thermoelectret formation (for unipolarly or bipolarly homocharged dielectrics, we refer to the extensive work by *Van Turnhout* [4.30]). The theory of the trapping model, more specific to materials polarized via photoelectret formation or charge injection, will be treated in Sect. 4.9.

A generalized theory of the charge carrier transport in dielectrics requires taking into consideration at least four basic processes, i.e., field-induced motion, diffusion-induced motion, recombination of carriers during transport and formation of new carriers by dissociation. Unfortunately, these processes are described by a set of nonlinear partial differential equations for which no general solution could be derived, not even for isothermal steady-state currents with single relaxation times. Obtaining analytical TSDC expressions is, therefore, dependent on a number of drastic approximations and previous assumptions concerning, among others, the respective contribution of each process, the spatial distribution of the excess charges and the blocking degree of the electrode contacts. At best, such expressions can be expected to predict qualitative behavior and will usually be of very limited practical use.

4.8.1 Nonuniform Space-Charge Distribution (Single Relaxation Time)

By neglecting diffusion, intrinsic conduction, recombination and generation of carriers, the continuity equations describing the field-induced transport of charge carriers along the spatial coordinate x $(0 < x < d)$ may be put in the form

$$\delta n(x, t)/\delta t = \mu_n(T)\delta n(x, t)F(x, t)/\delta x \tag{4.52}$$

$$\delta p(x, t)/\delta t = -\mu_p(T)\delta p(x, t)F(x, t)/\delta x, \tag{4.53}$$

where p and n are the positive and negative carrier densities and μ_n and μ_p their mobilities. The relation between the charge densities and the field is given by

Poisson's equation

$$\varepsilon_\infty \varepsilon_0 \delta F(x, t)/\delta x = p(x, t) - n(x, t).$$ (4.54)

On the other hand, during a TSDC experiment with shorted samples, the electric field satisfies the boundary condition

$$\int_0^d F(x, t)dx = 0$$ (4.55)

and the current density is [4.30, 141]

$$J_D(t) = \int_0^d [\mu_p(T)p(x, t) + \mu_n(T)n(x, t)]F(x, t)dx .$$ (4.56)

Numerical solutions of this complex function can only be found by assuming specific distributions to be initially present. The case of a distribution obeying the conditions

$$\begin{aligned}p(x, t) &= p_0 \cos(\pi x/2d) \\ n(x, t) &= n_0 \sin(\pi x/2d)\end{aligned} \qquad 0 \le x \le d$$ (4.57)

has been discussed by *Van Turnhout* [4.30]. To this end, the continuity and current equations are rewritten in a differential form by introducing the velocity of the zero-field point $x_0(t)$ defined by $F(x_0, t) = 0$ and assuming that $\delta p(x_0, t)/\delta t \simeq dp(x_0, t)/dt$ and $\delta n(x_0, t)/\delta t \simeq dn(x_0, t)/dt$. The evolution of the excess charge distribution calculated as a function of temperature during a TSDC experiment is represented in Fig. 4.24a and the corresponding charge and current release are shown in Fig. 4.24b. It is apparent that the ultimately released charge amounts to only a few percent of the initially stored charge, confirming the rather bad efficiency of the carrier drift process. In the absence of any information on the type of spatial distribution involved in a given material, such an approach is obviously rather formal and, consequently, great care should be taken in correlating it with experimental results.

If a high field is applied during the polarizing step, it is generally taken for granted that the thermally stimulated depolarization current resulting from the diffusion process will be much smaller than the drift current and can, consequently, be neglected. This will not be true, however, if large gradients are present, as commonly encountered in heteroelectrets made at high temperatures from strongly polar materials [4.30] or in monopolarly charged homoelectrets, in which the diffusion phenomenon may even be responsible for the occurrence of current reversals [4.30, 142]. The first theory of discharge currents caused by diffusion was developed by *Jaffe* [4.143], and *Jaffe* and *Lemay* [4.144] for the isothermal case. By neglecting the formation and

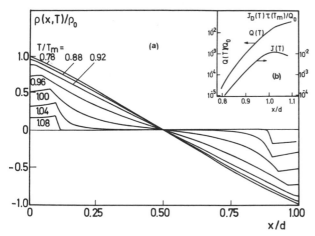

Fig. 4.24. Spatial distribution of the excess charge density at various temperatures during a TSDC experiment when the initial distribution is assumed to obey (4.57). Inset: Corresponding charge and current released by carrier drift ($E/kT_m = 20$) [4.30]

recombination of carriers and linearizing the internal field by assuming it to be zero, i.e., $F(x, t) = 0$, the $J(t)$ function was shown to obey

$$J(t) = 2v\mu F_p \sum_{j=0}^{\infty} G_j \exp(-z_{2j+1}^2 \zeta) \tag{4.58}$$

where $2v$ is the total equilibrium density of positive and negative charged carriers, ζ is a normalized time given by $\zeta = Dt/d^2$, D is the diffusion constant and G_j are constants depending on the polarizing field F_p, the mobility μ, the diffusion constant D, the blocking factor ψ, the sample thickness d and the eigenvalues z_{2j+1} of

$$\tan z = 2\theta z (z^2 - \theta^2)^{-1} \tag{4.59}$$

with $\theta = \psi d / D$. By assuming that the temperature dependence of μ and D can be described by the same shift factor $b^*(T)$ and ignoring the second and higher terms of the sum, (4.58) can easily be rewritten for nonisothermal conditions. Putting $\zeta = \zeta_r b^*(T)$, we obtain [4.10]

$$J_D(t) = 2v\mu F_p G_0 \exp\left[-z_1^2 \zeta_r \int_0^t b^*(T)dt\right] \tag{4.60}$$

which, for $b^*(T)$ of the Arrhenius type, leads to TSDC peaks similar in shape to those resulting from dipolar reorientation. Their other properties however will strongly differ from the latter in that the current will decrease for increasing sample thicknesses and will vary nonlinearly with the polarizing field. On the other hand, we should note that in normal heteroelectrets such a diffusion current will have the same direction as the drift current and, thus, will reinforce it [4.30]. This will not be the case in monopolar homoelectrets where carrier drift and diffusion occur in opposite directions.

Some general implications of a nonuniform space-charge distribution in a charge motion model including both carrier drift and diffusion (blocking conditions) were discussed by Kessler who postulated that only one kind of carrier is mobile [4.145]. Starting from the differential equation which describes the defect distribution in crystals [4.146], i.e.:

$$\frac{\delta n(x,t)}{\delta t} = \mu_n(T)\frac{\delta}{\delta x}\left[n(x,t)F(x,t) - \frac{kT}{e}\frac{\delta n(x,t)}{\delta x}\right], \tag{4.61}$$

where F is the polarization field in the sample as it follows from the space-charge distribution, the depolarization current density $J_D(t)$ and its integral $P_T = \int_0^\infty J_D(t)dt$ may be written in the general form

$$J_D(t) = -\mu_n(T)\frac{e}{d}\int_0^d dx \int_0^x \frac{\delta}{\delta x'}$$

$$\cdot \left[n(x',t)F(x',t) - \frac{kT}{e}\frac{\delta n(x',t)}{dx'}\right]dx' \tag{4.62}$$

$$P_T = -\frac{e}{d}\int_0^d dx \int_0^x [n(x',\infty) - n(x',0)]dx'. \tag{4.63}$$

No closed solution of these equations is possible but some useful qualitative predictions can be obtained. Both $J_D(T)$ and P_T depend on the shape of the charge distribution and on the charge density but, while a steeper distribution of a fixed quantity of accumulated charge carriers will reduce P_T, the initial current density $J_D(0)$ will be increased, at least in the case when the diffusion term in (4.62) dominates. Consequently, the maximum temperature of the peak T_m must be shifted towards lower temperatures. This is an important conclusion: Since a steeper distribution may be expected to occur when the polarizing temperature T_p or the polarizing time t_p is decreased (Fig. 4.25), T_m must be an increasing function of these two parameters. Furthermore, if a polarized sample is gradually depolarized, for example by the step-heating technique, P_T will decrease by each step less quickly than $J_D(0)$, also resulting in a progressive shift of T_m to higher temperatures. This behavior has been clearly observed in some materials, in particular in NH_4Cl, $NH_4Cl:NiCl_2$ and $CdF_2:NaF$ crystals [4.147, 148]. It must be emphasized, however, that these shifting properties cannot be considered as definitive arguments in favor of the existence of a space-charge polarization obeying the preceding model since they are also just those expected from a polarization process involving a distribution of relaxation times regardless of its microscopic origin (Sect. 4.5.4).

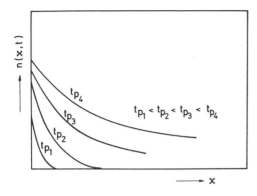

Fig. 4.25. Buildup of a space charge with time of polarization t_p for a charge motion model including carrier drift and diffusion (schematic) [4.145]

4.8.2 Uniform Space-Charge Distribution (Single Relaxation Time)

In order to obtain analytical solutions for the current resulting from a space-charge polarization, a number of authors have assumed that, besides neglecting diffusion and carrier recombination or generation, the excess charges can be considered as uniformly distributed and concentrated in narrow layers close to the electrodes (barrier-type polarization [4.149]), if at least one of the dielectric-electrode interfaces can be considered as fully blocking (insulating electrode). The thermally stimulated depolarization current, which is postulated to arise either from carrier drift or from intrinsic conduction (i.e., processes assumed to obey first-order kinetics), can then be calculated from the time variation of the space-charge density as a function of conductivity parameters or from equivalent model theories assuming that the polarized sample can be described by a set of RC circuits. Some authors have also discussed the case of a possible annihilation of charge carriers by bulk recombination. This process, which obeys second-order kinetics, is likely to occur only when very high concentrations of positive and negative charge carriers are created in the material. Whatever the case, the hypothesis of an initially uniform distribution is, a priori, rather unrealistic. Furthermore, the corresponding $J_D(T)$ functions are usually derived on the basis of a single relaxation time, although a number of experimental facts seem to show that the use of a distribution should often be much more appropriate. It follows that such expressions cannot be expected to lead to more than a crude approximation of the investigated phenomena.

The TSDC theory of barrier-type polarization obeying first-order kinetics was put forward, in particular, by *Bugrienko* et al. [4.150], *Nedetzka* et al. [4.12], *Muller* [4.19], and *Agarwal* [4.151]. In all cases, the $J_D(T)$ functions obtained are nearly equivalent to the *Bucci–Fieschi* equation and, thus, predict that the space-charge peaks will be characterized by the same typical asymmetrical shape as that outlined for dipolar peaks, i.e., that space-charge and dipolar processes will be analytically undistinguishable. This is not surprising in view of the adopted simplifying assumptions, and the primary interest of these theories will consist rather in emphasizing the essential role played in a

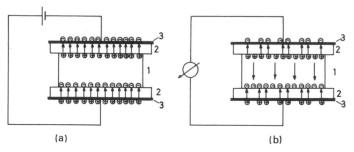

Fig. 4.26a and b. Charge distribution in Müller's model (**a**) during polarization and (**b**) during depolarization: (*1*) sample; (*2*) insulating foils; (*3*) metallic electrodes [4.19]

space-charge polarization by factors such as the nature of electrode material, the type of electrode–sample interface or the sample thickness and, thus, in providing arguments for differentiating space charge and dipolar peaks.

In the theory developed by *Muller* [4.19] a well-defined experimental arrangement, consisting of a sample that is fully insulated from the metallic electrodes by thin foils of nonconducting materials, is initially assumed (Fig. 4.26). Considering such a system as two parallel capacitors, the initial field F_1 inside the sample has the form

$$F_1 = \frac{P(T_0)}{\varepsilon_0 \varepsilon_1} \frac{2\varepsilon_1 d_2}{\varepsilon_2 d_1 + 2\varepsilon_1 d_2} \tag{4.64}$$

where ε_1 and d_1 are the relative permittivity and the thickness of the sample, ε_2 and d_2 are the corresponding parameters of the foils, $P(T_0)$ is the initial frozen-in polarization, i.e., $P(T_0) = Q_0/S$ with Q_0 the charge on the electrodes, and S the cross section of the samples. Assuming that the spatial extension of the accumulated charge is small compared with the thickness of sample and insulator and that the depolarization current is caused by intrinsic conduction rather than by redistribution of excess charges, i.e., that the mobility of these carriers is very small and that the number of free carriers in the bulk exceeds the number of those in the region of the accumulated charge which are free, the time variation of the polarization in short-circuited samples can then be written as

$$J(t) = -\frac{dP(t)}{dt} = \sigma(t) F(t) = \sigma(t) \frac{KP(t)}{\varepsilon_0 \varepsilon_1}, \tag{4.65}$$

where $K = 2\varepsilon_1 d_2/(\varepsilon_2 d_1 + 2\varepsilon_1 d_2)$ and $\sigma(t)$ is the material conductivity. Introducing an Arrhenius temperature dependence for the conductivity, i.e., $\sigma(T) = \sigma_0 \exp(-E/kT)$, and using a special reciprocal heating scheme[3] charac-

---

3 This heating scheme, also proposed by some authors in TSC and TL experiments [4.152], is in fact not very useful since the integral appearing in the $J_D(T)$ function presents little difficulty (Sect. 4.4.4). Furthermore, it is less easily reproduced and cannot be maintained over such a wide range as a linear temperature scan.

terized by $-d(1/T)/dt = T^{-2}(dT/dt) = q^*$, an analytical solution of (4.65) is directly obtained as

$$J_D(T) = \frac{K\sigma_0 P(T_0)}{\varepsilon_0 \varepsilon_1} \exp\left[-\frac{E}{kT} - \frac{K\sigma_0 k}{q^* E \varepsilon_0 \varepsilon_1} \exp\left(-\frac{E}{kT}\right)\right]. \tag{4.66}$$

As expected, this expression is similar to that derived from the dipolar theory after solving the integral [compare (4.66) and (4.14)], but it shows that a space-charge TSDC peak must depend on the nature of the dielectric–electrode interface (via the characteristics of the foils used as insulating electrodes) and, in addition, predicts a decrease of the current density with decreasing thicknesses (via the factor K). On the other hand, a linear dependence for the polarizing field is implicitly obtained. It is obvious, however, that a field-dependent distortion and broadening of the peaks can occur in the case of less defined conditions (e.g., direct contact of the sample with the electrodes or excess charges spatially distributed in the neighborhood to the electrodes).

The case of conduction-determined relaxation of space charges in semiconductors provided with a fully blocking contact and various electrode geometries has been extensively treated by *Fritzsche* and *Chandra* [4.153], and *Agarwal* [4.151]. For the simple arrangement depicted in Fig. 4.27, and using the equivalent circuit shown in Fig. 4.28, the current $i_D(T)$ obtained during the TSDC experiment is still nearly similar to the dipolar case and takes the form

$$i_D(T) = \frac{C_E^2 V_d}{(C_I + C_E)^2 R_0} \exp\left\{-y + \frac{E}{kq(C_I + C_E)R_0}\left[\frac{\exp(-y_0)}{y_0} - \frac{\exp(-y)}{y}\right.\right.$$
$$\left.\left. + \int_y^{y_0} \frac{\exp(-y)}{y} dy\right]\right\}, \tag{4.67}$$

where V_d is the polarizing voltage, C_E the insulator capacitance, C_I the sample capacitance, $y = E/kT$, $y_0 = E/kT_0$ and R_0 is defined by $R(T) = R_0 \exp(E/kT)$. More interesting are the cases where the samples are larger than the top metal electrode, an arrangement which is commonly used in the experimental practice. Let us first consider the extreme case depicted in Fig. 4.27b. When the field is applied, the capacitor formed by the MISM structure gets charged and can then discharge through the sample upon heating, producing a longitudinal relaxation current $i_D^*(T)$. Considering the equivalent circuit represented in Fig. 4.28b, where the sample length l is divided into small strips Δx having a capacitance $c\Delta x$ and resistance $r\Delta x$, this current is

$$i_D^*(T) = \frac{2V_d}{lr_0} \exp\left(-\frac{E}{kT}\right) \sum_{n=0}^{\infty} \exp\left[-\frac{(2n+1)^2 \pi^2}{4qr_0 C_E l^2} \int_0^T \exp\left(-\frac{E}{kT'}\right) dT'\right], \tag{4.68}$$

where r_0 is defined by $r = r_0 \exp(E/kT)$. This expression still leads to an asymmetrical TSDC peak, hardly distinguishable from the one represented by (4.67). Now, in the more realistic geometry of Fig. 4.27c, we are clearly dealing

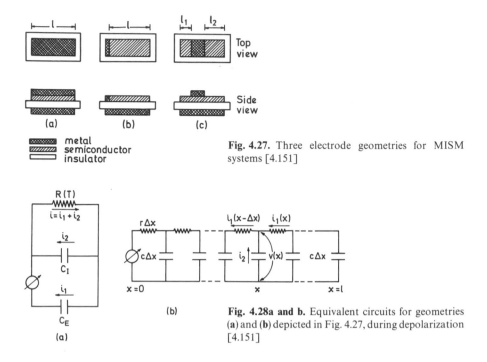

Fig. 4.27. Three electrode geometries for MISM systems [4.151]

Fig. 4.28a and b. Equivalent circuits for geometries (a) and (b) depicted in Fig. 4.27, during depolarization [4.151]

with a superposition of the two preceding cases and the total depolarization current will thus be obtained by the sum of (4.67, 68), resulting in the possible occurrence of two or even three peaks if the sample protruded beyond the top electrode more in one direction than in the other ($l_1 \neq l_2$). Obviously, the relative intensities of these peaks, and thus the more or less complicated structure of the TSDC spectrum, will depend on several parameters, including sample geometry, heating rate or activation energy. The effect of varying the semiconductor thickness d_S, for example, is shown in Fig. 4.29. With the chosen initial conditions (insulator thickness $d_I = 1\ \mu m$) a significant two-peak structure becomes apparent for $d_S \gtrsim d_I$. These conclusions are particularly important in pointing out that the electrode geometry can play a crucial role in TSDC experiments involving space-charge relaxation processes and that a complex spectrum is not necessarily to be interpreted in terms of multiple relaxation mechanisms or multiple trap centers (Sect. 4.9). This was experimentally confirmed by *Agarwal*, in particular in chalcogenide glass where the TSDC curves were shown to contain one or more peaks, the position and amplitude of which depended on the size and relative geometrical arrangement of the electrodes with respect to one another [4.151, 154, 155].

Some aspects of the TSDC theory for polarization processes obeying second-order kinetics have been discussed, in particular by *Podgorsak* and *Moran* [4.156], and *Vanderschueren* [4.157], in order to explain narrow symmetrical peaks appearing in the high-temperature region of certain po-

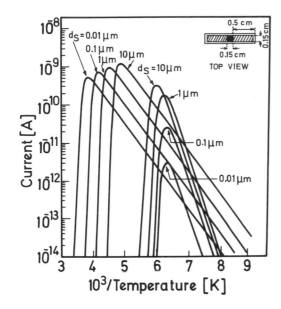

Fig. **4.29.** Calculated transversal and longitudinal currents (TSDC) for electrode geometry depicted in Fig. 4.27c $(l_1 = l_2)$ for various sample thicknesses d_S. The net current is obtained by adding the two peaks for a given d_S. Other parameters chosen are $d_1 = 1$ μm, $E = 0.5$ eV, $\sigma_0 = 9.2 \cdot 10^2 \, \Omega^{-1} \mathrm{cm}^{-1}$, $V_d = 150$ V, $q = 0.15$ K/s [4.151]

lymers [4.157] and CaF$_2$ single crystals [4.156]. On the other hand, peaks that are neither of first nor second order kinetics have been found experimentally by several investigators. In view of the complexity of the charge carrier accumulation in real dielectrics, this is expected and it seems questionable to analyze them in terms of general-order kinetics assuming an equation $J(t) = -\mathrm{const.}(T) \, P^n(t)$ to be valid.

4.8.3 Processes Characterized by a Distribution in Relaxation Time

Space-charge TSDC peaks observed in a number of materials show the characteristic properties of distributed processes, i.e., extension over a wide temperature range, large increasing shifts of the maximum temperature with increasing polarizing temperature, systematic variations in initial slope by applying the step-heating technique, etc. However, attempts to interpret these observations, even qualitatively, in terms of a distribution of conductivity parameters have been surprisingly rare and, up to now, no theoretical effort has been devoted to derive suitable expressions for the current density or to calculate the involved distribution functions. This is probably due to the fact that the observation of such a complex behavior, added to the inherent complexity of the space-charge process, has largely discouraged further detailed investigation.

Among the few significant cases mentioned in the literature as distributed processes are the high-temperature peaks appearing in heavily doped alkali halides [4.70] (distribution of the conductivity activation energy) and in phosphate glasses [4.66] (distribution of the pre-exponential factor).

4.8.4 Evaluation of the Conductivity Parameters

When a spatially uniform distribution of excess charges is involved and, provided that the depolarization process obeys first-order kinetics with a single relaxation time, most of the methods derived for analyzing dipolar processes remain valid (Sect. 4.6.2). This follows from the similarity of the basic $J_D(T)$ functions describing the two phenomena. Such an ideal case, unfortunately, is rarely encountered in practice.

For the other space-charge models the initial rise method is still directly applicable (with the restrictions already mentioned in Sect. 4.6.2). All other methods have to be appropriately modified, which is only possible in some specific cases. The graphical integration method, for example, has been extended by *Podgorsak* and *Moran* to single processes obeying a second-order kinetics [4.156] while adapting the same method to the more general case of a spatially nonuniform distribution has been proposed by *Wintle* [4.158], and *Van Turnhout* [4.30]. An equation valid over the whole temperature range can no longer be obtained, however, and only a partial linearization at low temperatures can be achieved by plotting, instead of $J(T)/\int_T^\infty j(T')dT'$, either the ratio of the released charge to the stored charge $\int_0^T j(T')dT'/\int_T^\infty j(T')dT'$ [4.158] or the quantity $T_m^2\int_0^T j(T')dT'/T^2\int_T^\infty j(T')dT'$ as a function of $1/T$ [4.30].

Owing to the close relationship existing between a space-charge peak and the normal dark conductivity, the TSDC characteristic parameters, activation energy E and pre-exponential factor σ_0, are expected to be identical with those governing the conductivity [4.27, 30, 151, 153, 159]. The TSDC method is therefore an alternative to the current–voltage–temperature measurements for studying the conduction mechanisms, particularly in cases where one blocking contact cannot, for some reason, be avoided. Furthermore, it is important to note that the TSDC measurements can be used to investigate the conduction type in semi-insulating thin films. This can be done either by establishing the influence of injected excess charge carriers on the TSDC peaks or by studying their possible shifts and distortions as a function of magnitude and polarity of the applied field [4.160, 161].

4.9 Theory of TSDC Involving Space-Charge Processes. The Trapping Model

In dielectrics polarized by the photoelectret effect or charge injection (via electrode injection, electronic bombardment or corona discharges), the long lifetime of carriers is generally assumed to result from trapping phenomena, and the space-charge peaks observed in TSDC experiments are usually interpreted in terms of trap-limited currents. However, as far as amorphous materials are concerned it must be remembered that the validity of a pure trapping model is still a moot point [4.162–166]. In such nonperiodic structures the immobilization of charge carriers for long periods does not necessarily indicate the presence of traps, a simple alternative explanation being that the carriers are slow moving to an extent which is determined by an inherently low-mobility process, such as hopping. Formally, this last mechanism can be considered as nearly equivalent to a gradual emptying of traps into a conduction or valence band but, physically, it implies that the interpretation of the observed phenomena will seldom be unambiguous. It underlines the need for great caution when calculating trapping parameters from TSDC data. Furthermore, before attributing TSDC peaks to special trap levels, it is also necessary to know to what extent equilibrium charge carriers take part in the investigated effects, which means, to what extent the peaks observed are caused or influenced by the bulk conductivity of the material [4.19, 151, 153]. Owing to its generality, this is a very important problem, and overlooking it has undoubtedly led to several erroneous interpretations of experimental results in terms of trapping parameters.

The basic equations needed for describing a thermal release from traps are similar to those of the charge motion model, except that the density of trapped charge must be introduced as an additional variable while recombination replaces intrinsic conduction as a neutralizing process. The resulting equations are intractable as well and manageable expressions for the depolarization current will only be obtained by means of several approximations regarding, among others, the kinetics of trapping, the nature of the metal–dielectric contact or the type of charge distribution involved. In view of the simplicity of the models usually adopted, the usefulness of such simplified expressions often seems questionable and their application to poorly characterized solids, in particular, must be attempted only with great caution. As for the charge motion model, we will restrict this section to a brief discussion of the main TSDC theories derived for heterocharged dielectrics and we refer again to the work by *Van Turnhout* for monopolarly and bipolarly homocharged materials [4.30].

4.9.1 Single Trap Level

A TSDC theory for materials previously polarized by the photoelectret effect was first proposed by *Zolotaryov* et al. considering a barrier-type polarization, in the quasi-equilibrium approximation (fast retrapping) [4.15]. Several authors working in the field of photoelectrets have further derived simplified theories applicable to other specific kinetics (mono- or bimolecular recombination, slow or fast retrapping [4.167–170]. The $J_D(T)$ functions obtained are generally of the *Bucci–Fieschi* type except for the bimolecular case which leads to more symmetrical peaks, the maximum temperatures of which are dependent on the initial concentration of trapped carriers and thus on the forming conditions [4.168].

Equations of the depolarization current in semiconductors previously charged by field effect and thermal cycling (thermally stimulated capacitor discharge method) have also been derived [4.171] while an extensive theory of thermally stimulated currents associated with various conditions of voltage bias applied to thin insulator films during the cooling and heating steps (stimulated dielectric relaxation currents method) has been given by *Simmons* and *Taylor* [4.23]. Considering that, for traps distributed throughout the material, the existence of blocking contacts involves a more or less severe distortion of the band edges (Schottky barriers) and establishing the theory of the subsequent nonequilibrium state between the external electric field and the excess charges, these authors have shown that the $J-V-T$ characteristics of the system must depend crucially on the relaxation properties of the depletion region. In a pure TSDC process, for example, the system relaxes from nonsteady state to steady state in two steps. The first one, associated with bulk properties, is a relaxation to a quasi-steady state arising from a redistribution of the excess charges between both depletion regions, while the second, associated with an electrode effect, is a relaxation from the quasi-steady state to equilibrium as a result of carrier injection from the electrodes into the insulator. Both mechanisms give rise to a TSDC peak having the familiar asymmetrical shape. The one appearing at lower temperature is the larger and corresponds to a release of charge in the external circuit given by

$$Q = \left(\frac{e\varepsilon N}{2}\right)^{1/2} \frac{SV_d}{(V_d + \Delta\phi_1)^{1/2} + \Delta\phi_2^{1/2}} \tag{4.69}$$

with N being the trap density, e the unit of electronic charge, ε the permittivity of the insulator, S the sample area, $\Delta\phi_1 \equiv (\phi_{m_1} - \phi_i)/e$ and $\Delta\phi_2 \equiv (\phi_{m_2} - \phi_i)/e$, where ϕ_{m_1}, ϕ_{m_2} and ϕ_i are the work functions of the lower and counter electrodes and of the insulator, respectively. This expression, which involves a quasi-linear variation of Q with V_d, remains valid until a threshold voltage V_T is reached for which an electrode-limited to bulk-limited transition occurs, indicating a sublinear behavior of Q vs V_d.

On the other hand, starting from a similar model but taking into account the variations in the internal electric field resulting from the gradual changes in the depletion region, *Gupta* and *Van Overstraeten* have shown by means of model calculations that the maximum temperature of the current peaks must be an increasing function of the occupied trap density and a decreasing function of the applied field [4.172].

Up to now, there has been little information on experimental verifications of these theories involving the depletion region. In thin film Al–CeF$_3$–Al samples the complex TSDC spectrum observed (Fig. 4.30) has been interpreted by *Simmons* and *Nadkarni* on the basis of a two-step process [4.173], but it was

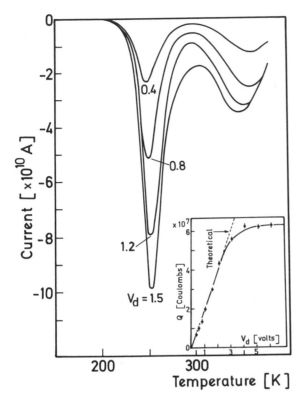

Fig. 4.30

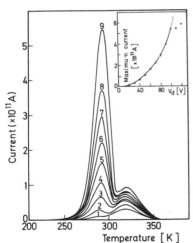

Fig. 4.30. TSDC spectra of thin film Al–CeF$_3$–Al samples for different values of polarizing voltage V_d. Inset: Plot of the released charge Q (main peak) vs V_d [4.173]

Fig. 4.31. TSDC spectra of Al-amorphous As$_2$S$_3$–Al samples ($d = 1$ mm) for different values of polarizing voltage V_d. Inset: Plot of the maximum intensity (main peak) vs V_d [4.175]

pointed out by *Muller* [4.19] that the second peak can also be predicted simply from a bulk-conductivity determined accumulation or depletion of charges in the surface region. On the other hand, in the system Al–As$_2$S$_3$–Al which, in principle, is particularly suitable to test the model because direct evidence for band bending exists [4.174] *Thurzo* and *Lezal* [4.175] showed that, contrary to theoretical predictions, the TSDC peak significantly shifts to higher temperatures with increasing voltage V_d and that its maximum intensity exhibits supralinear behavior vs V_d (Fig. 4.31). It should be noted, however, that these last experiments were performed on thick samples (1 mm) for which the theory of *Simmons* and *Taylor* can no longer be expected to be strictly valid.

4.9.2 Distribution of Trap Levels

Even though a distribution of localized states is commonly found in a wide range of insulators and semiconductors, particularly those existing in polycrystalline or amorphous forms, it has been postulated in only a few cases for explaining TSDC results [4.30, 173, 176] and no detailed theoretical treatment has been put forward. A good simplifying approach of the problem exists only for the analysis of thermally stimulated conductivity experiments in the high-field approximation [4.177].

4.9.3 Evaluation of Trapping Parameters

Provided that the motion of charge carriers in the conduction band is sufficiently rapid so as not to be rate limiting, the peak properties are essentially determined by the trapping parameters and, for the simplest models described by $J_D(T)$ functions similar to those encountered for dipolar or conductivity processes, the trap depth can be calculated in theory, by means of any one of the classical methods previously described (initial rise, heating rate, etc.). On the other hand, in absence of any previous argument in favor of a particular model (i.e., if the defect structure of the investigated solid is not reasonably well known beforehand), it is obvious that such procedures will never be very meaningful.

In other respects, e.g., in a picture involving a hopping process, the activation energy represents that of the carrier mobility only and, hence, is an inherent characteristic of the material, rather than of any defect.

4.10 Thermally Stimulated Depolarization Currents Involving Interfacial Processes

When structurally heterogeneous materials are investigated, one can expect that a field-induced ionic polarization will obey more closely an interfacial model of the Maxwell–Wagner–Sillars type than a space-charge model of the

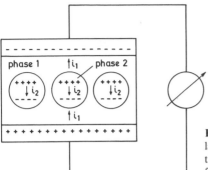

Fig. 4.32. Decay of the MWS polarization in a two-layer structure during a TSDC experiment (schematic); i_1 and i_2 are internal conduction currents flowing through the two phases

barrier type [4.38]. This will be the case, in particular, for the dielectrics used in electrotechnical applications since they are most often complicated mixtures, laminates or semicrystalline products. Obviously, for such complex compounds, a general theoretical treatment of the depolarization current appears to be intractable to analytical solutions and a number of more or less severe approximations are required to obtain a manageable $J_D(T)$ function. Furthermore, this type of polarization, like a space-charge polarization, does not usually have an efficiency of 100% when observed by the TSDC method. As a matter of fact, the charges piled up at the interfaces are neutralized by conduction currents in opposite directions (Fig. 4.32), which means that the observed charge release will be less than the initially stored charge, especially when the different phases have similar conductivities [4.30].

A simplified TSDC theory of interfacial polarization has been proposed by *Harasta* and *Thurzo* [4.178], and *Van Turnhout* [4.10, 30] by considering the model of a charged two-layer condenser and neglecting the possible dependences of the permittivities of the two layers on temperature. The expression for the current density is then obtained as [4.178]

$$J_D(T) = \frac{[\varepsilon_1\sigma_2(T)-\varepsilon_2\sigma_1(T)][\varepsilon_2\sigma_1(T_p)-\varepsilon_1\sigma_2(T_p)]d_1d_2}{(\varepsilon_1d_2+\varepsilon_2d_1)^2[\sigma_1(T_p)d_2+\sigma_2(T_p)d_1]} V_p \exp\left[-\int_0^T \frac{dT'}{q\tau^*(T')}\right],$$
(4.70)

where the subscripts 1 and 2 refer to the two layers and $\tau^*(T)$ is a relaxation time given by

$$\tau^*(T) = \frac{\varepsilon_0(\varepsilon_2d_1+\varepsilon_1d_2)}{d_1\sigma_2(T)+d_2\sigma_1(T)}.$$
(4.71)

Although the current is seen to depend on two varying parameters σ_1 and σ_2 and thus on two different activation energies E_1 and E_2, (4.70) will usually be represented by one asymmetrical TSDC peak, the maximum of which is determined by the fastest ohmic dissipation process and according to the expression

$$T_m = \left\{ \frac{\varepsilon_0 q(\varepsilon_1 d_2 + \varepsilon_2 d_1)[\varepsilon_1 \sigma_2(T_m)E_2 - \varepsilon_2 \sigma_1(T_m)E_1]}{k[\sigma_1(T_m)d_2 + \sigma_2(T_m)d_1][\varepsilon_1 \sigma_2(T_m) - \varepsilon_2 \sigma_1(T_m)]} \right\}^{1/2} \tag{4.72}$$

allowing us to predict a small shift of the peak with varying thicknesses of the layers. When the two functions $\sigma_1(T)$ and $\sigma_2(T)$ intersect, a second peak will appear and it will then be observed in a reverse direction [4.30]. Since the shape and location of the TSDC spectrum are affected by the properties of the two layers it is obvious that, except for the initial rise method, the classical techniques used for determining the conductivity parameters will lead to hybrid values.

In most real dielectrics, the interfacial and space-charge polarizations cannot be rigidly separated and a direct comparison between experiment and theory is, therefore, difficult. It follows that only a few experimental TSDC peaks have been successfully related to the existence of a pure Maxwell–Wagner–Sillars process. In polymers it was postulated by *Hedvig* [4.133, 179] that, owing to the unavoidable presence of structural inhomogeneities, this process could be the main cause of ionic polarization, but this question is still under discussion. In fact, significant results have essentially been obtained only in special systems, such as polycrystalline samples [4.180], polymer laminates [4.30] or ionic crystals provided with insulating surface layers [4.178]. In Fig. 4.33, for example, the TSDC spectrum of a laminate of teflon–FEP and polyimide is compared with that of a pure teflon–FEP sample. The peak appearing in the heterogeneous materials (ϱ_c) is clearly due to an interfacial polarization since its position and amplitude are markedly different from that of the classical space-charge peak (ϱ) observed in the homogeneous polymer.

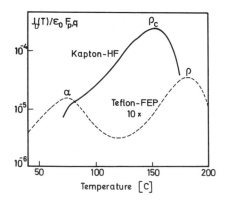

Fig. 4.33. TSDC spectra of pure teflon-FEP (dashed curve) and a laminate of teflon-FEP and polyimide (solid line) ($T_p = 473$ K, $F_p = 10^5$ V/cm, $t_p = 90$ min) [4.30]

4.11 Thermally Stimulated Polarization Currents

4.11.1 TSPC Involving Dipolar Processes

With the same assumptions as those used in deriving the *Bucci–Fieschi* formula and, provided that the temperature dependence of the equilibrium polarization is neglected, it is obvious that the $J_p(T)$ function associated with a gradual orientation of dipoles will be similar to (4.8). This implies that a TSPC peak should be characterized by the same position, height and shape as the corresponding TSDC peak, the only difference being that the polarization current is of the opposite sign [4.20, 88]

$$J_P(T) = -\frac{P_e(T_p)}{\tau_0}\exp\left(-\frac{E}{kT}\right)\exp\left[-\frac{1}{q\tau_0}\int_{T_0}^{T}\exp\left(-\frac{E}{kT'}\right)dT'\right]. \qquad (4.73)$$

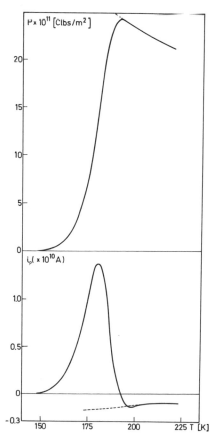

Fig. 4.34. Computed temperature dependence of polarization and TSPC for a dielectric obeying Fröhlich's model; typical parameters chosen are: $\tau_0 = 10^{-10}$ s, $E = 0.4$ eV, $N_d = 10^{20}$ m^{-3}, $p_\mu = 2$ Debye, $F_p = 10^3$ V/cm and $q = 1$ K/s. The dashed curves show the T^{-1} dependence of the equilibrium polarization and the T^{-2} dependence of the corresponding current

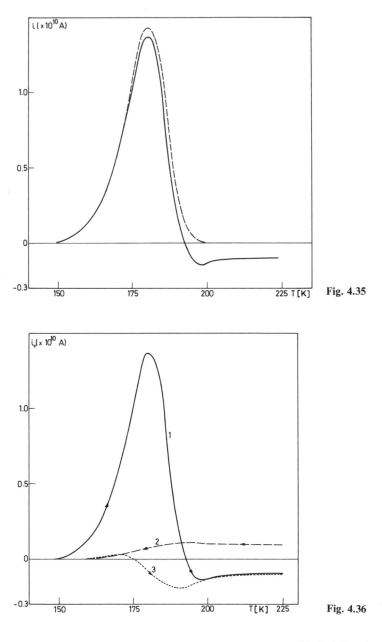

Fig. 4.35

Fig. 4.36

Fig. 4.35. Computed TSDC (dashed curve) and TSPC (solid line) for the same values of parameters as used in Fig. 4.34

Fig. 4.36. Computed theoretical temperature dependence of the current for successive thermal cycles under applied electric field (parameters used are the same as in Fig. 4.34); (*1*) first heating (TSPC); (*2*) cooling; (*3*) second heating

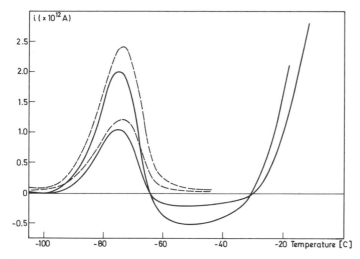

Fig. 4.37. TSDC (dashed curves) and TSPC (solid lines) observed for two field strengths (6 and 12 kV/cm) in a styrene-butadiene block copolymer. For convenience of comparison, the depolarization currents are plotted in the same direction as the main polarization peak

As shown by *Moran* and *Fields* however, this is no longer true if the $P_e(T)$ function is taken into account [4.14]. The current density then takes the form [4.181]

$$J_P(T)=\left\{\frac{1}{\tau(T)}\exp\left[-\int\limits_{T_0}^{T}\frac{dT}{q\tau(T)}\right]\right\}\int\limits_{T_0}^{T}\left[\frac{P_e(T')}{\tau(T')}\exp\int\limits_{T_0}^{T'}\frac{dT''}{q\tau(T'')}\right]dT'-\frac{P_e(T)}{\tau(T)}, \qquad (4.74)$$

which clearly shows a current reversal in the high temperature range. The occurrence of this phenomenon, which has been confirmed by numerical calculations based on the bistable model of *Frohlich* [4.93, 182] (Fig. 4.34), is an obvious consequence of the tangential convergence of the transient polarization $P(t)$ with its decreasing saturation limit $P_e(t)$.

Another consequence of the temperature dependence of the equilibrium polarization is that, owing to the polarization increase occurring during the cooling step of TSDC experiments (Sect. 4.4.5), the main TSPC peak will always be smaller than the TSDC peak obtained with the same field and heating rate (Fig. 4.35).

It thus appears that the comparison of TSDC and TSPC spectra allows us to obtain direct information on the importance and type of the temperature dependence characterizing the equilibrium polarization. It must be emphasized, however, that the TSPC technique is subject to several experimental difficulties resulting from the possible interference of conduction or injection phenomena. An injection of carriers, for example, can also lead to the appearance of current reversals and additional experiments are then required to distinguish between the two processes. This can be done either by varying experimental parameters such as magnitude and polarity of the applied field or the nature of electrodes (Sect. 4.12), or by carrying out a series of successive thermal polarization cycles including the following steps: I) Heating of the sample from T_0 to the temperature range where a current reversal occurs (normal TSPC), II) cooling from the maximum temperature reached in step 1 to T_0, III) repeating step I). As a matter of fact, model calculations show that, if the current reversal arising in step I) is of relaxational origin, a peak must appear during the second step with an amplitude depending on the cooling rate, while during the third step, a smaller peak followed by a current reversal must be apparent [4.182, 183] (Fig. 4.36).

Perhaps owing to the above-mentioned experimental difficulties, experimental data on TSPC of dipolar processes are rather scarce. In alkali halide crystals, which are doped with divalent metallic impurities, the technique was used by Mc Keever and Hughes to distinguish between dipolar and trapping mechanisms but no current reversal was observed [4.20]. In polymers, on the other hand, it has been shown by Mizutani et al. [4.184], and Vanderschueren et al. [4.93] that the TSPC spectra are often in good qualitative agreement with the theoretical predictions of a model characterized by a temperature-dependent equilibrium polarization (Fig. 4.37).

4.11.2 TSPC Involving Space-Charge Processes

Only a few simplified models of space-charge polarization predict the theoretical TSPC behavior. Starting from a charge motion model of a dielectric provided with fully blocking contacts, Muller has shown that thermally stimulated polarization and depolarization currents associated with a barrier-type polarization must obey similar laws [4.19], which obviously leads to the same conclusions as those obtained from a dipolar model characterized by a temperature-independent equilibrium polarization, i.e., the TSDC and TSPC peaks will be of opposite signs but strictly identical in magnitude, shape and position. According to the trapping model proposed by Simmons and Taylor [4.23] (Sect. 4.9.1) or Devaux and Schott [4.22], TSPC peaks without reversal can also be observed, provided that at least one of the electrodes is non-ohmic and they are similar not only to TSDC peaks but also to TSC peaks involving the same trap levels. In the model advocated by Gupta and Van Overstraeten [4.172] (Sect. 4.9.1), theoretical TSPC characteristics have been extensively discussed by means of model calculations of the trapping parameters but no comparison has been made with the TSDC process.

Due to the small number of experimental results reported in the literature and in view of the complex behavior expected in real dielectrics, it is still difficult to judge the validity of these theories.

4.12 Experimental Means for Differentiating Between the Various Polarization Processes Involved in FITSC Experiments

Perhaps the fundamental difficulty of the FITSC method lies in the absence of selectivity towards the various possible polarization processes. In view of the complexity of electrical conduction and the charge storage mechanisms in dielectrics on one hand, and the uncertainty associated with the nature of electrode–sample interface and the physico–chemical structure of the materials investigated on the other, it is advisable to be extremely cautious in giving interpretations of measured current curves in terms of simplified models and in calculating the relaxation parameters involved. In fact, analyses of data performed before knowing the exact origin of the underlying microscopic process remain phenomenological. There are many examples in the literature where unrealistic values of physical parameters have been chosen to fit the experimental data without looking for other possibilities. Furthermore, we have seen in the preceding chapters that most of the models usually adopted for describing dipolar, ionic or electronic processes predict similar functional relationships (and consequently lead to current peaks which look deceptively similar) and that their predictions regarding the change of peak properties with a variation of external or internal characteristic parameters are not often valid beyond the particular assumptions of the model.

All these facts imply that an unequivocal distinction between the various polarization mechanisms (which can be responsible *a priori* for the peaks appearing in a given material) will be a virtually impossible task if we start from an analytical basis. The inescapable conclusion is that, at the present state of our knowledge, isolated FITSC measurements will be rarely self-consistent and carry with them relatively little information content. It is particularly significant that for many of the investigated materials the interpretation of data is still the subject of much controversy. Most of the cases where a satisfactory physical explanation has been achieved have in common that the current measurements were complemented by other physical or chemical techniques.

In the following we will attempt, through some characteristic examples, to devise some critical experiments that allow us to obtain information on the nature of charge carriers and the type of polarization mechanism involved in typical FITSC measurements. However we must make clear that none of the measurements on its own leads to a direct identification, particularly when several processes coexist and interfere with each other and that firm conclusions may be reached only by using them in conjunction with others.

Ideally, the general procedure for interpreting a given current spectrum should include, as a first stage, a careful testing of the experimental arrangement and operating process to ensure that a part of or the total observed current is not induced or influenced by spurious effects (poor metal–dielectric

contacts, thermal gradients, electrochemical processes, triboelectric phenomena, etc.) (Sect. 4.12.1). Subsequently, a systematic variation is required of
those experimental parameters which are expected to affect differently the
various possible polarization mechanisms (electrode material, sample thickness, intensity of electric field, thermal cycling, etc.) (Sect. 4.12.2) and, as a
necessary confirmative stage, correlation studies should be carried out with all
available techniques concerned with thermal relaxation effects (Sect. 4.12.3). In
addition, it is obvious that the interpretation of these experiments will be
greatly facilitated if such relevant factors as chemical and physical structure,
doping content and thermal and radiative history are reasonably well known
by preliminary investigations of, e.g., infrared spectroscopy, optical absorption,
X-ray scattering or chemical analysis. We must not forget that, if the presence
of a particular type of charge carrier sometimes appears to be the most likely
because of the chemical structure of the material under study, the additional
presence of other carriers should never be rejected. It is well known, for
example, that electronic carriers may arise both intrinsically and from easily
ionized impurities or electrode injection, that the presence of mobile ionic
species need not necessarily be an intrinsic property of the material but may
arise from chemical degradation or from impurities such as additives, dissolved
gases or incorporated water and that, even in nonpolar structures, dipoles may
exist as a result of polar impurities or can be created by oxidation, annealing or
irradiation. Finally, it must also be borne in mind that in many materials
surface states and absorbed layers may be responsible for polarization processes and can play an important role in the exchange properties at the metal–
dielectric interface.

4.12.1 Parasitic Currents and Spontaneous Polarization

Surface Currents and Electrode Geometry

It is a well-known fact in conventional permittivity and conductivity measurements that the bulk current and capacitance of a material can be markedly
affected by surface currents in certain atmospheres and temperature ranges
[4.29]. Usually however, the problem is not too severe since the collection of
these surface currents in the measuring circuit can be prevented by using a
three-terminal arrangement (guarded electrodes). Although often overlooked,
the role of surface conductivity is obviously also of primary importance in
FITSC measurements performed without a guard ring and it is not unusual
that such experiments lead to additional background currents several orders of
magnitudes larger than the bulk current [4.185] or to one or more additional
current peaks, particularly when the samples or the atmosphere are not
previously carefully dried [4.27, 186] (Fig. 4.38).

   In MISM structures, on the other hand, it must also be recalled (Sect. 4.8.2)
that the position, shape and number of the conduction-induced TSDC peaks
may depend on the sample–electrode geometry.

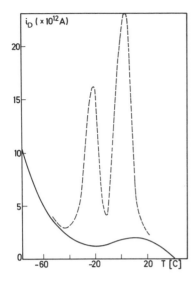

Fig. 4.38. Spurious peaks (dashed curve) appearing in polyethyleneterephtalate during TSDC measurements performed in a humid atmosphere without guard ring. The solid line represents the normal spectrum ($T_p = 295$ K, $F_p = 26$ kV/cm, $q = 5$ K/min)

Spurious Currents Related to Electrode–Dielectric Contact

Ideally, the electrode–dielectric contact should be electrically "invisible" for investigating intrinsic relaxation properties. This is obviously not possible and the problem of applying electrodes and their adhesion to samples plays, therefore, a key role in the determination of electrical properties of solids. This is particularly true in FITSC measurements where generally large changes in temperature are involved. Poor contacts will lead to irreproducible measurements, high noise level, poor agreement between results from different samples of the same material and sometimes additional current peaks due, in particular, to homocharge formation (Fig. 4.39). As already emphasized in Sect. 4.3, the most convenient way to avoid these difficulties is to use metal films carefully evaporated onto cleaned surfaces. Without this cleaning procedure large parasitic currents may be observed in certain materials [4.185]. This is illustrated in Fig. 4.40 for aluminum oxide samples provided with aluminum electrodes.

A further complication related to contact problems can result from the often large differences in thermal expansion between the electrodes and the sample. Fracture and peeling phenomena can occur during the cooling or heating steps, leading to abnormally high resistivity and thus to misleading results [4.21].

Finally, reversible parasitic currents may also appear in nonpolarized samples as a result of electrochemical processes between the metallic electrodes and the sample (oxidation–reduction reactions, for example). Such currents, increasing at high temperatures in the same way as true conduction currents, have been clearly observed particularly in ferroelectric ceramics [4.187] and polymers, even in samples provided with identical electrodes [4.27, 30, 188].

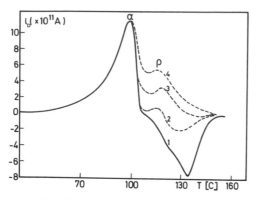

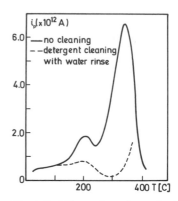

Fig. 4.39. Effect of a progressive improvement of the electrode-dielectric contact on TSDC of non-metallized samples of poly(methyl methacrylate) (the pressure exerted by electrodes on the sample gradually increases from 4 to 1)

Fig. 4.40. Effect of surface cleaning prior to electrode evaporation on TSPC observed in aluminum oxide samples for $F_p = 6580$ V/cm [4.185]

No simple means for solving these contact problems exist, except to remember their existence and to minimize their effects.

Spontaneous Polarization

Aside from these spurious effects it is also essential, before any attempt of interpretation of FITSC data, to ascertain that one or more of the observed peaks do not result from some previous spontaneous polarization of the sample. Such peaks, easily identified since they must also be obtained during heating without any previously applied voltage, have been reported in a number of materials, even nonferroelectrics (Fig. 4.41) and several explanations have been given based on the nature of the material investigated and the method of sample preparation: Molecular orientation produced during film fabrication either as a direct consequence of extrusion, drawing or molding processes (especially in polymers [4.27, 30, 189]) or, for solvent-cast materials, as a result of charge gradient created during the unavoidably asymmetric

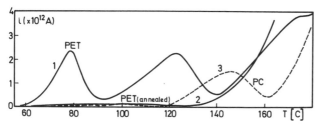

Fig. 4.41. Short-circuit currents obtained by heating nonpolarized commercial samples of poly-ethyleneterephthalate (*1*) and polycarbonate (*3*) (spontaneous polarization). Curve *2* shows a spectrum obtained after prolonged annealing at 450 K

evaporation process [4.190]; charge separation due to a gradual evaporation of volatile impurities [4.21]; production of carriers during thermally induced chemical degradation taking place asymmetrically in regions near opposite surface of films (especially in polymers [4.191]); injection of defects formed under the influence of moisture on the sample surface (ionic crystals [4.192]); thermally generated electromotive forces arising from a thermal gradient across the sample [4.193], etc. Most, but not all, of these mechanisms are irreproducible and can be eliminated relatively easily by appropriately conditioning the sample, e.g., by repeatedly heating and cooling the material under vacuum until reproducible current spectra are obtained (see curves 1 and 2 in Fig. 4.41).

4.12.2 Variation in Experimental Conditions

Preliminary Remarks

As stated above, even if we assume that sufficient care has been taken in sample and electrode preparation and that any spurious or spontaneous current has been either eliminated or identified, the simple examination and analysis of an isolated FITSC spectrum cannot be expected to yield the origin of the peaks. In certain favorable cases, however, some general indications can be obtained, based on charge calculations, current direction or peak position. The calculation of the released charge, for example, can lead to unrealistic values if a particular type of polarization is assumed (in ionic crystals, the number of dipoles deduced from the area of the high temperature TSDC peaks often exceeds the number of impurity atoms added [4.70]) or, conversely, to a close agreement with the previously known concentration of a particular type of charge (in polyethylene, a good correspondence between the absolute $C=O$ concentration as determined by TSDC and IR spectroscopy has been demonstrated [4.69] and in a series of polar–nonpolar ethylene–vinyl acetate copolymers, it has been shown that the ratio of the height of the low-temperature TSDC peaks closely agrees with the ratio of the dipole concentration [4.184]).

The direction of the TSDC peaks, on the other hand, is often an important, but not definitive argument for separating heterocharge processes (dipole orientation, ionic drift, electronic space charge of internal origin) and homocharge processes (electronic or ionic injection from the electrodes or surrounding atmosphere). Provided no significant diffusion occurs, heterocharge peaks always appear in a direction opposite to that of the normal charging current, while the direction of the external current generated by release of homocharge depends on the nature of the contact between dielectric and electrodes. For blocking contacts the homocharge decays by internal conduction and the current is opposite to the heterocharge current (complete preceding depolarization supposed) (Fig. 4.1). For nonblocking electrodes, the homocharge decays by conduction through the dielectric–electrode interface and the current has the same direction as the heterocharge current but its amplitude is smaller [4.10] (in the extreme case of an interface with negligible thickness, the homocharge decay can be observed by the noise resulting from internal breakdown phenomena [4.194]). Note that the situation can be further complicated in TSPC measurements where reversals can arise not only from diffusion and homocharge processes but also from the temperature dependence of the equilibrium dipole polarization (Sect. 4.11.1).

Finally, the temperature range where the peaks appear can also provide information. We may expect, for example, that conduction-induced peaks will be the last to occur in the spectrum owing to the macroscopic displacements involved. This was shown to be the case in a number of materials, in particular in ionic solids, where the space-charge peaks always appear at temperatures higher than room temperature [4.70] and in polymers, where they are observed only after the glass transition has occurred [4.27, 30].

Whatever the case, any further differentiation between the various types of carriers and polarization mechanisms requires a series of measurements in a number of different experimental conditions.

Polarization and Depolarization Conditions (T_p, t_p, F_p, q)

As far as a distinction between various heterocharge mechanisms is concerned, the study of the properties of a given TSDC peak as a function of polarizing time and temperature is usually not very meaningful. As a matter of fact, a systematic variation in shape and position could be attributed to dipolar processes (with a distribution in relaxation times) as well as to ionic or electronic space-charge phenomena (involving a spatially nonuniform charge distribution or a distribution in activation energy) while, with other starting assumptions (single relaxation time, barrier-type space-charge polarization), the absence of such variations could also be explained by the same basic mechanisms.

On the other hand, it may be useful to know the influence of the polarizing temperature T_p for identifying homocharge processes occurring in certain temperature ranges, especially when they result from injected carriers deeply trapped in the material. In this case, one or more peaks may appear at temperatures much higher than T_p, which is never possible according to the usual heterocharge mechanisms. This is illustrated in Fig. 4.42, where TSDC obtained by the *thermal sampling technique* (Sect. 4.5.4) for Al-polyethylene-terephtalate-Au structures are shown. For forming temperatures below $-40\,^{\circ}\text{C}$, each curve shows only one peak at a temperature a little higher than T_p which can be governed for example by distributed dipolar processes. Above $-40\,^{\circ}\text{C}$, each spectrum shows additional broad peaks, extending over several tenths of degrees beyond T_p, suggesting they originate from carriers injected into the bulk.

One of the arguments advocated for distinguishing a dipolar polarization (or, more generally speaking, a uniform polarization) from a space-charge polarization is based on the field dependence of the TSDC properties. This is so because TSDC peaks corresponding to the usual dipole models are character-

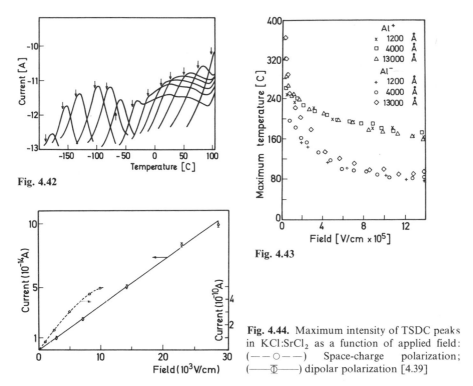

Fig. 4.42

Fig. 4.43

Fig. 4.44. Maximum intensity of TSDC peaks in KCl:SrCl$_2$ as a function of applied field: (— — ○ — —) Space-charge polarization; (——— ⌀ ———) dipolar polarization [4.39]

Fig. 4.42. TSDC obtained by thermal sampling technique for Al-polyethyleneterephtalate-Au systems. The arrows show the values of T_p corresponding to each TSDC curve [4.195]

Fig. 4.43. Dependence of T_m of the α peak on field for Al–SiO$_2$–Si samples ($3 \cdot 10^{13}$ Na$^+$/cm^2) with different oxide thicknesses [4.196]

ized by field-independent maximum temperatures and amplitudes (or areas) that are strictly proportional to the field strength until electrical saturation is reached (i.e., for field intensities less than a few hundred thousand V cm^{-1})[4] while more complex behavior is obtained from processes involving space-charge polarization, since generally the buildup, release and equilibrium spatial distribution of the charge are strongly dependent on the applied voltage. In the case of ionic migration, for example, the peak position is usually field dependent [4.196] (Fig. 4.43) and saturation (i.e., polarization increasing less than linearly with increasing field strength) may be expected to be reached much sooner than for dipolar processes, since the gradual accumulation of carriers near the electrode rapidly decreases the effective directing field (Fig. 4.44). When carrier injection occurs, with or without subsequent trapping,

---

4 This approximate value is that for which a measurable saturation effect occurs in liquids [4.43]. It could be less in solids, but it is difficult to predict to what extent because experimental data related to this phenomenon are very scarce and the exact theory is extremely complicated.

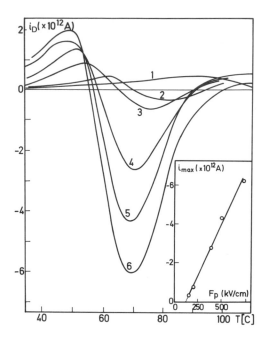

Fig. 4.45. TSDC spectra of polyethylene for $T_p = 358$ K. Curves *1–6* correspond to polarizing voltages of 350, 500, 750, 1400, 1700, and 2500 V, respectively ($d = 35\,\mu m$). Inset: Plot of the maximum current (main peak) vs polarizing field [4.197]

one may also generally predict a field-dependent peak position and a more or less complicated non-ohmic behavior, e.g., due to inefficient transport, potential-dependent injection (Schottky barrier lowering) or potential-dependent rate of detrapping (Poole–Frenkel effect). Furthermore, injection processes will be often associated with the observation of a treshold voltage as well (Fig. 4.45). This will be the case, in particular, when injection directly results from the progressive increase in the field induced by the accumulation of previously trapped internal space charges. The field-dependence argument should, however, be viewed with caution for a number of reasons. First of all, in the most commonly investigated region of low to moderate fields (10^2–10^5 V cm^{-1}), space-charge peaks often behave only slightly nonlinearly and an ohmic behavior is then detected in first approximation. Certain space-charge models, on the other hand, predict a linear dependence of the depolarization or polarization currents on the field strength (barrier-type polarization [4.19]). Conversely, the behavior of peaks resulting from bulk polarization can be more complex than expected [4.198] either due to the simultaneous presence of space charges modifying the internal field [4.30] or because the polarization is governed by special mechanisms such as two-site asymmetrical hopping processes [4.95]. Whatever the case, it is often questionable to consider an experimentally observed field dependence as a convincing confirmation of a simple theoretical model.

Finally, a dependence of the charge released on the heating rate can be expected only in some special cases. When diffusion is present, for example, a space-charge peak may vary as a consequence of unequal temperature de-

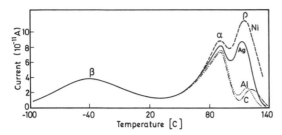

Fig. 4.46. TSDC spectra of poly(methyl methacrylate) obtained with various electrode materials ($T_p = 403$ K, $F_p = 20$ kV/cm, $q = 3$ K/min)

pendences of internal conduction and diffusion [4.199]. A similar effect may occur in heterogeneous dielectrics (interfacial polarization) when the temperature dependences of the conductivities of the components differ [4.30].

Electrode Material

The study of the influence of electrode material is undoubtedly one of the most reliable ways for distinguishing between dipolar and space-charge processes in FITSC measurements because the various metals used as contacting electrodes in independent experiments do not obviously affect a dipolar polarization (unless coexisting with an electrode-dependent space-charge polarization), but they may lead to quite important variations in formation and release of excess charges both of external and internal origin, owing to the differences in work function and blocking factor. An illustration of this behavior in polymers is given in Fig. 4.46. The amplitude and position of the space-charge ϱ peak are seen to be markedly dependent on the electrode material, contrary to those of the dipolar α peak. Other significant examples have been reported by a number of authors for electronic as well as ionic processes [4.27, 30, 195, 196], although cases where the electrodes did not play an important role have also been mentioned [4.176]. In carrying out these experiments, on the other hand, one must be aware that the results may also be affected by a possible diffusion of the metal or metallic ions into the sample.

A further attempt of differentiation can possibly be reached by means of more sophisticated systems such as M_1–I–M_2 structures (insulator sandwiched between two different metallic electrodes) or samples provided with an air gap or insulating electrodes (i.e., electrodes insulated from the sample by thin foils of high-resistivity materials). The use of insulating electrodes in comparison with contact electrodes can demonstrate the existence of possible injection phenomena. In using this procedure, it is important to previously ensure that the foils are thermally inert (no measurable current is observed in the range of temperatures investigated). By performing TSDC experiments with one-sided or two-sided metallized samples provided with an air gap, the conduction-induced space charge peaks are readily identified from the current direction and then correlated with the classical spectrum obtained with shorted samples [4.30, 200, 201]. Finally, the combined use of asymmetrical electrodes and change in forming field polarity can also be useful for identifying injection

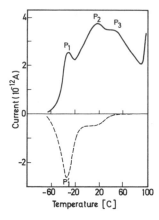

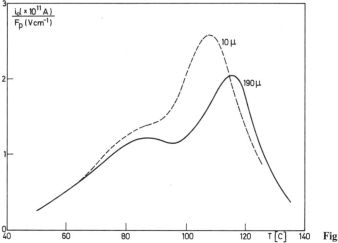

Fig. 4.47. Effect of the polarity of the polarizing field on the TSDC spectra of Al-polyethyleneterephtalate-Au systems ($T_p = 233$ K, $F_p = 10^6$ V/cm) Upper curve: Al−; lower curve: Al+ [4.195]

Fig. 4.48. Effect of sample thickness on the TSDC spectrum of polyethyleneterephtalate ($T_p = 403$ K, $F_p = 26$ kV/cm, $q = 3$ K/min

Fig. 4.48

processes. Such an experiment is shown in Fig. 4.47, where the TSDC spectrum of Al-polyethyleneterephtalate-Au structures is seen to be strongly affected by the field polarity [4.195]. The P_1 and P_1' peaks, which are similarly located close to the polarizing temperature and have polarity-independent magnitudes, are presumably of dipolar origin, while the P_2 and P_3 high-temperature peaks, which are only significantly observed in case of negative forming field applied to the Al electrode (characterized by a low work function) can reasonably be ascribed to injected electronic carriers. Other sandwich structures, including the asymmetrical MOS and MIS systems, have been studied with the same goal of isolating electronic or even ionic mechanisms [4.160, 196].

Sample Thickness

The sample thickness is also an important parameter to be considered for separating volume, space-charge and contact-induced processes. In theory, depending on the type of behavior observed for a given applied voltage, the following conclusions can be drawn: I) Thickness-independent results will imply that the phenomenon is associated with the contacts, II) a linear dependence of the current on sample thickness will be indicative of a uniform bulk polarization, and III) more complex functions will involve the probable presence of excess charges. When ion migration takes place, for example, it is often found that for a given field strength the peak amplitude decreases and the maximum temperature increases with increasing thickness (Fig. 4.48), probably because the excess charges have to move a larger distance to become neutralized at the electrodes [4.30].

It is obvious, however, that these conclusions cannot be considered as definite arguments and that all the limiting conditions already emphasized in discussing the influence of the field strength are also to be taken into consideration here.

The Sectioning Technique

The sectioning technique consists of measuring the charge released during reheating sections of variable thickness from different regions of the previously polarized material [4.202]. The existence of a uniform bulk polarization is ascertained if this charge is found to be independent of section thickness and position in the material, while a space-charge polarization will be apparent from a drastic decrease of the charge when the external layers are removed. However, this method, only feasible in certain materials, does have its pitfalls because of possible changes in the polarization or possible formation or annihilation of charge during sectioning and deposition of new electrodes.

Doping and Aging Processes

It is well known that impurities as well as the thermal, electrical and radiation history markedly affect the electrical properties of materials and this is especially true in FITSC measurements, which by their nature involve the application of high field strengths as well as accentuated annealing and quenching of samples. Each polarization mechanism is to some extent sensitive to these effects and it is for this reason that only in certain favorable cases a distinction can be attempted on this basis.

A polarization resulting from orientation of molecular dipoles is the only one which can be expected to be relatively little affected by contaminants and, therefore, can lead to nearly identical FITSC spectra for differently doped materials or samples of various origins (Fig. 4.49a). This is obviously no longer true when ionic dipoles such as impurity–vacancy complexes in alkali halides are involved. On the contrary, until the limit of solubility is reached, a close correlation must exist between the released charge and the concentration of ionic impurities responsible for the polarization (Fig. 4.50).

Whatever the material, the properties of space-charge peaks may be expected to be strongly dependent on concentration and type of impurity or

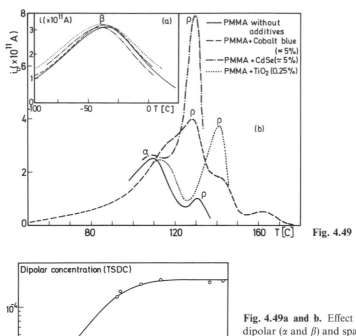

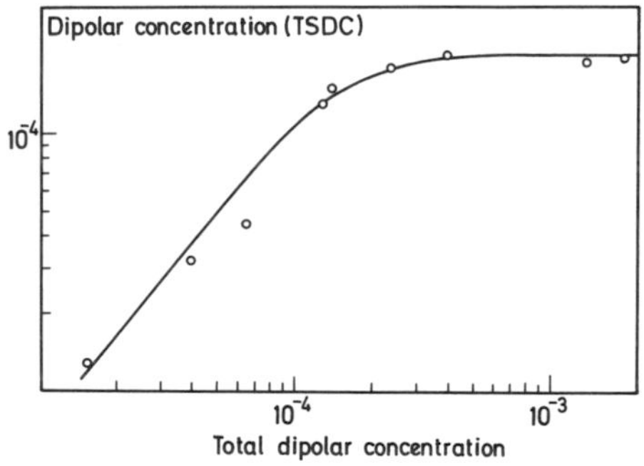

Fig. 4.49a and b. Effect of additives on the dipolar (α and β) and space-charge (ϱ) TSDC peaks of poly(methyl methacrylate). $T_p = 413$ K, $F_p = 10$ kV/cm, $q = 5$ K/min (**a**) and 3 K/min (**b**)

Fig. 4.50. Dipolar concentration, determined by TSDC, vs total dipole concentration in LiF:Mg^{++} single crystals [4.42]

dopant, since they are usually governed by many factors that are not typical of the chemical composition of the investigated sample. A variation in the number of extrinsic ionic carriers, as a result of doping, for example, will affect the initial distribution of excess charges and, consequently, the shape, position and amplitude of the corresponding TSDC peaks (Fig. 4.49b). On the other hand, impurities such as water will noticeably increase the bulk conductivity by which most of a previously stored space charge can be dissipated [4.27, 30, 203] (Fig. 4.51). A somewhat similar neutralization phenomenon can occur when extrinsic ions are directly injected or diffused into the bulk of the sample [4.204].

Annealing and quenching are inherent procedures in the FITSC method and their influence must therefore be accurately known before interpreting experimental curves observed as a result of a particular polarization–cooling–heating sequence. Prolonged thermal treatments often have pronounced effects on the current spectra, not only when spontaneous polarization processes are initially present, but also because they can lead in certain materials to important modifications of the physical structure of the sample, by which the

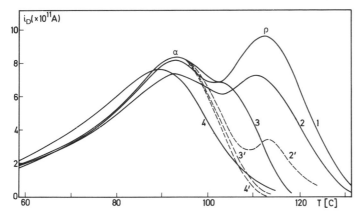

Fig. 4.51. TSDC observed in poly(methyl methacrylate) after exposure of polarized samples ($T_p = 408$ K, $T_0 = 295$ K, $F_p = 20$ kV/cm) to water vapor at 100 % RH (solid lines) for times $t_h = 0$ (*1*), 2.5 h (*2*), 6 h (*3*), 17 h (*4*), and to γ radiation (dashed curves) at doses of $4.5 \cdot 10^4$ (*2'*), $4.3 \cdot 10^6$ (*3'*) and $2 \cdot 10^7$ rads (*4'*) [4.205, 206]

various polarization mechanisms will obviously be affected. In partially crystalline films, in particular, a slow cooling may completely change the crystalline–amorphous ratio as compared to that of rapidly cooled samples, involving marked variations in trapping properties, space-charge formation and dipole orientation, while in ionic crystals, annealing can produce precipitation or dissolution of dipolar agglomerates, strongly altering the number of relaxing dipoles [4.40]. It is obvious that poor reproducibility of FITSC measurements in certain temperature ranges can result from such effects. This is often the case for the high-temperature TSDC peaks appearing in molecular compounds such as polymers, in particular when excess ionic charges are involved [4.27, 30, 207].

The variations in TSDC properties following low-temperature irradiation (IR, uv, X-rays or γ rays) on previously polarized samples have sometimes been used as a criterion for identifying specific polarization mechanisms, but these studies must be carried out with great caution because the effects of such a stimulation are usually manifold: Variation in intrinsic conductivity, modification of electrode–dielectric contacts, production of free radicals, new traps or new carriers migrating in internal fields, molecular degradation, photochemical reactions, etc. Three typical examples are shown in Figs. 4.51, 52, 53. Figure 4.52 illustrates the influence of ultraviolet illumination upon the TSDC spectrum of silicon monoxide doped with metallic impurities [4.208]. Peak H, attributed to electronic carriers, is noticeably lowered, while peak I, associated with ionic charge, is in no way affected. As shown in Fig. 4.53, similar uv irradiation can also have a marked effect on ionic dipoles, presumably as a consequence of photochemical reactions which decrease the effective number of dipoles [4.209]. Finally, the difference in behavior between dipolar and ionic space-charge TSDC peaks of polymers towards γ rays is represented in

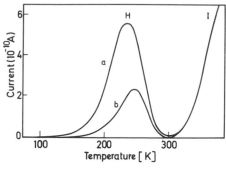

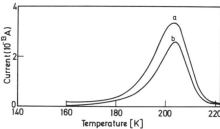

Fig. 4.52. Effect of ultraviolet illumination on H and I TSDC peaks in SiO ($V_p = 8$ V, $T_p = 400$ K, $T_0 = 78$ K). (**a**) without illumination; (**b**) with illumination for 5 min at 78 K and $V_p = 0$ [4.208]

Fig. 4.53. Effect of ultraviolet illumination on the dipolar peak in KI:S^{--} crystals ($F_p = 3 \cdot 10^4$ V/cm, $T_p = 210$ K, $T_0 = 179$ K). (**a**) without illumination; (**b**) with illumination in the 368 nm absorption band for 5 min at 179 K and $V_p = 0$ [4.209]

Fig. 4.51. The dipole polarization (α peak) is seen to be little affected while the ionic process is drastically decreased, probably as a result of either an increase in material conductivity or reactions of charge carriers with free radicals produced by irradiation [4.206].

Thermal Cycling of Samples Through Various Polarization–Depolarization Sequences

It was shown in Sect. 4.11.1 that dipolar processes associated with a temperature-dependent equilibrium polarization can be, in theory, identified from a comparison between TSDC and TSPC spectra. These are characterized by current reversals which can be distinguished from diffusion or injection currents by means of repeated thermal cycles under appropriate bias (Fig. 4.36). Provided that diffusion is negligible, on the other hand, the TSDC–TSPC comparison can also be a useful tool for separating conduction-induced and trap-limited current peaks in samples supplied with blocking contacts, since only the first mechanism should produce spectra identical in structure and opposite in sign (Sect. 4.11.2). Using asymmetrical contacts as a further stage, we can even identify in certain cases the nature and sign of the carriers involved [4.160]. It must be remarked, however, that a close analogy between trap-controlled polarization and depolarization currents can also be obtained in thin film MIM systems when Schottky barriers exist at the metal–insulator interfaces [4.23].

Unfortunately, as already emphasized in Sect. 4.11, a significant comparison between the TSDC and TSPC results is often a difficult task for a number of reasons: Possible overlapping diffusion or injection phenomena (masking or giving rise to parasitic reversals), possible superposition of true conduction currents and higher noise level in TSPC spectra, buildup during the polarizing period of ionic space charge reducing the forming field and hence leading to dipolar TSDC peaks lower than expected, etc.

4.12.3 Correlation Studies of FITSC with Other Physical Methods

The methods available for investigating the various forms of relaxational and electrical behavior in dielectrics are numerous, each of which being more or less specific to one or several particular mechanisms, but only a few are self-consistent. When combined, they can often serve as a firm basis for their common understanding and thus for an unequivocal determination of the microscopic origin of the involved processes. It is obvious that for interpreting FITSC data, such cross-checking experiments will be necessary in view of the diversity and complexity of the associated relaxation phenomena. As a first step, the results should be correlated with those obtained, simultaneously wherever possible, from other nonisothermal methods, since they are mainly concerned with the nonequilibrium state of the material: Differential thermal analysis (DTA), differential scanning calorimetry (DSC), thermally stimulated exoelectron emission (TSEE), TL, TSC, etc. Complementary data should then be obtained by studying the temperature dependence of a number of material parameters such as optical absorption, Hall coefficient, capacitance, dielectrical and mechanical losses, NMR, ESR, etc.

It is obviously not possible in the framework of this article to review extensively all these methods. In the following, we will therefore give only some brief illustrations of typical correlated studies involving FITSC measurements with particular references to electron trapping, ionic drift and dipolar mechanisms.

The principle of a general procedure for isolating trap-controlled processes by the combined use of TSPC and TSC measurements is shown in Fig. 4.54, where the following assumptions have been made: i) The sample contains both trapped charge and dipoles, ii) the dipolar relaxation occurs at a lower temperature than the charge release from traps, and iii) the dipolar equilibrium polarization is temperature independent [4.20]. Such a procedure, however, is only valid for identifying dipolar (or ionic) processes appearing as parasitic phenomena in the normal TSC measurements and the distinction between trap-controlled peaks and peaks induced by other mechanisms appearing in FITSC spectra is usually best performed, either by a direct comparison with TSC, TL or TSEE spectra or by correlated studies of the temperature variations of optical absorption and ESR. Figure 4.55, for example, shows typical TL and TSPC curves observed in high purity CaF_2 [4.210]. The peaks appearing at

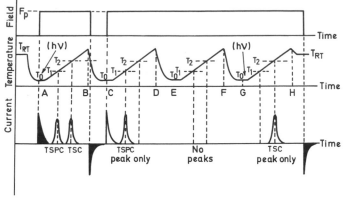

Fig. 4.54

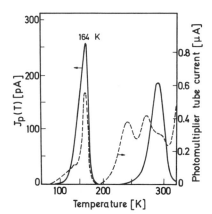

Fig. 4.54. General procedure for obtaining and isolating TSC and TSPC in dielectrics (schematic) [4.20]

Fig. 4.55. Comparison of TSPC and TL spectra of CaF_2 in the low temperature range (exposure 100 R at 77 K, $F_p = 6 \cdot 10^3$ V/cm, $d = 0.8$ mm) [4.210]

164 K in both experiments have similar properties and are also closely related to variations in the ESR signal, probably attributable to thermally activated motion of self-trapped holes (V_k centers). The interpretation of such experiments is not always so easy, however, because it must be borne in mind that a lack of correlation is not a sufficient argument to rule out any possibility of trapping processes in FITSC measurements (the trap involved may be different and the detrapping process is not necessarily accompanied by radiative emission) and, conversely, that the observation of a close correspondence between FITSC, TSC and TL peaks does not necessarily imply identical origins. In polymers, for example, the dipolar TSDC peaks and the trap-controlled TL or TSC peaks are often similarly located because, since the dipolar groups act as trap centers, charge release from traps is associated with the onset of dipolar motion [4.211].

Few methods exist which provide evidence for an ionic space-charge polarization in solid dielectrics. In some favorable cases the presence of ionic carriers can be demonstrated by spectroscopic techniques such as reflection infrared spectroscopy [4.191] or by chemical or chromatographic analysis which show mass transfer or gas evolution resulting from electrolytic processes [4.212]. Usually even this is not possible and it is particularly significant that in most dielectrics the nature of carriers involved in conductivity measurements remains a much discussed question. In certain systems such as MOS structures, however, ion migration can sometimes be identified by studying the shift of capacitance–voltage curves following a heating cycle under voltage bias [4.213]. Using this method in conjunction with the TSDC and TSPC techniques, *Manifacier* ct al. have been able to distinguish between ionic polarization and electron trapping in $Mo–SiO_2–Si$ devices [4.214].

Dipole polarization, especially when associated with large-scale molecular motions, is perhaps the easiest one to identify by cross-checking of various methods. This mechanism is related, both in theory and experiment, to all the relaxation effects involving the strain of a volume element of the material due to an applied stress, which can be thermal (as in DTA and DSC), mechanical or thermomechanical (as in creep, thermostimulated creep [4.215], torsion, compression or dynamic mechanical measurements), electrical (as in isothermal current decay or dielectric capacitance and loss methods), electrical and mechanical (as in the piezostimulated current method [4.216]), magnetic (as in NMR), etc. A characteristic example of some of these possible correlations is

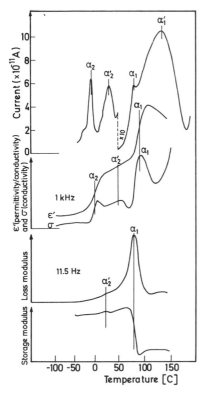

Fig. 4.56. Comparison between TSDC, ac dielectric measurements, mechanical loss and storage modulus for a poly(vinyl chloride)/rubber blend [4.217]

given in Fig. 4.56 where the TSDC spectrum of a poly (vinyl chloride)/rubber blend is compared with the corresponding dielectric (1 kHz) and mechanical (11.5 Hz) relaxation spectra [4.217]. It is seen that the α_1 transition (glass transition of the polyvinylchloride phase) and α'_2 transition (glass transition of the elastomeric phase) manifest themselves to a greater or lesser extent in all the curves while the α_2 transition (presumably due to more local molecular motions) is only visible by using the TSDC and dielectric methods. The large α'_1 TSDC peak, on the other hand, probably resulting from ionic polarization, appears neither in mechanical nor in dielectric measurements where, due to the higher measuring frequency, it occurs at such high temperatures that it merges into the conduction losses.

4.13 Conclusions

In principle, the FITSC technique offers an important key to the comprehension of the fundamental mechanisms for charge storage and release in dielectrics and semiconductors and is also considered as a very sensitive probe of kinetic transitions and molecular relaxation processes in polar materials. As such, it appears to have several basic advantages, compared with the more conventional step response or loss measurements: It is inherently more sensitive, allowing us for example to detect dipole concentrations of less than 0.1 ppm [4.40] or carrier concentrations of about 10^8 to 10^9 cm^{-3} [4.218] and it is characterized by a high resolving power, enabling us to resolve relaxation processes arising from sets of dipoles with only slightly different energies (0.01 eV for comparable concentrations [4.40]).

The FITSC measurements have not yet become standard methods, however, because, due to the variety and complexity of interfering mechanisms and the number and uncertainty of the physical parameters involved, the pertinent theoretical equations are usually intractable and, thus, it is extremely difficult to correlate theory with experiment. Most of the various simplified models adopted lead to similar functional relationships and the interpretation of data in terms of these models is therefore seldom very meaningful and still the subject of much controversy for most of the materials investigated. Among the major problems that remain to be solved is that of unambiguously sorting out the microscopic mechanisms responsible for the space-charge polarization occurring in a given dielectric. To this end, additional theoretical studies are needed. However the emphasis should now be on experimental studies including a more complete characterization of the test specimens and determination of the effect exercised by systematically varying carefully controlled experimental conditions. This should include more detailed investigations of the correlations between the FITSC results and the data obtained with comparable, well-characterized samples by all the other available physical methods that are sensitive, to some extent, to relaxation or conduction properties. In fact, it is now very clear that the fundamental development and future importance of these FITSC measurements will closely be subordinated to their strong interrelation with the other branches of solid-state physics.

From a more practical viewpoint, however, it must be emphasized that the FITSC method remains a very useful and sensitive tool for investigating a number of physico–chemical phenomena such as aggregation and precipitation of ionic dipoles due to annealing or quenching procedures [4.40], physical aging and moisture effects in polymeric materials [4.133, 205, 206, 219], formation or annihilation of color centers in irradiated crystals [4.220, 221]) as well as for assessing the variations of morphology or microscopic structure, especially in polymers (curing of epoxy systems [4.133, 219], crosslinking processes [4.133, 222], changes in tacticity [4.27, 223], heterogeneities of blends or block copolymers [4.133, 207], etc.). Finally, it is also important to note that the method can be applied in research directed towards more technical applications, for example by allowing to predict the potential usefulness of electrets [4.30, 224] or to monitor the production of dielectrics with desired electrical properties.

Acknowledgements
We are greatly indebted to Prof. V. Desreux (Belgium) and J. P. Fillard (France) for stimulating discussions and interest in our own work in the field of FITSC measurements. One of us (J.V.) expresses also his gratitude to the Fonds National de la Recherche Scientifique (Belgium) for a research grant. Particularly we acknowledge the help of J. Moers, R. Remy and Mrs. B. Manifacier in the final preparation of the manuscript.

List of Symbols

u^*	shift factor of the relaxation time	F_p	polarizing field strength [V·cm^{-1}]
b^*	shift factor of ζ	$g(E)$	normalized distribution function of E
C_E, C_I, c	capacitances [F]		
d	thickness of sample [cm]	H	function characterizing the filling state of a polarized sample
d_g	thickness of air gap [cm]		
D	diffusion constant [cm^2 s^{-1}]	i	current [A]
e	elementary charge [C]	i_D	thermally stimulated depolarization current [A]
E	activation energy [eV]		
E_m	most probable value of E for distributed processes [eV]	i_D^*	longitudinal thermally stimulated depolarization current in Agarwal's model [A]
E_W	apparent activation energy in WLF equation [eV]		
$f(\tau), f(\tau_0)$	normalized distribution functions of τ and τ_0	i_P	thermally stimulated polarization current [A]
$f'(\tau), f'(\tau_0)$	logarithmic distribution functions of τ and τ_0	J	current density [A·cm^{-2}]
f_{eq}, f'_{eq}	equivalent ac frequency of TSDC [Hz]	J_D	thermally stimulated depolarization current density [A·cm^{-2}]
F	electric field in general [V·cm^{-1}]	J_P	thermally stimulated polarization current density [A·cm^{-2}]
		k	Boltzmann's constant [eV·K^{-1}]

K	constant	U_1, U_2	constants of the WLF equation ($U_1 = 40$ and $U_2 = 325$ K for amorphous polymers)
l	length of sample [cm]		
n	density of negative excess charge carriers [cm^{-3}]	V_d	polarizing voltage [V]
N	trap density [cm^{-3}]	w	general distribution parameter for symmetrical distributions
N_d	dipolar concentration [cm^{-3}]		
$N(E)$	distribution function of activation energy in term of dipolar concentration [cm^{-3}]	x	depth below the sample surface ($0 < x < d$) [cm]
		y, y_0, y_m	$= E/kT$; E/kT_0; E/kT_m
p	density of positive excess charge carriers [cm^{-3}]	$\bar\alpha$	distribution parameter in the Wagner distribution
p_μ	dipole moment [Debye]	$\bar\beta$	distribution parameter in the Cole-Cole distribution
P	polarization [C·cm^{-2}]		
P_e	equilibrium polarization [C·cm^{-2}]	$\bar\gamma$	distribution parameter in the Fuoss-Kirkwood distribution
		δ	dielectric loss angle
$P(E)$	distribution function of activation energy in term of dipolar polarization [C·cm^{-2}]	$\bar\delta$	distribution parameter in the Davidson-Cole distribution
		ε'	real part of dielectric constant
q	heating rate [K·s^{-1}]	ε''	imaginary part of dielectric constant or dielectric loss factor
q^*	quadratic heating rate [K^{-1}s^{-1}]	ε_0	dielectric constant of free space
		ε_s	static dielectric constant
Q	charge released during a TSDC experiment [C]	ε_∞	high frequency dielectric constant
Q_0	charge initially stored [C]	$\bar\varepsilon, \bar\eta$	distribution parameters in the Havriliak–Negami distribution
R, r	resistances [Ohm]		
r_0	pre-exponential factor in r [Ohm]	θ	$= \psi d/D$
		ζ	$= Dt/d^2$
R_0	pre-exponential factor in R [Ohm]	κ	local field correcting factor in Langevin's function
s	geometrical factor in Langevin's function	μ_n	mobility of negative charge carriers [cm^2 V^{-1} s^{-1}]
S	area of the sample [cm^2]	μ_p	mobility of positive charge carriers [cm^2 V^{-1} s^{-1}]
t	time [s]		
t_p	polarizing time [s]	ξ	$= (kT^2/qE)\exp(-E/kT)$
t_0	storage time at T_0 [s]	ϕ_i	work function of sample [eV]
T	temperature [K]	ϕ_m	work function of electrode [eV]
T_p	polarizing temperature [K]	ψ	blocking factor
T_0	initial temperature [K]	σ	electrical conductivity [Ω cm^{-1}]
T_g	glass transition temperature [K]		
T_m	temperature of TSDC maximum [K]	σ_0	pre-exponential factor in σ [Ω cm^{-1}]
T_1	temperature at half maximum of a TSDC peak (low temperature side) [K]	τ	relaxation time [s]
		τ_0	pre-exponential factor in τ [s]
T_∞	characteristic temperature in WLF equation [K]	τ_m	most probable value of τ for distributed processes [s]
u	$= \ln(\tau/\tau_m)$	τ_{0_m}	most probable value of τ_0 for distributed processes [s]
		ω	angular frequency

References

4.1   A.Servini, A.K.Jonscher: Thin Solid Films **3**, 341 (1969)
4.2   G.M.Sessler (ed.): *Electrets*, Topics in Applied Physics, Vol. 33 (Springer, Berlin, Heidelberg, New York 1979)
4.3   H.Frei, G.Groetzinger: Phys. Z. **37**, 720 (1936)
4.4   B.Gross: J. Chem. Phys. **17**, 866 (1949)
4.5   S.Wikstroem: Ericsson Tech. **9**, 225 (1953)
4.6   A.N.Gubkin, B.N.Matsonashvili: Fiz. Tverd. Tela **4**, 1196 (1962)
4.7   P.V.Murphy: J. Phys. Chem. Sol. **24**, 329 (1963)
4.8   C.Bucci, R.Fieschi: Phys. Rev. Lett. **12**, 16 (1964)
4.9   T.Takamatsu, E.Fukada: Polym. J. **1**, 101 (1970)
4.10   J.Van Turnhout: Polym. J. **2**, 173 (1971)
4.11   R.A.Creswell, M.M.Perlman: J. Appl. Phys. **41**, 2365 (1970)
4.12   T.Nedetzka, M.Reichle, A.Mayer, H.Vogel: J. Phys. Chem. **74**, 2652 (1970)
4.13   B.T.Kolomietz, V.M.Lyubin, V.L.Averyanov: Mat. Res. Bull. **5**, 655 (1970)
4.14   P.R.Moran, D.E.Fields: J. Appl. Phys. **45**, 3266 (1974)
4.15   V.F.Zolotaryov, D.G.Semak, D.V.Chepur: Phys. Status Solidi **21**, 437 (1967)
4.16   J.Chiu: J. Polym. Sci. C**8**, 27 (1965)
4.17   M.I.Pope: Polymer **8**, 49 (1967)
4.18   B.K.Shim: J. Polym. Sci. C**17**, 221 (1967)
4.19   P.Müller: Phys. Status Solidi A**23**, 165 (1974)
4.20   S.W.S.Mc Keever, D.M.Hughes: J. Phys. D**8**, 1520 (1975)
4.21   D.A.Seanor: "Electrothermal Analysis of Polymers", in *Techniques and Methods of Polymer Evaluation*, Vol. 2, ed. by P.E.Slade,Jr., L.T.Jenkins (M.Dekker, New York 1970) pp. 293–337
4.22   P.Devaux, M.Schott: Phys. Status Solidi **20**, 301 (1967)
4.23   J.G.Simmons, G.W.Taylor: Phys. Rev. B**6**, 4804 (1972)
4.24   R.Gerson, J.M.Rohrbaugh: J. Chem. Phys. **23**, 2381 (1955)
4.25   H.J.Wintle: J. Non-Cryst. Sol. **15**, 471 (1974)
4.26   D.K.Das Gupta, K.Joyner: J. Phys. D**8**, 829 (1976)
4.27   J.Vanderschueren: Thesis (Liège 1974) (unpublished)
4.28   T.J.Gray: "Detailed Electronic Structure by Measurements of Thermally Stimulated Electron Currents and Photoelectrets", in *Electrets, Charge Storage and Transport in Dielectrics*, ed. by M.M.Perlman (The Electrochemical Society, Princeton 1973) pp. 75–83
4.29   Amer. Soc. for Testing and Materials: "Electrical Resistance of Insulating Materials", in 1965 *Book of ASTM Standards*, pt. 27, D 257–61 (ASTM, Philadelphia, 1965) pp. 78–100
4.30   J.Van Turnhout: *Thermally Stimulated Discharge of Polymer Electrets* (Elsevier, Amsterdam 1975)
4.31   P.Dansas, S.Mounier, P.Sixou: C. R. Acad. Sci. Paris, B**267**, 1223 (1968)
4.32   B.Gross: J. Electrochem. Soc. **115**, 376 (1968)
4.33   T.A.Vassilev, D.D.Christosov: Natura (Plovdiv) **4**, 21 (1971)
4.34   T.A.Vassilev, T.A.Terziiski, I.I.Popov, D.B.Mindeva: C. R. Acad. Bulg. Sci. **29**, 1421 (1976)
4.35   C.Solunov, C.Ponevsky: J. Polym. Sci., Polym. Phys. Ed. **14**, 1801 (1976)
4.36   C.Solunov, C.Ponevsky: J. Polym. Sci., Polym. Phys. Ed. **15**, 969 (1977)
4.37   P.H.Ong, J.Van Turnhout: "TSD of Polymer Electrets having a Distributed Dipole Polarization", in *Electrets, Charge Storage and Transport in Dielectrics*, ed. by M.M.Perlman (The Electrochemical Society, Princeton 1973) pp. 213–229
4.38   V.V.Daniel: *Dielectric Relaxation* (Academic Press, London/New York 1967)
4.39   C.Bucci, R.Fieschi, G.Guidi: Phys. Rev. **148**, 816 (1966)
4.40   R.Capelletti, R.Fieschi: "Ionic Thermoconductivity: A Method for the Study of the Temperature Dependent Ionic Polarization in Condensed Matter", in *Electrets, Charge Storage and Transport in Dielectrics*, ed. by M.M.Perlman (The Electrochemical Society, Princeton 1973) pp. 1–14

4.41   A.Brun, P.Dansas, P.Sixou: Solid State Commun. **8**, 613 (1970)
4.42   C.Laj, P.Berge: J. Phys. **28**, 821 (1967)
4.43   C.J.F.Böttcher: *Theory of Electric Polarisation* (Elsevier, Amsterdam 1952)
4.44   T.A.Vassilev, P.M.Mihailov, P.G.Todorov: C. R. Acad. Bulg. Sci. **24**, 27 (1971)
4.45   I.Kunze, P.Müller: Phys. Status Solidi **33**, 91 (1969)
4.46   N.G.McCrum, B.E.Read, G.Williams: *Anelastic and Dielectric Effects in Polymeric Solids* (Wiley, London 1967)
4.47   B.Gross: J. Polym. Sci., Polym. Phys. Ed. **10**, 1941 (1972)
4.48   P.Varotsos, D.Kostopoulos, S.Mourikis, S.Kouremenou: Solid State Commun. **21**, 831 (1977)
4.49   H.Eyring: J. Chem. Phys. **4**, 283 (1936)
4.50   W.Kauzmann: Rev. Mod. Phys. **14**, 12 (1942)
4.51   S.Glasstone, K.J.Laidler, H.Eyring: *The Theory of Rate Processes* (McGraw-Hill, New York 1941)
4.52   M.L.Williams, R.F.Landel, J.D.Ferry: J. Am. Chem. Soc. **77**, 3701 (1955)
4.53   M.H.Cohen, D.Turnbull: J. Chem. Phys. **31**, 1164 (1959)
4.54   D.Turnbull, M.H.Cohen: J. Chem. Phys. **34**, 120 (1961)
4.55   G.Adam, J.H.Gibbs: J. Chem. Phys. **43**, 139 (1965)
4.56   P.B.Macedo, T.A.Litovitz: J. Chem. Phys. **42**, 245 (1965)
4.57   K.C.Rush: J. Macromol. Sci., Phys. B**2**, 179 (1968)
4.58   I.J.Saunders: Br. J. Appl. Phys. **18**, 1219 (1967)
4.59   G.A.Dussel, R.H.Bube: Phys. Rev. **155**, 764 (1967)
4.60   T.A.T.Cowell, J.Woods: Br. J. Appl. Phys. **18**, 1045 (1967)
4.61   V.Harasta: Fyz. Casop. (Czech) **19**, 232 (1969)
4.62   H.Solunov, T.Vassilev, P.Hedvig: Univ. Plovdiv (Bulg.), Scientif. Pap. Physics **12**, 47 (1974)
4.63   G.Caserta, A.Serra: J. Appl. Phys. **42**, 3778 (1971)
4.64   R.D.Shelley, G.R.Miller: J. Solid State Chem. **1**, 218 (1970)
4.65   I.Thurzo, E.Mariani, D.Barancok: Czech. J. Phys. B**24**, 203 (1974)
4.66   I.Thurzo, E.Mariani, H.Heks: Czech. J. Phys. B**23**, 1241 (1973)
4.67   J.P.Stott, J.H.Crawford, Jr.: Phys. Rev. B**6**, 4660 (1972)
4.68   B.Gross: "Persistent Internal Polarization and Distribution of Activation Energy", in *Electrophotography*, ed. by J.H.Howard, Applied optics, Suppl. 3 (Amer. Inst. of Physics, New York 1969) pp. 176–179
4.69   P.Fischer, P.Röhl: J. Polym. Sci., Polym. Phys. Ed. **14**, 543 (1976)
4.70   C.Bucci, S.C.Riva: J. Phys. Chem. Sol. **26**, 363 (1965)
4.71   K.W.Wagner: Ann. Phys. **40**, 817 (1913)
4.72   B.Gross: *Mathematical Structure of the Theories of Viscoelasticity* (Herman and Cie, Paris 1953)
4.73   A.Van Roggen: "Distribution of Relaxation Times" in IEEE Trans. EI-**5**(2), 1 (1970)
4.74   M.Gevers: Philips Res. Rep. **1**, 447 (1946)
4.75   H.Fröhlich: *Theory of Dielectrics* (Oxford University Press, London 1958)
4.76   K.S.Cole, R.H.Cole: J. Chem. Phys. **9**, 341 (1941)
4.77   R.M.Fuoss, J.G.Kirkwood: J. Am. Chem. Soc. **63**, 385 (1941)
4.78   D.W.Davidson, R.H.Cole: J. Chem. Phys. **19**, 1484 (1951)
4.79   S.Havriliak, S.Negami: Polymer **8**, 161 (1967)
4.80   E.J.Hennelly, W.M.Heston, Jr., C.P.Smyth: J. Am. Chem. Soc. **70**, 4102 (1948)
4.81   P.Van de Walle, P.Sixou, P.Dansas: C. R. Acad. Sc. Paris C**264**, 469 (1967)
4.82   S.Havriliak, S.Negami: J. Polym. Sci. C**14**, 99 (1966)
4.83   E.Ikada, T.Watanabe: J. Polym. Sci., Polym. Chem. Ed. **10**, 3457 (1972)
4.84   C.J.Knauss, R.R.Myers, P.S.Smith: J. Polym. Sci., Polym. Lett. **10**, 737 (1972)
4.85   J.Mrazek: "Relaxation Polarization and Thermoelectrets", in *Electrets, Charge Storage and Transport in Dielectrics*, ed. by M.M.Perlman (The Electrochemical Society, Princeton 1973) pp. 260–268
4.86   J.Mrazek: Acta Technica CSAV **4**, 402 (1972)
4.87   B.Heller, J.Mrazek: Acta Technica CSAV **6**, 515 (1973)

4.88 M. Abkowitz, P.J.Luca, G.Pfister, W.M.Prest,Jr.: Proc. Piezoelectric and Pyroelectric Symposium-Workshop (Nat. Bur. Stand. U. S., Interagency Report, NBSIR 75–760, 1975) pp. 96–119

4.89 J.Vanderschueren: Appl. Phys. Lett. **25**, 270 (1974)

4.90 J.Vanderschueren: 1974 Ann. Rep. Conf. Electrical Insulation and Dielectric Phenomena (Nat. Acad. Sci., Washington D. C. 1975) pp. 339–346

4.91 M.Kryszewski, M.Zielinski, S.Sapieha: Polymer **17**, 212 (1976)

4.92 T.Terziiski, T.Vassilev: Univ. Plovdiv (Bulg.) Scient. Pap. Physics **12**, 75 (1974)

4.93 J.Vanderschueren, J.Gasiot, J.P.Fillard, A.Linkens, P.Parot: Preprints of the European Symposium on Electric Phenomena in Polymer Science (Pisa, 1978) pp. 143–148

4.94 H.Adachi, Y.Shibata: Jpn. J. Appl. Phys. **13**, 1479 (1974)

4.95 H.Adachi, Y.Shibata: J. Phys. D**8**, 1120 (1975)

4.96 J.Vanderschueren: J. Polymer Sci., Polymer Phys. Ed. **15**, 873 (1977)

4.97 T.Hino: Jpn. J. Appl. Phys. **12**, 611 (1973)

4.98 P.Fischer, P.Röhl: J. Polym. Sci., Polym. Phys. Ed. **14**, 531 (1976)

4.99 C.Ponevski, C.Solunov: J. Polym. Sci., Polym. Phys. Ed. **13**, 1467 (1975)

4.100 M.Zielinski, M.Kryszewski: Phys. Status Solidi A**42**, 305 (1977)

4.101 D.Chatain, C.Lacabanne, M.Maitrot, G.Seytre, J.F.May: Phys. Status Solidi A**16**, 225 (1973)

4.102 D.Chatain, C.Lacabanne, M.Maitrot: Phys. Status Solidi A**13**, 303 (1972)

4.103 D.Chatain, P.Gautier, C.Lacabanne: J. Polym. Sci., Polym. Phys. Ed., **11**, 1631 (1973)

4.104 A.K.Jonscher: J. Non-Crystal. Solids **8–10**, 293 (1972)

4.105 A.K.Jonscher: J. Phys. C**6**, L235 (1973)

4.106 A.K.Jonscher: Nature (London) **253**, 717 (1975)

4.107 A.K.Jonscher: J. Electrostatics **3**, 53 (1977)

4.108 J.Vanderschueren, A.Linkens: J. Electrostatics **3**, 155 (1977)

4.109 J.T.Randall, M.H.F.Wilkins: Proc. R. Soc. London A**184**, 366 (1945)

4.110 P.Bräunlich: "Thermoluminescence and Thermally Stimulated Current – Tools for the Determination of Trapping Parameters", in *Thermoluminescence of Geological Materials*, ed. by D.J.Mc Dougall (Academic Press, New York 1968) pp. 61–88

4.111 P.Kivits, H.J.L.Hagebeuk: J. Lumin. **15**, 1 (1977)

4.112 P.Kelly, M.J.Laubitz, P.Bräunlich: Phys. Rev. B**4**, 1960 (1971)

4.113 J.P.Fillard, J.Gasiot, J.C.Manifacier: Phys. Rev. B **18**, 4497 (1978)

4.114 J.Hill, P.Schwed: J. Chem. Phys. **23**, 652 (1955)

4.115 W.Hoogenstraaten: Philips Res. Rep. **13**, 515 (1958)

4.116 H.Gobrecht, D.Hofman: J. Phys. Chem. Sol. **27**, 509 (1966)

4.117 L.Heijne: Philips Res. Rep., Suppl. 499 (1961)

4.118 G.F.J.Garlick, A.F.Gibson: Proc. Phys. Soc. London A**60**, 574 (1948)

4.119 C.H.Haake: J. Opt. Soc. Am. **47**, 649 (1957)

4.120 P.L.Land: J. Phys. Chem. Sol. **30**, 1693 (1969)

4.121 M.M.Perlman: J. Appl. Phys. **42**, 2645 (1971)

4.122 C.Laj, P.Berge: C.R.Acad. Sci. Paris B**263**, 380 (1966)

4.123 L.I.Grossweiner: J. Appl. Phys. **24**, 1306 (1953)

4.124 A.H.Booth: Can. J. Chem. **32**, 214 (1954)

4.125 A.Bohun: Czech. J. Phys. **4**, 91 (1954)

4.126 A.P.Kulshreshtha, V.A.Goryunov: Sov. Phys.-Solid State **8**, 1540 (1966)

4.127 L.R.Krumberg: Fiz. Tekh. Poluprovodn. **1**, 1456 (1967)

4.128 C.Lacabanne, D.Chatain: J. Polym. Sci., Polym. Phys. Ed. **11**, 2315 (1973)

4.129 F.R.Schwarzl, A.J.Staverman: Physica **18**, 791 (1952)

4.130 T.Hino, K.Suzuki, K.Yamashita: Jpn. J. Appl. Phys. **12**, 651 (1973)

4.131 S.Ikeda, K.Matsuda: Jpn. J. Appl. Phys. **15**, 963 (1976)

4.132 W.Primak: J. Appl. Phys. **31**, 1542 (1960)

4.133 P.Hedvig: *Dielectric Spectroscopy of Polymers* (A.Hilger, Bristol 1977)

4.134 V.G.Luydskanov, T.A.Vassilev, Y.V.Zelenev: Vysokomol. Soyed A**14**, 161 (1972)

4.135 G.A.Lushcheikin, L.I.Voiteshonok: Vysokomol. Soyed A**17**, 429 (1975)

4.136 H. Solunov, T. A. Vassilev: J. Polym. Sci., Polym. Phys. Ed. **12**, 1273 (1974)
4.137 P. C. Mehendru, K. Jain, V. K. Chopra, P. Mehendru: J. Phys. D**8**, 305 (1975)
4.138 B. V. Hamon: Proc. Inst. Electr. Eng. (Monograph $n°$ 27) **99**, 151 (1952)
4.139 F. R. Schwarzl, L. C. E. Struik: Adv. Molecular Relaxation Processes **1**, 201 (1967–68)
4.140 M. M. Perlman, S. Unger: J. Appl. Phys. **45**, 2389 (1974)
4.141 J. Van Turnhout: "Current and Charge TSD of Polymer Electrets resulting from the Motion of Excess Charges", in *Electrets, Charge Storage and Transport in Dielectrics*, ed. by M. M. Perlman (The Electrochemical Society, Princeton 1973) pp. 230–251
4.142 B. Gross: "Persistent Polarization and Space Charge Layers in Dielectrics", Lecture at Northern Electric, Canada, Report T0144 (1970)
4.143 G. Jaffé: Ann. Phys. **16**, 217 (1933)
4.144 G. Jaffé, C. Z. Lemay: J. Chem. Phys. **21**, 920 (1953)
4.145 A. Kessler: J. Electrochem. Soc. **123**, 1236 (1976)
4.146 A. B. Lidiard: *Encyclopedia of Physics*, Vol. XXII (Springer, Berlin 1957)
4.147 P. Berteit, A. Kessler, T. List: Z. Phys. B**24**, 15 (1976)
4.148 A. Kessler: J. Phys. C**6**, 1594 (1973)
4.149 I. R. Freeman, H. P. Kallman, M. Silver: Rev. Mod. Phys. **33**, 553 (1961)
4.150 V. I. Bugrienko, V. K. Marinchik, V. M. Belous: Sov. Phys.-Solid State **12**, 36 (1970)
4.151 S. C. Agarwal: Phys. Rev. B**10**, 4340 (1974)
4.152 P. J. Kelly, M. J. Laubitz: Can. J. Phys. **45**, 311 (1967)
4.153 H. Fritzsche, S. Chandra: Proc. Symp. on Thermal and Photostimulated Currents in Insulators, ed. by D. M. Smyth (The Electrochemical Society, Princeton 1976) pp. 105–117
4.154 S. C. Agarwal, H. Fritzsche: Phys. Rev. B**10**, 4351 (1974)
4.155 S. C. Agarwal, H. Fritzsche: Bull. Am. Phys. Soc. **19**, 213 (1974)
4.156 E. B. Podgorsak, P. R. Moran: Phys. Rev. B**8**, 3405 (1973)
4.157 J. Vanderschueren: J. Polym. Sci., Polym. Lett. **10**, 543 (1972)
4.158 H. J. Wintle: J. Appl. Phys. **42**, 4724 (1971)
4.159 A. Kessler, J. E. Caffyn: J. Phys. C**5**, 1134 (1972)
4.160 P. Müller: Phys. Status Solidi A**28**, 521 (1975)
4.161 P. Müller: Phys. Status Solidi A**33**, 543 (1976)
4.162 H. Bauser: Kunststoffe **62**, 192 (1972)
4.163 E. H. Martin, J. Hirsch: J. Appl. Phys. **43**, 1001 (1972)
4.164 K. C. Frisch, A. Patsis (eds.): *Electrical Properties of Polymers* (Technomic, Westport 1972)
4.165 W. L. Mc Cubbin: Chem. Phys. Lett. **8**, 507 (1971)
4.166 G. Caserta, B. Rispoli, A. Serra: Phys. Status Solidi **35**, 237 (1969)
4.167 G. A. Bordovskii, V. G. Boytsov, B. A. Demidov: Fiz. Tekh. Poluprovodn. **8**, 1918 (1974)
4.168 G. A. Bordovskii: Phys. Status Solidi A**29**, K183 (1975)
4.169 D. G. Semak, A. A. Kikineshi: Phys. Status Solidi A**9**, K141 (1972)
4.170 N. Kashukeev, A. Antonov, G. Zadorzhnii: C. R. Acad. Bulg. Sci. **14**, 447 (1961)
4.171 A. G. Zhdan, V. B. Sandomirskii, A. D. Ozheredov: Solid State Electron. **11**, 505 (1968)
4.172 H. M. Gupta, R. J. Van Overstraeten: J. Phys. C**7**, 3560 (1974)
4.173 J. G. Simmons, G. S. Nadkarni: Phys. Rev. B**6**, 4815 (1972)
4.174 P. Nielsen: Solid State Commun. **9**, 1745 (1971)
4.175 I. Thurzo, D. Lezal: J. Phys. C**9**, L163 (1976)
4.176 I. Thurzo, J. Doupovec: J. Non-Crystal. Solids **22**, 205 (1976)
4.177 J. G. Simmons, G. W. Taylor, M. C. Tam: "Direct Determination of Trap Distributions from High Field Thermally Stimulated Currents", in *Electrets, Charge Storage and Transport in Dielectrics*, ed. by M. M. Perlman (The Electrochemical Society, Princeton 1973) pp. 202–212
4.178 V. Harasta, I. Thurzo: Fyz. Casop. (Czech) **20**, 148 (1970)
4.179 P. Hedvig: 1974 Ann. Rep. Conf. Electrical Insulation and Dielectric Phenomena (Nat. Acad. Sci., Washington DC 1975) pp. 3–12
4.180 C. Bucci: RC (65A) Riun. Assoc. Electrotec. Ital. Fasc. III, paper 87 (Palermo, 1964) pp. 1–5
4.181 J. C. Manifacier, J. Gasiot, P. Parot, J. P. Fillard: J. Phys. C **II**, 1011 (1978)
4.182 A. Linkens, P. Parot, J. Vanderschueren, J. Gasiot: Comput. Phys. Commun. **13**, 411 (1978)
4.183 J. Vanderschueren, J. Gasiot, J. P. Fillard, P. Parot, A. Linkens: Unpublished results

4.184 T. Mizutani, Y. Suzuoki, M. Ieda: J. Appl. Phys. **48**, 2408 (1977)

4.185 G. D. Fullerton, J. R. Cameron, P. R. Moran: "Thermocurrent Dosimetry with High Purity Aluminum Oxide"; USAEC Tech. Rpt. C00-1105-217 (1974)

4.186 P. Devaux: Thesis (Paris 1969) (unpublished)

4.187 J. W. Northrip: J. Appl. Phys. **31**, 2293 (1960)

4.188 T. Furukawa, Y. Uematsu, K. Asakawa, Y. Wada: J. Appl. Polym. Sci. **12**, 2675 (1968)

4.189 E. Sacher: J. Macromol. Sci., Phys. B**4**, 449 (1970)

4.190 S. I. Stupp, S. H. Carr: J. Appl. Phys. **46**, 4120 (1975)

4.191 S. I. Stupp, S. H. Carr: J. Polym. Sci., Polym. Phys. Ed. **15**, 485 (1977)

4.192 A. Kessler: J. Electrochem. Soc. **123**, 1239 (1976)

4.193 A. R. Mc Ghie, G. Mc Gibbon, A. Sharples, E. J. Stanley: Polymer **13**, 371 (1972)

4.194 S. Kojima, K. Kato: J. Phys. Soc. Jpn. **6**, 207 (1951)

4.195 K. Kojima, A. Maeda, M. Ieda: Jpn. J. Appl. Phys. **15**, 2457 (1976)

4.196 T. W. Hickmott: J. Appl. Phys. **46**, 2583 (1975)

4.197 T. Hashimoto, M. Shiraki, T. Sakai: J. Polym. Sci., Polym. Phys. Ed. **13**, 2401 (1975)

4.198 G. Seve, L. Lassabatere: Thin Solid Films **15**, 285 (1973)

4.199 B. Gross: J. Electrochem. Soc. **119**, 855 (1972)

4.200 Y. Asano, T. Suzuki: Jpn. J. Appl. Phys. **11**, 1139 (1972)

4.201 T. Takamatsu, E. Fukada: "Surface Charge, Depolarization and Piezoelectricity in Electrets", in *Electrets, Charge Storage and Transport in Dielectrics*, ed. by M. M. Perlman (The Electrochemical Society, Princeton 1973) pp. 128–140

4.202 B. Gross, R. J. De Moraes: J. Chem. Phys. **37**, 710 (1962)

4.203 P. K. C. Pillai, K. Jain, V. K. Jain: Nuovo Cimento B**11**, 339 (1972)

4.204 G. Zuther, H. Prandtke, M. Schmidt: Phys. Status Solidi A**20**, K123 (1973)

4.205 J. Vanderschueren: J. Polym. Sci., Polym. Phys. Ed. **12**, 991 (1974)

4.206 J. Vanderschueren, A. Linkens: J. Polym. Sci., Polym. Phys. Ed. **16**, 223 (1978)

4.207 P. Alexandrovich, F. E. Karasz, W. J. Mc Knight: J. Appl. Phys. **47**, 4251 (1976)

4.208 J. Pinguet, S. S. Minn: Phys. Status Solidi A**35**, 431 (1976)

4.209 J. Prakash, F. Fischer: Phys. Status Solidi A**39**, 499 (1977)

4.210 E. B. Podgorsak, G. E. Fuller, P. R. Moran: Radiat. Res. **59**, 446 (1974)

4.211 R. H. Partridge: "Thermoluminescence in Polymers", in *The Radiation Chemistry of Macromolecules*, Vol. 1, ed. by M. Dole (Academic Press, New York 1972) pp. 193–222

4.212 D. A. Seanor: J. Polym. Sci. A-**2**, **6**, 463 (1968)

4.213 E. H. Snow, A. S. Grove, B. E. Deal, C. T. Sah: J. Appl. Phys. **36**, 1664 (1965)

4.214 J. C. Manifacier, P. Parot, J. P. Fillard: J. Electrostatics **3**, 203 (1977)

4.215 D. Chatain, C. Lacabanne, J. C. Monpagens: Makromol. Chem. **178**, 583 (1977)

4.216 Bui Ai, P. Destruel, Hoang The Giam, R. Loussier: Phys. Rev. Lett. **34**, 84 (1975)

4.217 S. A. Yalof, P. Hedvig: Thermochem. Acta **17**, 301 (1976)

4.218 M. Campos, G. L. Ferreira, S. Mascarenhas: J. Electrochem. Soc. **115**, 388 (1968)

4.219 J. B. Woodard: J. Electronic Mater. **6**, 145 (1977)

4.220 C. Bucci, R. Capelletti, L. Pirola: Phys. Rev. **143**, 619 (1966)

4.221 C. Bucci: Phys. Rev. **152**, 833 (1966)

4.222 P. K. C. Pillai, P. K. Nair, R. Nath: Polym. **17**, 921 (1976)

4.223 C. Linder, I. F. Miller: J. Polym. Sci., Polym. Chem. Ed. **11**, 1119 (1973)

4.224 B. Gross: Endeavour **30**, 115 (1971)

5. Exoemission

H. Glaefeke

With 24 Figures

5.1 Background

The phenomenon of exoemission (EE), defined below, has been observed as early as 1902 by *McLennan* [5.1]. However, in the subsequent decades reports were published only sporadically, e.g., [5.2, 3] dealing with EE from metals after mechanical pretreatment and electron bombardment. In 1949 and 1950, *Kramer* [5.4, 5] published an extended study of the phenomenon. Pioneering work in the field of exoemission was also reported by most of the participants of the First International Symposium on Exoelectrons, held at Innsbruck, Austria, in 1956 [5.6]. The main concepts of EE were already discussed there. EE was recognized to be a structure-dependent effect.

Especially the analogy to luminescence phenomena was outlined and correlations of EE with sorption and transport phenomena were pointed out. The symposium decided to adopt the presently only rarely used "Kramer effect" for all exoemission phenomena. The prefix "exo" was originally chosen to suggest that exothermal processes may be the cause of EE. Some review articles on EE are listed in [5.7–10].

The phenomenon of exoemission was found to be a very complex one belonging to the group of emission phenomena which occur during relaxation of perturbations p_i^* of thermodynamic equilibria in the bulk or at the surface of a solid, according to

$$\frac{\partial p_i^*}{\partial t} = - \frac{p_i}{\tau_{i\,\text{relax}}}. \tag{5.1}$$

In this chapter, only those relaxation processes are discussed which are accompanied by EE. Originally, the use of the term exoemission was restricted to the emission of electrons only (frequently referred to as EEE). However, recently EE is being used not only for electron emission but for ion emission as well.

The emission process is correlated, causally and temporally, with the relaxation process. In contrast to such stationary effects as photoemission (PE) and thermionic emission (TE), it is

    I) nonstationary,

    II) present only after generation of a perturbation p_i^*,

    III) governed by the relaxation of p_i^*.

The emission process may occur spontaneously, even during ongoing perturbation, or it may require external stimulation in the form of increased temperature or optical photons in order to reduce the relaxation time $\tau_{i\,relax}$ to a value at which monitoring the relaxation process is experimentally feasible. Thermally stimulated EE may be observed isothermally or nonisothermally. The latter is called thermally stimulated exoemission (TSEE).

Optical stimulated exoemission (OSEE) makes use of externally supplied photons to accelerate the relaxation process. A special case of OSEE is observed when the perturbation p_i^* relaxes without emitting charged particles, but instead results in the emission of photons through a radiative transition (e.g., thermoluminescence). The generated luminescence photons in turn may stimulate some other relaxation process that yields observable OSEE.

The physical or chemical nature of perturbation p_i^* leading to EE is multifaceted and its experimental identification is sometimes difficult because several p_i^* may be superimposed.

In this chapter we discuss the reversible and irreversible perturbations of atomic structure and bond state and of physico-chemical equilibria (adsorption–desorption, adsorption–chemisorption, chemical conversions), the reversible perturbations of the statistical equilibrium distribution of electrons and holes over the energy states of the solid and of the charge equilibrium.

The perturbations are generated by mechanical deformation, exposure to gases, irradiation with electromagnetic radiation (uv, X-, γ-rays) and particle radiation (α- and β-rays, neutrons and protons, slow electrons and ions), etc.

The elementary microscopic as well as the macroscopic processes of the generation of perturbations and the elementary mechanism of the liberation and the escape of emitted particles are subjects of present research. In some instances TSEE and OSEE may be interpreted as thermionic emission or photoemission from shallow trap levels which are not occupied according to thermodynamic equilibrium. For special cases different emission mechanisms have been proposed based on well-known elementary interactions. Essential experimental findings and interpretations concerned with these problems are treated in Sect. 5.3.

Two theoretical concepts have been established.

I) The surface concept deals with EE as a physico-chemical reaction or an effect caused by ionization.

II) The volume concept is based on the assumption that electrons may escape from thin surface layers, the thickness of which is determined by a field-strength-dependent escape depth.

A single microscopic theory of EE does not exist because of the multifaceted nature of EE. The common theoretical aspect linking EE phenomena appears to be a thermodynamic one. However, it has not yet been treated in any detail. Aspects of microscopic theories and kinetics of stimulated relaxation are discussed in Sect. 5.4.

A large number of more or less successful attempts have been undertaken to apply EE phenomena. They are dealt with in Sect. 5.5. Certain EE phenomena may serve as a basis for experimental methods to characterize bulk and/or surface properties of solids, especially insulators. However, up to this point in time EE phenomena have at best been shown to augment modern surface analytical techniques such as Auger spectroscopy, residual gas analysis, ESCA, SIMS and so on. In turn, a combination of these techniques with experiments on EE appears to be the best approach to attain the kind of data urgently needed to unravel the mechanism of a large number of unexplained or only poorly understood exoemission phenomena. This notwithstanding, exo-emission has been successfully applied to radiation dosimetry and a new type of surface imaging, where the emission phenomenon is exploited without detailed understanding of its mechanism.

5.2 The Exoemission Effect

5.2.1 General Aspects

As mentioned before, EE is a concomitance of the relaxation of perturbations of various thermodynamic equilibria [5.11]. The temporal behavior is displayed in Figs. 5.1, 2. The relaxation of the perturbations often takes place by overcoming an energy barrier E_r. Its temperature dependence is described by

$$\frac{1}{\tau_r} = \frac{1}{\tau_{r0}} e^{-E_r/kT} . \tag{5.2}$$

Since EE is only an indirect relaxation phenomenon (Chap. 1), it is generally governed by a separate $\alpha_e \equiv 1/\tau_e$,

$$\alpha_e = \nu_e e^{-E_e/kT} . \tag{5.3}$$

Figures 5.1, 2 show the simple case $\tau_r = \tau_e$. In general, several perturbations p_i^* may overlap. The EE-current density is given by

$$J_e \propto \frac{\partial p^*(t)}{\partial t} , \tag{5.4}$$

with

$$p^*(t) = \sum_{i=1} p_{0i}^* e^{-t/\tau_{ri}} . \tag{5.5}$$

An example is the simultaneous generation of vacancies and dislocations [5.12]. The diffusion of vacancies toward the surface and the recovery of dislocations control the EE process. The τ_{ri} of diffusion and recovery are EE rate determining.

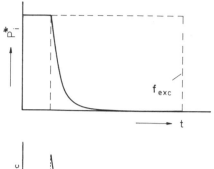

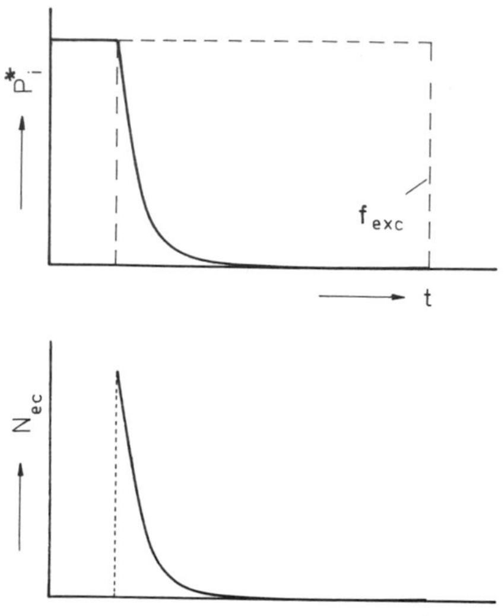

Fig. 5.1. Schematic representation of the temporal correspondence between perturbations p_i^* and spontaneous EE. f_{exc}=external excitation function, N_{ec}=exoemission count rate

Fig. 5.1

Fig. 5.2a–d. Schematic representation of the temporal correspondence between perturbations p_i^* and stimulated coemission (2.1) and postemission (2.2) f_{exc}=external excitation function, f_{st}=external stimulation function, N_{ec}=exoemission count rate

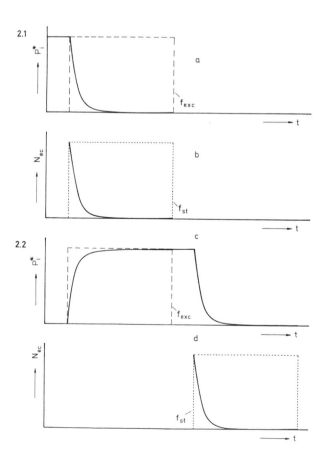

Fig. 5.2

The relaxation time τ_r of a single perturbation which relaxes through several "channels" may disperse into as many τ_i as channels available. For example, the perturbed Fermi distribution may relax via the observed "channels" conductivity, luminescence and EE. τ_r disperses experimentally into τ_L for luminescence, τ_C for conductivity and τ_e for EE. The connection between τ_r and the independently observed τ_C, τ_L and τ_e is no longer trivial [5.13].

Figure 5.1 represents a situation that has so far received only limited attention. Spontaneous EE in the dark was found during chemisorption of O_2 and H_2O on clean Mg [5.14] and Al surfaces [5.15] and during mechanical deformation of Al [5.16, 17]. In the case of chemisorption of O_2 and H_2O on Mg, the external excitation function f_{exc} in Fig. 5.1 represents the gas pressure in a given time interval. The perturbation p represents the concentration of centers capable of sorption as a function of time. EE only occurs as long as gases react with the surface.

Numerous reports on externally stimulated EE have been published (Fig. 5.2). The external stimulation function f_{st} may be superimposed to the external excitation function (Fig. 5.2a) without influencing the relaxation process. Tensile straining of metals under simultaneous illumination is an example. Illumination merely enables electrons to escape from the solid. The relations between strain, work function and OSEE are given, e.g., in [5.18].

Besides this "coemission," "postemission" has been measured most frequently (Fig. 5.2b). The external excitation function is determined, e.g., by the parameters of the irradiation with electromagnetic or particle radiation.

The spatial region may be selected by proper choice of the energy of exciting electrons. Electrons in the eV range will excite only the surface layer [5.19], while keV electrons will perturb the bulk as well as the surface. The possible electronic transition in a solid may also be predetermined by choice of proper wavelength or kinetic energy of the exciting radiation [5.20]. Electrons and light quanta of varying energy have been widely used for the excitation of EE. Postemission also occurs as a consequence of gross perturbations of atomic structure or physico-chemical equilibria.

The particular external stimulation function f_{st} in Fig. 5.2 corresponds to radiation of constant energy and intensity. The excited perturbation either decays isothermally without additional external stimulation [5.21] or it may relax only under the action of light (photochemical process, photodepopulation of traps). In any case, the measured OSEE will have the shape of a decay curve as long as T and $h\nu$ remain constant. If $h\nu$ is varied continuously from the near infrared to band-gap light of the material under investigation (wide-band semiconductor, insulator), the measured curve will behave in a twofold manner [5.22]. The nonselective OSEE curve will join the normal photoemission curve without exhibiting maxima on the frequency scale. Selective OSEE will show maxima which coincide with optical absorption [5.23]. Selective OSEE is assumed to be a photothermal process.

In the case of thermal stimulation, f_{st} represents the temperature. A simple decay curve is obtained with T fixed. Commonly, two different types of heating programs, $f_{st} = T(t)$, are employed. The first,

$$T = T_0 + qt, \tag{5.6}$$

produces normal "glow" curves; the second,

$$T = T_0 + qt + \Delta T_0 \cdot f(t), \tag{5.7}$$

the fractional "glow curve". Here $q = dT/dt$ (at $\Delta T_0 = 0$) is the heating rate and $f(t)$ a periodic function (ramp voltage) which modulates the increase of T.

In principle, a complete description of EE is possible only if the physical or chemical processes of the excitation of perturbations and their relaxation are known. The encountered complexity of this problem stems from the possibility that several different superimposed perturbations may relax via different channels. Since EE techniques monitor only one of these channels and provide, therefore, only indirect information on the relaxation process, exclusive observation of EE will, in general, not yield sufficient data to completely characterize the relaxation kinetics. Only in the most simple cases, e. g., electron emission from adsorbed molecules, is the EE procedure a direct indicator of the ongoing relaxation and only then is it possible to extract meaningful values for the kinetic parameters from EE measurements. Notwithstanding these complexities, a number of practical applications have been explored which are based solely on the existence of EE phenomena.

The elementary process responsible for EE has always been subject to controversy. None of the hypotheses for the emission mechanism discussed hitherto seem to have general validity. At present, some progress has been achieved by employing a well-defined excitation and stimulation procedure, combined with modern surface analytical methods and careful sample selection (use of otherwise well-characterized materials).

In general, the intensity of TSEE or OSEE and the energy and angular distribution of emitted species are measured. In investigating EE, two aspects are essential, the first one dealing with the rate-determining processes and the second one looking for the escape mechanism of electrons and ions. The second aspect has recently attracted special attention. With few exceptions, it has been found that the kinetic energy of thermally stimulated exoelectrons is "over-thermal" and that of optically stimulated exoelectrons exceeds the values predicted by the Einstein relation for photoemission.

5.2.2 Experimental Procedure

The EE effect can be observed in ambient air. Excitation is brought about by α-, β-, γ-, X-rays or uv light. Heat or visible light is used for stimulation. The emitted particles are detected with the aid of special open point counters [5.24] or commercial gas flow counters. This procedure is simple and inexpensive and is mainly used in commercial EE dosimetry, e. g., in [5.25].

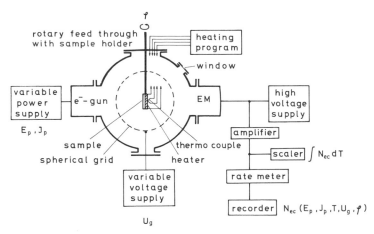

Fig. 5.3. Schematic diagram of a typical TSEE arrangement

Since EE is a phenomenon occurring on the surface of the solid, research concerning its mechanism requires an ultrahigh vacuum (UHV) facility and samples with well-characterized surfaces. This entails careful structural and chemical analysis of the surface as well as knowledge of the composition of residual gases in the vessel. The EE facility should therefore be equipped at least with an Auger spectrometer and a quadrupole mass analyzer. In addition, a capability to measure the work function (e.g., via the Kelvin or Fowler method) is desirable as well as a LEED capability for characterization of the surface structure. A quartz window should be provided since optical measurements are often needed. A combination of EE, luminescence, optical absorption and conductivity measurements has proved to be quite valuable [5.26].

In principle, an apparatus for measuring EE consists of an excitation unit, a stimulation unit and a detection unit. A TSEE facility for excitation with electrons is shown schematically in Fig. 5.3. With the aid of a rotary feedthrough the sample can be scanned by an electron beam with arbitrary angle of incidence. Energy E_p and current density of incident primary electrons J_p are variable. E_p determines the thickness of the surface layer to be excited and the product $J_p \cdot t_i$ the degree of excitation (perturbation); $p^*(t_i)$ can be controlled via the assumed proportionality

$$p^*(t_i) \propto \int_0^\infty N_{ei} dT, \tag{5.8}$$

where N_{ei} is the rate of emitted electrons due to perturbation $p^*(t_i)$.

During excitation the analyzer voltage U_g of the spherical grid analyzer controls surface charging of the sample and, especially, the formation of space charge within the insulating material (see Sect. 5.4). During stimulation by heating, U_g is used to determine the energy distribution of emitted particles. In

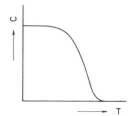

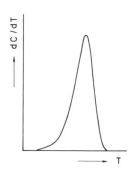

Fig. 5.4. Relative concentration C^* of excited TSEE-active centers and corresponding normal TSEE glow curve (dC^*/dT vs T)

Fig. 5.3 the sample is adjusted for electron bombardment. It is turned around 180° when EE is detected by the electron multiplier (EM). In addition to measuring the energy distribution, the spherical grid analyzer can also be used to measure the angular distribution of emitted charged particles, by simply turning the sample around on its axis.

The electronically controlled heating program determines the shape of the measured curve $N_{ec}(T)$, where N_{ec} is the rate of counted electrons. Using $T = T_0 + qt$ yields the normal glow curve dC^*/dT (Fig. 5.4) with

$$C^*(T_i) = \frac{\int\limits_{T_i}^{\infty} N_{ec}dT}{\int\limits_{0}^{\infty} N_{ec}dT}, \tag{5.9}$$

the relative concentration of excited TSEE-active centers. In the simplest case of first-order kinetics the kinetic parameters E_e and v_e (5.3) can be approximately evaluated from the following expressions [5.27],

$$\frac{d\ln(-\ln C^*)}{d(1/T)} \approx -\frac{E_e}{k} - 2T, \tag{5.10}$$

and

$$\ln(-\ln C^*) \approx -\frac{E_e}{kT} - 2\ln\frac{E_e}{kT} + \ln\frac{v_e E_e}{qk}. \tag{5.11}$$

A TSEE facility with precise T measurement is described in [5.28]. Application of the simple approximations (5.10, 11) will yield satisfactory results if single peaks due to a single activation energy E_e are analyzed. The presence of several E_{ei} can be tested by applying the operator [5.29]

$$\Omega = kT^2 \frac{d}{dT} \ln\left(-\frac{d}{dT} \ln ...\right)$$ (5.12)

to $C^*(T)$. The general result

$$E_e(T) = \Omega C^*(T)$$ (5.13)

indicates a single activation energy by $E_e(T) =$ const. The case of a continuous energy distribution $N(E_e)$ has to be treated by the fractional glow technique (FGT) at the higher expense of more elaborate experimental techniques and electronic data processing [5.30]. The heating program employed for this purpose is represented by (5.7). The FGT is, in principle, based on the initial rise method [5.31]

$$\frac{d \ln N_e(T)}{d(1/T)} \approx -\frac{E_e}{k},$$ (5.14)

which is a good approximation and can be applied during heating as well as during cooling of the sample in each heating cycle. Assumptions made are:

I) the number of liberated electrons per heating cycle is small compared to the total number emitted in a peak,

II) the frequency factor v_e is independent of T in first approximation,

III) retrapping is negligible [5.32].

The first and third assumption can be checked experimentally. Each cycle will yield two E_{ei}-values; E_{e1} upon heating and E_{e2} upon cooling. E_{ei} is a mean activation energy of the emptied preferential traps at T. Averaging for each heating cycle results in $\langle E_e \rangle$ with $E_{e1} < \langle E_e \rangle < E_{e2}$. E_{ei} follows (Fig. 5.5) from

$$E_{ei} = k \tan \vartheta_i.$$ (5.15)

From the sequence of heating cycle results $\langle E_e(T) \rangle$ is obtained. A plot of measured pulse sums per heating cycle versus $E_e(T)$ represents the trap occupation function $f(E)N(E)$, reflecting the relative distribution of electrons over the energy levels in the trap level distribution. The FGT method has been applied successfully to TSEE from Al_2O_3 [5.33]. A fully automated HV apparatus is described in [5.34]. FGT is applicable if the TSEE rate N_{ec} is at least of the order of 10^2 counts/s cm^2. A difficult problem is the interpretation of E_e [5.35]. Care has to be taken to choose the proper evaluating procedure [5.36].

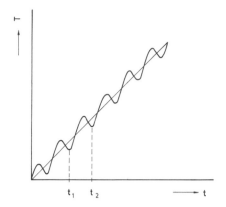

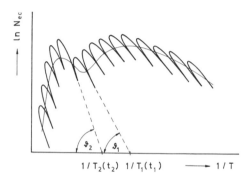

Fig. 5.5. Heating program
$T = T_0 + qt + \Delta T_0 f(t)$
and resulting fractional glow curve
($\ln N_{ec}$ vs $1/T$)

Special attention has been paid to energy and angular distribution measurements. They reflect the interaction of electrons with the solid and thus give an indication of the emission mechanism. Various energy analyzers [5.37–40] have been employed. However, angular distributions have only occasionally been reported in the literature [5.39, 40].

Frequently plane retarding field analyzers were used, since a high luminosity is needed. More advantageous is a spherical grid analyzer, since it permits the measurement of angular and energy distributions of the emitted charged particles [5.41]. Cylindrical mirror analyzers and other special arrangements have also been employed [5.42, 43].

An OSEE facility is equipped with special uv windows or light sources. Short wave uv or X-rays excite EE, and uv or visible light will stimulate the exoemission [5.44]. If the temperature of the sample and the flux of the stimulating light quanta are kept constant, one can measure the "excitation spectra" of OSEE as a function of energy of the exciting quanta. The design of an OSEE apparatus is described in [5.46]. A quick survey of the behavior of the stimulation spectrum can be made using graded interference filters. If high resolution is desired, quartz double-monochromators or modern tunable light sources are required.

Single decay curves at a constant wavelength are discussed in [5.45], the procedure for evaluating them in [5.46–48]. On the basis of a most simple model, the decay is exponential with time according to

$$N_{ec}(t) = N_{ec}(t_0) e^{-\alpha_e t}. \qquad (5.16)$$

OSEE excitation and stimulation spectra have been discussed in connection with optical absorption. If the selective OSEE maximum and the optical absorption maximum coincide, both observations are attributed to the same trap level, e.g., F centers, just as is the case when TSEE and thermal bleaching of optical absorption peak at the same temperature [5.49]. A correlation of TSEE and OSEE has been attempted by considering the thermal and optical activation energy in terms of the Franck–Condon ratio $E_{opt}/E_{therm} \geqq 1$ [5.28]. If perturbations p_i^* are thermally and optically annealable, one may anneal thermally a small fraction of p_i^* and observe the remaining OSEE intensity or, vice versa, one may anneal optically a small fraction of p_i^* and observe the remaining TSEE intensity.

OSEE in connection with mechanical deformation is measured with arrangements similar to those shown in Fig. 5.3. During tensile deformation the sample is exposed to light of different wavelengths and intensities $I(hv)$ and the resulting OSEE is detected with the aid of channel or dynode multipliers.

5.2.3 Correlated Effects

The interpretation of EE results is based on the knowledge of the detailed processes occurring during excitation and stimulation. They are called "correlated" if they influence the EE intensity and/or the decay constant α_e. Hitherto existing observations seem to show a considerable number of effects being involved.

OSEE is sometimes correlated with a time-dependent change of the work function ϕ which, in turn, is affected by gross structure perturbations of clean surfaces, mechanical removal of adsorbates and subsequent readsorption, change of the adsorption–desorption equilibrium, diffusion of bulk defects to the surface and rearrangement of the surface structure, and charging of cleavage planes of solids with polar bonds. Optical excitation and stimulation of EE may cause photoadsorption, photodesorption, photochemical reactions and dissociation. Ultraviolet light even generates bulk defects in alkali halides. These defects may be the sites of electron emission.

TSEE was found to correlate with ϕ as well. Thermally stimulated diffusion in the bulk or at the surface may lead to recombination of vacancies and interstitials or rearrangement of the surface structure. The energy released in the process is often assumed to be transferred to the electrons which may be emitted by overcoming the work function barrier. Thermally stimulated desorption frequently accompanies exoelectron emission. The variation of the energy of exciting electrons is of considerable influence on the structure and intensity of TSEE and OSEE spectra. Extremely low energy electrons (eV range) are attached to the solid surface, sometimes by dissociative attachment, and the surface becomes negatively charged. Intermediate energy electrons (of the order 10^2 to 10^3 eV) affect TSEE and OSEE spectra by electron-stimulated desorption (ESD). The ESD threshold of insulator–adsorbate systems is about 20 eV but depends on the kind of solids and adsorbed species. Electrons of several hundred eV generate defects in alkali halides and even desorb lattice atoms. Oxides are known to be decomposed by 10^3 eV electrons. Moderate energy electrons generate space charges inside the insulators. The buildup of this space-charge structure proceeds in a complicated manner and results in a positive–negative double layer. Secondary electron emission kinetics play a decisive role. Electron bombardment and field-enhanced ion drift occurs in the space-charge region. Irradiation of insulators with uv light and X-rays can cause a positive charge at the surface. High energy radiation (α, β, γ, n) produces displacement effects in the whole sample. The pyroelectric effect is accompanied by high internal fields and voltages and seems to be responsible for emitted high-energy electrons upon heating of pyroelectrics. First- and second-order phase changes, e.g., at the Curie and Neel point, are detectable with TSEE. This effect is attributed to the structure damage in the oxide covering the material.

Besides this category of "correlated" effects there are the simultaneous effects being observable during excitation and stimulation. Observation of these effects augment the "EE picture" of a solid rather than explain the EE phenomenon. Effects of this kind are, for example, luminescence occurring during excitation with light quanta, electrons or during structure deformation; photo- and electron-induced conductivity; optical absorption; and, finally, TSC, TSL, TSDC and TSD.

5.3 Experimental Findings and Interpretations

A selection of recent work – in the author's opinion representative and typical of current topics in the field of EE – has been chosen to display the present state of development. It was not intended to give a historical review. Thus, some important early work is not mentioned here. The reader may consult the cited review articles [5.7–10]. The subheadings of this section refer to the perturbations. EE is treated as a consequence of their relaxation.

5.3.1 Mechanically Deformed Structure – Atomically Clean Surface

Mints et al. [5.50] tried to correlate the process of recovery and recrystallization of metals after plastic deformation with exoemission. The studies of *Kortov* et al. [5.51] were aimed at the same goal. The subject of their work was the plastic deformation of Al in ultrahigh vacuum. The possibility of EE occurring from atomically clean metal surfaces was dismissed by many authors. However, it has found some support by recent results. The mechanical deformation generates a high defect concentration which causes a change of the density distribution of electronic states, resulting in a small shift ΔE_F of the level. This is of the order $10\,meV$ for a change of defect concentration of 1% [5.52–54]. The thermally enhanced annealing of defects leads, via $\Delta E_F(T, t)$, to a small change of the photoemission (PE) from the metal. This change in PE may be called OSEE.

During the relaxation of structure perturbations, defects can diffuse to the surface and rearrange there exothermally. The localized release of energy and a simultaneous interaction with a photon could result in a diffusion-controlled OSEE component. Clean metal surfaces of high roughness display a lowered work function ϕ in comparison to smooth metal surfaces. The lowering of ϕ may be as much as a few hundred meV [5.55, 56].

Buck et al. [5.57] explain this effect by a roughness-enhanced coupling of incident light with surface charge oscillations which are effective sources of photoemission. A thermally stimulated annealing of roughness should cause a change in the wavelength dependence of the photoyield.

After electron bombardment and/or exposure to X-ray irradiation, no OSEE or TSEE was measured from atomically clean metal surfaces.

5.3.2 Adsorption – Initially Clean Surface

During adsorption of reactive gases EE may arise from the surface of metals, even in the dark. At constant T the adsorption process is described by the isotherm of adsorption. The adsorption–desorption equilibrium is reached rapidly without activation in the case of physisorption. The heat of physisorption is of the order of $0.1\,eV$ per species. The adsorption–desorption equilibrium shifts with increasing T toward desorption as determined by the Le Chatelier–Braun principle. Physisorption is reversible and forms multimolecular layers.

Chemisorption requires an activation energy and is preceded by physisorption. The heat of chemisorption is of the order of several eV per species. Chemisorption is partially irreversible and ceases at monolayer coverage. In the case of oxidation species may diffuse through the growing oxide layer. As long as the chemical reaction velocity $K = K_0 \exp(-E/kT)$ is rate determining,

$$\frac{dc_A}{dt} = Kc_A c_B = \frac{1}{\tau} c_{AB} \tag{5.17}$$

is valid. When the diffusion process becomes rate limiting, the oxidation process is governed by Fick's first law

$$\frac{dn_A}{dt} = -Dq\frac{dc_B}{dx} \tag{5.18}$$

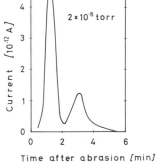

Current [10^{-12} A]

Time after abrasion [min]

2×10^{-8} torr

Fig. 5.6. Time development of the spontaneous EE current during exposure of freshly abraded magnesium to oxygen [5.14]

where c is the concentration and n_A the number of reacting atoms A. $D = D_0 \exp(-Q/kT)$ is the diffusion constant and dc_B/dx the concentration gradient of the reaction partner B within the oxide layer of area q.

Gesell et al. [5.14] reported on the emission of electrons from clean Mg surfaces exposed to 5×10^{-6} Torr O_2 or H_2O in the dark. At the beginning of chemisorption the emission increases steeply and then passes through two maxima (Fig. 5.6). After a few minutes the emission has decayed. The maxima of EE are attributed to different adsorbing species.

The dependence of the work function on surface coverage during ongoing chemisorption was measured as well. The maxima of EE correspond to the minima of $\phi(t)$. The emission is attributed to two processes. The first one is the lowering of the surface potential barrier by adsorption of molecules with a dipole moment favoring emission. The second one is the release of adsorption heat and its transfer to electrons which then can escape because of the lowered work function. Physisorption is not expected to cause emission since the heat of physisorption is too low. The emission is controlled by thermally stimulated chemisorption.

Born and *Linke* [5.58] found a considerable decrease in the work function of Al during adsorption of water. Adsorption of O_2, N_2 and H_2 does not affect ϕ for partial gas pressure of 10^{-10} to 10^{-2} Torr. ϕ was determined using the Fowler method. At 10^{-10} Torr, a work function of $\phi_{Al} = 4.25 \pm 0.05$ eV was measured. After exposure to water vapor, it reduces to $\phi_{Al} = 1.3$ eV. A mixture of O_2/H_2O yields a smaller reduction in ϕ. Again, the minimum of $\phi(t)$ coincides with the maximum of "dark EE". This kind of emission turns out to be dependent on pressure, dose and kind of the residual gas. The intensity of the emission is correlated with the lowering of ϕ by adsorption. The process of chemisorption is assumed to generate surface plasmons by the heat of chemisorption. As surface plasmons annihilate, they transfer their energy to the electrons enabling them to escape from the solid surface. A known surface plasmon peak in Al is positioned at about 10 eV. The emission probability per incident molecule is reported to be about 10^{-9}.

5.3.3 Desorption – Adsorbate-Covered Surface

Thermally stimulated desorption (TSD) is sometimes accompanied by exo-emission. TSD is governed by the equation

$$\frac{d\theta}{dt} = -A\theta^n e^{-E/kT} \tag{5.19}$$

where θ is the degree of coverage, n the order of reaction and $1/\tau = A\exp(-E/kT)$ the reciprocal relaxation time. *Krylova*'s explanation of TSEE [5.59] is based on a deformation of the adsorption layer provoked by the desorption of species. The nonequilibrium state of the surface (active surface) relaxes, being thermally stimulated by recombination of active species at the surface with sites which have become vacant through desorption. The released recombination energy is consumed by the escape of the electrons. If the recombination proceeds spontaneously, only $d\theta/dt$ will be rate limiting for EE. If it requires an activation energy, the desorption and recombination probability will determine the rate of emission. If, in addition, surface diffusion is involved, there are three different relaxation times τ affecting the course of $N_{ec}(t)$. *Krylova* [5.10] found that highly hydrated oxides (Al$_2$O$_3$, MgO, etc.) and alkali halides (NaCl, etc.) show a strong gas evolution accompanied by electron emission. Measurements were carried out in ultrahigh vacuum (5×10^{-9} Torr). The concentration of residual gases was monitored by mass analysis. The combined effect of EE and TSD was observed after immersing metals and semiconductors, and immersing alkali halides in water and alcohol. Maxima of TSD were observed at about 140 to 160, 180, and 360° C. They have been identified as being due to water desorption. The number of emitted exoelectrons per desorbed molecule has been estimated to be 10^{-6}. Electron emission occurs without any preliminary excitation by electrons or X-rays. This kind of emission is called "self-excited" and has already been reported by *Bohun* [5.60].

5.3.4 Physico-Chemical Changes in Adsorbates

In addition to adsorption and desorption, excitation of EE is obtained following physico-chemical changes induced by irradiation or mechanical deformation. This physico-chemical concept, especially promoted by *Krylova* [5.10], is characterized by the following scheme of the stimulation process: Heating – phase conversion in adsorption compounds – formation of a mobile adsorption layer – thermally stimulated recombination of ions and radicals – emission of electrons, ions and neutrals.

It is assumed that on the surfaces of different substances similar reactions will occur. This assumption is in accordance with experimental findings displaying nearly the same peak position on TSEE from various substances, thus indicating the involvement of material-independent emission centers. In the subsequent example X-ray induced physico-chemical changes of adsorbed water are assumed:

I) Excitation of emission (generation of ions and radicals)

$$H_2O \xrightarrow{h\nu} 2H^+_{ads} + O^{2-}_{ads}$$

$$H_2O \xrightarrow{h\nu} H^+_{ads} + OH^-_{ads}$$

$$H_2O \xrightarrow{h\nu} H_{ads} + OH_{ads}$$

II) Stimulation of emission (recombination of ions and radicals)

$$O^{2-}_{ads} + H_{ads} \xrightarrow{Q^*} OH^-_{ads} + Q_{rec} + e^- \uparrow$$

$$OH^-_{ads} + H_{ads} \xrightarrow{Q^*} H_2O^-_{ads} + Q_{rec} \rightarrow H_2O^- \uparrow$$

$$OH_{ads} + H_{ads} \xrightarrow{Q^*} H_2O_{ads} + Q_{rec} \rightarrow H_2O \uparrow .$$

Q^* represents the supplied amount of thermal energy during heating. Q_{rec} is needed for the emission process. The emission of ions and neutrals occurs also upon exciting the EE by irradiation or mechanical deformation of the surface. While being excited with electrons [5.61], ESD of H_2 and CO was observed from SiO_2, Al_2O_3, MgO and NiO. In the case of SiO_2, ESD-induced O^+_2 was detected [5.62]. A subsequent TSEE measurement exhibited an ESD-rate-dependent TSEE intensity. During mechanical deformation a simultaneous emission of electrons, ions, neutrals, and photons can be observed from oxide-covered metals [5.63].

Besides these physico-chemical changes, a simple negative ionization of adsorbed species by electron attachment was observed. Electrons of several eV kinetic energy, impinging on the surface, will convert an adsorbate complex via Franck–Condon transitions from the neutral ground state to the single negatively ionized state [5.64]. *Krylova* [5.65], and *Euler* et al. [5.66] discussed negatively ionized states of adsorbed oxygen on oxides. *Euler* et al. proved that oxygen is involved in the emission process from BeO. The surface was monitored by AES and could be cleaned with the aid of argon–ion bombardment. Similar results were obtained by *Jakowski* et al. on SiO_2 [5.67]. Electron-beam induced desorption of O^+_2 caused a considerable decrease of the TSEE signal. A uniform description of the involved excitation and stimulation processes and an interpretation of experimental results is given by *Glaefeke* et al. [5.64]. Within the framework of a potential curve model, both the threshold energy and resonance energy of excitation can be interpreted. For the first time

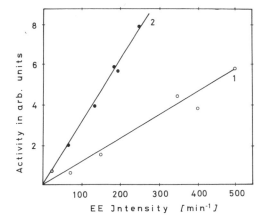

Fig. 5.7. Relation between EE intensity and catalytic activity of (1) Pt/SiO$_2$ with different Pt contents and (2) ZnS/Cu with different copper contents [5.10]

Hiernaut et al. [5.68] reported the observation of excitation resonances. Besides oxygen, OH$^-$ is assumed to be an EE-active center as proposed by *Kriegseis* et al. [5.69].

A detailed discussion of the surface concept in terms of simple ionization is presented by *Huster* [5.70]. The emission of exoelectrons from the adsorbate is affected by a "perturbed" electron affinity. Adsorbed molecules at the surface of a solid reside within the region of strong change of the undisturbed image potential. Whether emission is favored or not depends on the orientation of an assumed dipole moment of the molecule and the interaction with the image potential. The frequently used assumption of negligible influence of the image potential on electron emission from adsorbed species, which has often been taken for granted, is dubious. During mechanical deformation, point defects are created which diffuse toward the surface. They perturb the oxide at the surface and appear to control EE by their diffusion, as discussed in Sect. 5.3.9.

5.3.5 Catalytic Reactions

Heterogeneous catalytic reactions take place at centers where the electron exchange in the catalyst-substrate system is favored. EE also arises from centers at the surface where electrons are most weakly bound. Since TSEE occurs in the same range of temperatures (20 to 500° C) which is typical for heterogeneous catalysis and since it has been observed during catalytic reactions, catalysis and TSEE seem to be correlated. Indeed, a reproducible relation has been found between catalytic activity and TSEE capacity in different groups of material (Fig. 5.7) [5.10]. The catalytic oxidation of hydrogen and carbon monoxide at NiO, ZnO and platinum black was found to be accompanied by EE. Simultaneous emission of electrons and ions has been observed [5.10]. The catalyst platinum black has been activated in an oxygen atmosphere at 400° C and thereafter exposed to hydrogen. During hydrogen adsorption, decaying EE could be detected which was controlled by two relaxation times. A subsequent heating at a constant rate resulted in TSEE peaks at 130 to 150° C, 180 to 190° C and 240° C to 260° C. The analysis of TSD conducted under equivalent conditions proved that a catalytic reaction takes place.

In heterogeneous catalysis, oxygen is active in the adsorbed form O_{ads}^- and O_{ads}^{2-} stemming from O_{2ads} and O_{2ads}^-. The valence transformation

$$O_{2ads} + e^- \overset{exc}{\underset{stim}{\rightleftarrows}} O_{2ads}^-$$

is assumed to be exoemission active. In addition, the reactions

$$O^{2-} \rightarrow O^- + e^-$$

and

$$O^- \rightarrow O + e^-$$

are discussed. Thus, certain correlations between catalysis and EE appear to be established. Similar results were found during catalytic decomposition of methanol [5.10]. The energy needed by electrons and ions to escape from the solid surface is gained from the recombination process of species on the surface.

5.3.6 Phase Transformations

EE has been observed during ongoing changes of the state of aggregation or modification (first-order phase changes) and during changes of orientation, e.g., of microregions of ferroelectrics (second-order phase changes). In first-order phase changes, macroscopic properties and thermodynamic state variables (e.g., density, molar volume, enthalpy) change steplike with temperature. In second-order phase changes properties and state variables change in an extended range of temperature.

When oxide covered metals and alloys are heated, stress is generated within the oxide since the molar volume changes at the phase change temperature. Simultaneous illumination creates EE in the T range of the phase change. A survey of this subject was given by *Sujak* et al. [5.71].

EE has been detected during the transition from the hexagonal to the cubic configuration of thallium and during changes of the modification of hexagonal chromium and selenium. Solidification of aqueous ammonia solutions, water, melts of lead, tin and lead-tin eutectic alloy is accompanied by EE as well as melting of lead, tin and bismuth. EE during second-order phase transitions has been found in $Ni + NiO$ and $Cr + Cr_2O_3$. The Neel temperature of the oxides and the Curie point of Ni controls the EE. All these samples were heated or cooled and illuminated simultaneously. Austenite steel exhibits EE induced by first- and second-order phase changes. The maxima of EE appear during heating and cooling at different T. First-order phase changes are attributed to the transition $\varepsilon \to \gamma$, second-order phase changes to the transition antiferromagnetic state $\to$ paramagnetic state.

It is assumed that the EE intensity is a measure of the degree of completion of the phase change. The EE intensity is assumed to be proportional to the number of defects created in the oxide.

However, unequivocal results have not been obtained so far. Sometimes EE maxima are shifted against the temperature of phase change. A widely quoted interpretation of these results was given by *Sujak* et al. [5.72]. The change of macroscopic properties of the oxide-covered material generates stress within the oxide. Perturbations of the oxide structure favor physico-chemical processes at the oxide surface, lowering the work function ϕ. Thus OSEE becomes possible. The phase change is thermally stimulated and controls the generation of structure defects. The relaxation of these defects is displayed by OSEE.

Earlier, *Hanle* et al. [5.73] investigated the thermally stimulated decomposition of hydrated sulphates, chlorides and carbonates. Negative ions were emitted during heating of these substances. The TSEE peaks (negative ions) correspond exactly to the temperatures at which the hydrates of different configuration decompose.

5.3.7 Pyroelectricity

Second-order phase changes in crystalline pyroelectrics are accompanied by EE of high-energy particles. At the Curie point the domain structure of pyroelectrics changes. This change is correlated with the generation of high internal electric fields which cause the EE.

Rosenblum et al. [5.74] correlated the domain structure change of lithium niobate with EE. The emission mechanism was recognized as a thermally stimulated field effect (TSFE). The source of the electric field is the thermally induced change of the regions of spontaneous polarization at the surface. The TSFE occurs if the rate of change of spontaneous polarization is not compensated by free charge carriers and the polarization field. The TSFE curves were obtained in UHV without preceding excitation. After electron bombardment OSEE is observed but no TSEE.

Syslo [5.75] observed the emission of negatively charged particles and photons. The particles were detected by an open point counter. Results are interpreted in terms of electrical discharges at the surface. The discharge is provoked by thermally induced polarization changes of adjoining domains. The discharge is the source of the emitted negative particles and photons.

Brunsmann et al. [5.76] reported a similar behavior of TSEE and TSL from BeO crystals. The emission occurred without previous excitation during heating as well as during cooling. The kinetic energy of the emitted electrons exceeds 3 keV. This effect is not observed with polycrystalline BeO samples.

5.3.8 Mechanically Deformed Oxide-Covered Structure: Dark Emission

This kind of emission accompanies the mechanical deformation of oxide-covered metals and does not require previous excitation or additional stimulation. *Sujak* et al. [5.16] reported first results obtained with anodically oxidized Al. Subjecting Al to the tensile strain, two EE maxima appeared at strains of 1 to 2% and 3 to 4%. Light did not affect the emission. It was shown that the intensity of the dark emission depends on the oxide thickness, the electrical field accelerating the emitted electrons, the rate of deformation and the preliminary treatment of the sample. This kind of EE is interpreted in terms of field emission from microcracks propagating in the oxide layer during tensile deformation of the sample. It is assumed that the opposing surfaces of propagating cracks become charged. This "electrified fissure model" was reinvestigated by *Ramsey* et al. [5.17]. They restrict its applicability to polar substances and propose to interpret results from nonpolar substances by rearrangement of dangling bonds created when cracks propagate while straining the sample. An estimate of the total energy released during fracture of the oxide indicated the feasibility of this mechanism.

Recent results obtained by *Rosenblum* et al. [5.77] point to a different interpretation. Oxide-covered Al, Ni and Ti were shown to emit simultaneously electrons, negative and positive ions, and photons. Emission occurs in bursts on straining the sample. The tips of the propagating cracks are the origin of the emission. Calculations [5.63] lend strong support to the idea that the release of strain energy in the vicinity of a propagating crack results in local heating to a temperature sufficient to cause thermionic emission from the walls. Electrical fields are assumed to participate in the emission process. The observed ion emission is restricted to Al and Ni. The temperature of the crack tip was estimated to reach up to 3000 K. Photon emission is attributed to the relaxation of excited neutrals or ionized impurities that evaporate from the freshly formed surfaces of the cracks.

5.3.9 Mechanically Deformed Oxide-Covered Structure: OSEE

OSEE from Al_2O_3 coated Al wires was investigated quantitatively by *Brotzen* et al. [5.78, 79]. The samples were subjected to plastic strain and illuminated simultaneously with light of wavelength $\lambda \geq 3000$ Å. The strain generates point defects which diffuse toward the surface. Diffusion is controlled by the diffusion constant $D = D_0 \exp(-Q^*/kT)$. The OSEE rate was proportional to the decay of EE active oxide centers introduced into the oxide by point defect diffusion. The decay kinetics was found to be of first order. The detected $N_{ec}(t)$ increased rapidly with strain and went through a maximum after the strain was removed. This anomalous decay was attributed to time-delayed appearance of the diffusion maximum at the surface. Deformation at low T ($-145°$ C) and subsequent heating at a constant rate confirmed that the diffusion process is

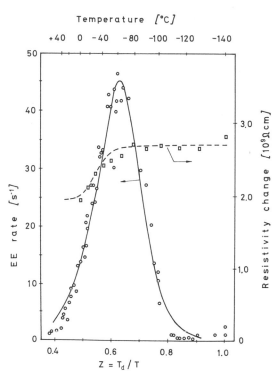

Fig. 5.8. Comparison of experimental with computed emission rates and electrical resistivities for annealing of oxide-covered Al wires at a constant T rate after tensile strain of 7.5 % at $-140\,°C$; ○ EE, experimental; □ resitivity change $\Delta\varrho$, experimental; —— EE, computed; $-\,-\,-\,\Delta\varrho$, computed; T_d = deformation temperature [5.79]

thermally activated. Experimental determination of the activation energy Q^* yielded a value of 0.44 eV, in fairly good agreement with activation energies obtained by independent methods. The emission was stimulated optically. The "glow curve" as shown in Fig. 5.8 is the thermally stimulated diffusion controlled OSEE. The change in resistivity, $\Delta\varrho$, is plotted as well. It is indicative of the proposed change in defect concentration, which occurs within the T region of the OSEE maximum. This dependence of ϱ on temperature corroborates the assumed hypothesis of a diffusion controlled OSEE.

The rate-limiting aspect of OSEE has been treated quantitatively by *Shorshorov* et al. [5.12]. Plastic deformation of metals was correlated with the generation of dislocations and vacancies. The behavior of these defects has been studied by OSEE and metallographic methods (electron microscopy, X-ray analysis). OSEE was found to consist of two components, the first one reflecting the relaxation of dislocations and the second being determined by diffusion of vacancies toward the surface. This has been verified in molybdenum, which exhibits no diffusion but relaxation of dislocations, and in α iron. The latter showed a considerable flow of vacancies toward the surface. Calculated relaxation and diffusion curves agreed well with the measured OSEE curves. Thus, OSEE was interpreted to be limited by both recovery and by diffusion.

5.3.10 Perturbations Induced by UV and X-Rays

Exposure of alkali halides to uv light is known to form color centers. The reverse process, annealing, is accompanied by EE if the electron affinity χ at the surface is sufficiently small. When the EE spectra are recorded as a function of the energy of exciting uv quanta under otherwise constant stimulation conditions, one obtains the so-called "creation spectra." If, on the other hand, energy and intensity of the exciting uv light are kept constant and the wavelength of stimulating light is varied, "stimulation spectra" are measured. The emission mechanism of exoelectrons from color centers in low affinity alkali halides can be a two-step photothermal process. The first step is the optical transition from the color center to the conduction band. The second step is the thermionic emission from the conduction band into vacuum according to $N_e \propto \exp(-\chi/kT)$.

Bichevin et al. [5.20] simultaneously investigated the optical absorption and the exoelectron emission from NaCl and KCl after excitation with uv light (7 to 11 eV). The samples were heated at a constant rate and simultaneously illuminated with pulsed F-band or K-band light. The OSEE-creation spectra indicate that stable F centers in KCl are generated at 300 K by light quanta of energy $h\nu = 7.7$ eV or $h\nu = 8.4$ eV. Excitons e^0 and electron–hole pairs $e^- + e^+$ are involved in the creation process.

The obtained TSEE creation spectra indicated that excitons and electron–hole pairs are equally effective in the F-center creation in KCl. This statement is valid for a 400 K TSEE peak. The 450 K peak is attributed to a photon-excited transition from the valence band to the d-conduction band resulting in F-center creation. Various photon energies lead to the formation of F centers. It was revealed that excitons and electron–hole pairs form F centers in regular lattice sites. The presence of defects is not a necessary condition.

F-center formation is described by

$$e^0 \rightarrow F + H \quad \text{and} \quad e^- + e^+ \rightarrow e^0,$$

where F is the F center and H the interstitial halogen atom (H center). The essential conditions of defect creation are

$$E_{exc} > E_d \quad \text{and} \quad \tau_{exc} > \tau_{vib}.$$

If these conditions are fulfilled in a solid, a low radiation resistance appears to result ($E_{exc} =$ energy of exciting $h\nu$, $E_d =$ energy needed for defect formation, $\tau_{exc} =$ lifetime of e^0 and $e^- + e^+$, $\tau_{vib} =$ period of vibrational frequency of lattice ions).

In other crystals the conditions

$$E_{exc} < E_d, \tau_{exc} > \tau_{vib},$$

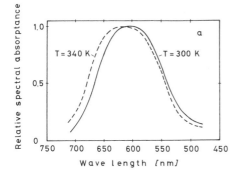

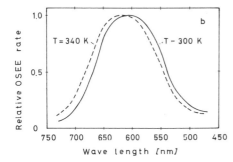

Fig. 5.9a and b. Spectral distribution of optical absorption (**a**) and selective OSEE (**b**) of CsCl [5.23]

or

$$E_{exc} > E_d, \tau_{exc} < \tau_{vib}$$

may hold. The latter one implies high resistance of the material to radiation. In examining EE from alkali halides, the uv energy of the stimulating photons has been extended to soft X-rays (40 to 160 eV). While the absorption spectrum and the photoelectron spectrum show the same structure as a function of energy, the OSEE creation spectrum is rather flat and less structured. This flattening is tentatively explained by a compensation effect [5.80].

Stimulation spectra have been studied extensively. *Nink* and *Holzapfel* [5.23] examined F centers in CsCl by OSEE and optical absorption. The crystal was colored by X-rays. The peak of optical absorption was found at 605 nm. Raising the temperature from 300 to 340 K shifts the absorption peak to 620 nm (Fig. 5.9). Heating up to 450 K bleaches the F centers completely.

OSEE exhibited a selective behavior. The wavelength dependence of the OSEE intensity peaks at 605 nm. As found for absorption, the OSEE peak shifts to 620 nm if T is raised from 300 to 340 K (Fig. 5.9). Temperature-dependent F-band broadening was found in both effects. Heating of the colored CsCl sample yields TSEE. Maxima are positioned at 330, 360 and 400 K. The additional supply of F-band light results in a glow curve superimposed by a decay curve. Photodesorption was assumed during illumination. The activation energy was determined from thermal and optical measurements. The electron affinity χ of CsCl is extremely small; thus, the thermally released electrons are directly emitted and the thermal activation energy should reflect the trap depth. A thermal activation energy E_{th} was found to be 0.95 eV. From the OSEE curves E_{opt} was estimated to be 2.30 eV. The ratio E_{opt}/E_{th} is interpreted in terms of the Franck–Condon principle. The experimentally determined ratio of 2.3/0.95 is in reasonable agreement with theory.

Attachment and detachment of an electron from an F center is accompanied by a repositioning of nuclei. F-center creation and annihilation comprises the production of an interstitial halogen (H center) and the association of the F^0 and e^- and a recombination of F and H (F^0 = vacancy, $F^0 + e^- = F$). The excitation of EE with extremely low energy electrons has revealed a quite similar behavior.

Nonselective OSEE has been measured as well. A simple decay curve at constant T and $h\nu$ can give an indication of the number of elementary processes involved by plotting the intensity on a semilogarithmic scale [5.46]. The emission mechanism is a direct transition of electrons from traps into the continuum above the vacuum level.

The photothermal emission process which governs selective OSEE was observed preferentially in alkali chlorides and fluorides ($\chi < 0.8$ eV) but not in iodides and bromides ($\chi > 0.8$ eV) [5.81].

An obvious and impressive correlation of TSL and TSEE after excitation with X-rays has been presented by *Holzapfel* and *Krystek* (Fig. 5.10) [5.82]. Undoped $BaSO_4$ and $SrSO_4$ exhibit no TSL but only TSEE. Doping with Eu yields both TSL and TSEE without mutual influence. The kinetics is described by a simple picture: Electrons are thermally liberated from shallow traps belonging to intrinsic defects. The small χ of some 1/10 eV favors thermionic emission. The radiative recombination into Eu activator levels is responsible for TSL. From the analysis of the kinetics a first-order reaction is derived. However, a large frequency factor points to strong retrapping. This contradiction is resolved by assuming a potential barrier for retrapping. The evaluated parameters of the kinetics for TSL and TSEE agree extremely well. The agreement of activation energies, frequency factors, peak position and shape is strongly indicative of an identical origin of emitted and recombined electrons. The missing mutual influence is due to the volume character of TSL in contrast to the surface-specific nature of TSEE.

In general, the connection between TSL and TSEE is not that unequivocal. *Tomita* et al. [5.83], for example, found only a partial correlation of TSL and TSEE from LiF after X-ray excitation, and this seems to be the rule.

Oxides, irradiated with X-rays, have been studied, among others, by *Maenhout–Van der Vorst* [5.84]. The relative glow peak intensities and the measured activation energies appeared to be specific to the type of oxide. A change of the residual gas pressure from 10^{-5} to 10^{-9} Torr did not affect the

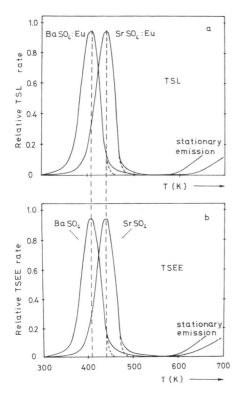

Fig. 5.10a, b. TSL and TSEE of BaSO$_4$ and SrSO$_4$ [5.82]

results. LEED pattern confirmed, e.g., that V$_2$O$_5$ is converted above 100° C to V$_4$O$_9$. In this process oxygen is released. One TSEE peak is attributed to the oxygen release. These investigations seem to elucidate that volume states, typical of the material, and adsorbate states, specific of adsorbed species only, determine the glow curve spectrum.

In most cases a direct liberation of electrons from traps is assumed when EE is studied. *Bichevin* and *Käämbre* [5.85] confirmed such a process which had already been proposed earlier. NaCl and KCl were excited with X-rays. The thermal annealing process was followed by TSEE and ESR. It was found that the annealing of V$_K$ centers (self-trapped holes) corresponds to a TSEE peak. The assumed transition of the emitted electrons is an Auger-like process. Trapped holes are thermally released from V$_K$ centers and migrate to F′ centers. One of the two electrons in the F′ center recombines with the hole, whereby the released energy is transferred to the second electron of the F′ center. If the amount of transferred energy is sufficient, the second electron is emitted. Additional experiments with optical redistribution of electrons over available trap levels seem to confirm these ideas. The probability of these transitions has also been examined theoretically by these workers.

5.3.11 Perturbations Induced by Electron Beams

Atomically clean metal surfaces were never found to emit exoelectrons after irradiation. The presence of oxide or adsorbed species is a necessary condition. Apparently this is also true for semiconductors. In insulators EE has been observed from clean as well as adsorbate-covered surfaces.

Excitation of EE with low-energy (0.2 to 30 eV) electrons shows some similarity to uv-light induced "creation spectra". Moderate energy electrons (several 10^2 to 10^3 eV) generate TSEE glow-curve spectra which resemble those obtained after X-ray excitation.

Above the ESD threshold (about 16 to 20 eV for insulators) EE spectra change with ongoing desorption. Electrons of several hundred eV kinetic energy may remove lattice ions from alkali halides [5.86]. *Carriere* and *Lang* [5.87] reported decomposition of oxides during AES analysis with 1000 eV electrons. The correlation between the created defects and EE has not yet been investigated in detail.

The role of the kinetic energy of exciting electrons has to be considered from another point of view. Very low energy electrons only excite adsorbate or surface states and are deposited in the upper surface layers, whereby the solid charges up negatively. In addition, moderate-energy electrons excite volume states. A complicated space-charge structure is generated inside insulators depending on the primary energy E_p, the deposited total charge, and the thickness of the affected layer [5.88].

Low-Energy Electrons

ESD of positive O_2^+ ions above the desorption threshold of 16 to 20 eV was studied in SiO_2 [5.62]. However, desorption of neutrals is a much stronger effect. ESD can influence TSEE spectra considerably [5.67]. Both the intensity as well as the position of the TSEE peaks are affected. Therefore, electron-induced creation or excitation spectra are studied below the ESD threshold.

Hiernaut et al. [5.68] reported the first experimentally obtained excitation spectra (Fig. 5.11). The TSEE intensity exhibits a maximum positioned at an energy E_p of the exciting electrons of about 3 eV in NiO, ZnO and CuO.

Several resonances where found in SiO_2 [5.89]. Peaks at 2.5, 7, and 9 eV were tentatively attributed to various states of negatively ionized adsorbed oxygen molecules. It is of interest to note that the 2.5 and 7 eV resonance coincide with those measured for electron impact production of negative oxygen ions in gases. A threshold for excitation exists. Further, nitrogen compounds and hydroxyl groups are hypothetically assumed to form EE-active centers. Both TSEE and OSEE intensities increase with the rate of adsorbed oxygen. In the case of SiO_2, water adsorption suppresses the emission. Adsorption of argon does not influence EE, whereas N_2 reduces the intensity.

Only wet oxidized BeO shows TSEE or OSEE. *Euler* and *Scharmann* [5.90] found a resonance at $E_p = 3.5$ eV for the 400° C TSEE maximum. They

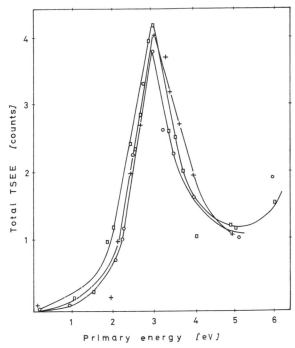

Fig. 5.11. Total number of TSEE counts obtained from NiO (+, factor 10^3), ZnO (O, factor 10^2) and CuO (□, factor 10^5) as a function of the primary energy of the exciting electrons [5.68]

postulate the creation of EE-active centers via

$$O_2 + e^- \rightarrow O_2^{-*} \rightarrow O_2^-,$$

together with the dissociative attachment

$$O_2 + e^- + E_{kin} \rightarrow O + O^- + E'_{kin}.$$

Glaefeke et al. [5.64] found an anisotropic cross section which is assumed to correspond to the 7 eV resonance of the $^2\Pi_u$ state of O_2^-. The thermally stimulated transition into the $^3\Sigma_g^-$ ground state of O_2 is indicated by the 160° C TSEE peak.

Euler and *Scharmann* [5.91] monitored the BeO surface by means of AES. In the 400° C TSEE maximum of BeO a phase transition was found involving the recombination of mobile O with adsorption sites at the surface. In BeO *Euler* and *Scharmann* [5.92] observed disappearance spectra (DAPS) complementary to ELS. Very slow electrons which are elastically backscattered from BeO are analyzed with lock-in techniques. All electrons which are inelastically scattered disappear from the spectrum. A disappearance peak near 3.5 eV might be correlated with excitation resonance of TSEE. *Brunsmann* and *Scharmann* [5.37] continued applying DAPS and TSEE excitation spectra and compared the results with ELS. A DAPS peak at 10.5 eV in NaF correlated with the exciton creation in NaF (Fig. 5.12). The combination of TSEE excitation spectra, DAPS and ELS seems to be promising.

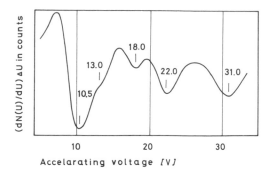

Fig. 5.12. Differential DAPS spectrum of a clean NaF surface. The scale of U has been adjusted expecting a correspondence of the first minimum with exciton absorption [5.37]

Moderate-Energy Electrons

Excitation of EE with several hundred to several thousand eV electrons generates a positive–negative space-charge structure inside of highly insulating and radiation-resistant materials. The temporal buildup of this space-charge structure has been demonstrated by a self-consistent computer simulation [5.88] comprising the calculation of surface potential shifts and secondary electron coefficients which could be confirmed experimentally. The agreement between experimental and theoretical parameters is both a necessary condition and a check of the reliability of the computer simulation. Space-charge-generated electric fields inside the insulator of the order of 10^6 V cm^{-1} appear possible. Surface potential shifts in the order of one to 10 V have been found. After the electron bombardment the space charge is frozen in and the resulting high electric fields will cause a transport of liberated electrons, holes, and ions together with polarization effects. In addition to charge depth profiles there will arise a lateral nonuniform charge resulting in patch fields. Between the first and second crossover of the secondary emission coefficient $\sigma = \delta + \eta$, the resulting total charge of the insulator is positive, outside of this range negative.

Brunsmann and *Scharmann* [5.37] attributed EE, measured in NaF, to volume traps. The crystals were excited with electrons in the keV range. Infrared spectra indicated no OH$^-$ bands. AES monitoring prior to and after each TSEE cycle revealed a clean surface. Prior to measurements the samples were heated for cleaning in the 10^{-10} Torr pressure range. After electron bombardment a strong emission appeared. The fact that no adsorbed species were detected led to the interpretation in terms of a volume trap concept. These thermally liberated electrons are assumed to stem from the high-energy tail (exceeding the electron affinity) of the Maxwellian energy distribution of electrons in the conduction band. Careful measurements of the energy distribution of emitted electrons do not contradict these assumptions. Strong support is given to the validity of the volume concept of EE in alkali halides, which has already been established by *Bohun* in the early fifties [5.93].

Kriegseis and *Scharmann* [5.69] studied ZnO. Contrary to NaF, clean ZnO surfaces, monitored by AES and cleaned by argon–ion bombardment, do not

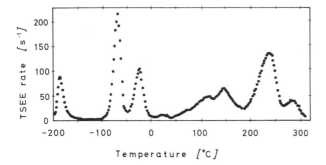

Fig. 5.13. Electron beam excited TSEE obtained from an oxidized and gas contaminated Al surface; $p \leq 10^{-8}$ Torr, heating rate $q = 0.1$ K s^{-1} [5.94]

emit after electron excitation. However, TSEE was obtained after adsorption of water and oxygen. N_2, CO, H_2 and CO_2 do not affect EE from this material. Residual gases were monitored by a quadrupole mass spectrometer. During electron bombardment, water and oxygen supposedly dissociate into the adsorbed species H, H^+, H^-, H_2^+, OH, OH^-, OH^+, O^+, O^-. After electron bombardment some of these species recombine if no activation energy is required for recombination. Upon heating the remaining species recombine at temperatures which coincide with the observed TSEE peaks. The reaction

$$O^- + H_2O + e^- \underset{E}{\rightleftharpoons} 2OH^-$$

demonstrated this in principle. The reverse reaction requires an activation energy and an electron is released. *Euler* and *Scharmann* [5.91] confirmed that dry oxidized BeO is not TSEE active. Wet oxidized BeO emits exoelectrons as long as T does not exceed 650° C and $\chi = 2.5$ eV. Thermal desorption during TSEE measurement was not observed. Excitation with 2.5 keV electrons in the temperature range from 400 to 600° C causes TSEE to disappear, indicating at thermally assisted ESD. AES measurements in the same T interval indicated chemical changes at the surface. Therefore, TSEE is attributed to adsorbed species. The high-temperature maximum of BeO is interpreted as a result of recombining O atoms. The low-temperature maxima are correlated with different but not specified emission mechanisms.

Schlenk [5.94] reported that Al_2O_3 does not exhibit EE after argon–ion bombardment. Prior to ion "cleaning" the TSEE intensities from different oxides peak at nearly the same temperature (Fig. 5.13). Adsorption of water on Al_2O_3 affects the intensity of TSEE but causes no additional maxima. A thermionic emission concept is ruled out, since the kinetic energy of the emitted electrons is incompatible with the thermal equilibrium of the lattice. The mean energy of emitted exoelectrons decreases with increasing T.

Glaefeke et al. [5.95] studied SiO_2 layers which were thermally grown in wet oxygen. Different TSEE spectra were recorded for undoped and doped SiO_2. Dopants (P, B, OH) cause stable peaks. Investigation of the layer structure indicates that, besides surface emission, a field-assisted TSEE component from the conduction band is observed. Electrons are thermally liberated in the bulk where they are subjected to accelerating electric fields (10^6 V cm^{-1}) caused by prior electron bombardment. The field-dependent escape depth enables electrons to escape from a depth of several thousand Å. The field-determining negative space charge in the bulk is mainly formed by occupied deep traps not involved in the emission process. The positive space charge near the surface is caused by trapped holes and ionized donors. Measured angular distribution of the emitted electrons corroborates this model.

Krylova [5.65] claimed that TSEE from oxides is caused by physico-chemical reactions at the surface. Al_2O_3, SiO_2, MgO, ZnO, TiO_2, ZrO_2, Cu_2O and NiO were investigated. Most oxides show a desorption maximum (TSD) which is assumed to be correlated with TSEE. Chemisorption of H_2 or H_2O on Al_2O_3 raises the TSEE intensity. The 140 to 160°C peak, typical of all oxides, grows with O_2 adsorption as confirmed in zinc, silicon, and magnesium oxide. The excitation and emission mechanism is explained as follows:

By electron bombardment (radiolysis) free radicals are produced in the water layer at the surface. Thermally stimulated recombination of radicals generates desorption and electron emission. The 140 to 160°C maximum is attributed to valence transformations of oxygen according to

$$O^{2-}_{Lattice} \rightarrow O^-_{ads} + e^- \uparrow,$$

i.e., upon excitation oxygen from lattice sites is transported to adsorption sites. During heating TSEE is produced according to

$$O^-_{ads} \rightarrow O_{ads} + e^- \uparrow, \quad \text{and TSD by} \quad 2O_{ads} \rightarrow O_2 \uparrow.$$

5.3.12 Energy and Angular Distributions

Particular attention was paid to energy and angular distribution measurements since they reflect the interaction of emitted species with the solid. The emission mechanism has been discussed in terms of the observed energy and angular distribution. In special cases kinetic energies up to 120 keV were reported [5.96]. In general, the kinetic energies of the emitted charged particles are larger than thermal energies and the measured distribution is broader than a Maxwellian distribution. Kinetic energies far in excess of thermionic energies are due to accelerating electric fields inside and outside the emitting solid. Measurements of angular distribution are rare.

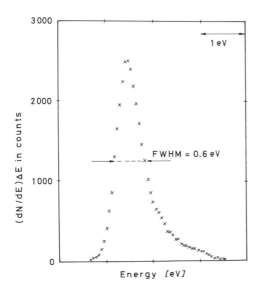

Fig.. 5.14. Energy distribution of TSEE from in situ evaporated LiF irradiated with $2 \cdot 10^{13}$ one keV electrons per cm^2 [5.43]

One expects the adsorption coverage to influence the electron energy spectra (EES). *Samuelsson* [5.43] studied LiF layers deposited in situ. Shape and full width at half maximum (FWHM) of the EES (Fig. 5.14) do not differ appreciably when recorded under high vacuum as compared to ultrahigh vacuum conditions. He concludes that the sorption layer does not measurably influence EES. In contrast to this, *Brunsmann* and *Scharmann* [5.97] found that sorption layers on K_2SO_4 strongly affect EES. In a separate paper *Brunsmann* et al. [5.98] studied different EE glow peaks obtained from LiF. They state that the EES are independent of the excitation dose and emission intensity in the range examined. The EES below 350° C may be described by a Maxwellian distribution. However, above 350° C this is no longer possible, which demonstrates that the thermionic emission concept is not generally valid in LiF.

Hayakawa and *Oda* [5.99] noted the effect on EES of charging the insulators during excitation. In LiF powder this charging of the sample disappears for $T > 500°$ C. *Brunsmann* and *Scharmann* [5.37] measured the energy distribution of electrons emitted from NaF under well-controlled conditions (residual gas analysis and AES monitoring of the surface in UHV). The observed energy spectra distinctly deviate from Maxwellian distributions. The mean energy decreases with decreasing energy of the exciting electrons and increasing temperature. The TSEE intensity does not affect the EES from NaF. Auger-like transitions causing the liberation of electrons were not evident in these energy distributions, neither was broadening as a consequence of electron–phonon interaction at elevated temperatures. The "overthermal" width of the distribution was traced back to patch fields along the surface.

Schlenk [5.100] found that the energy spectra obtained from oxide layers on aluminum can be approximated by a Maxwellian distribution, but the mean energy decreased with increasing T.

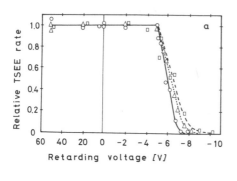

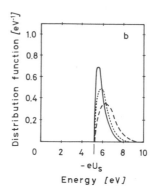

Fig. 5.15

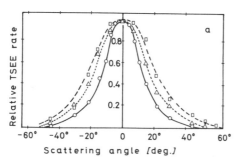

Fig. 5.16

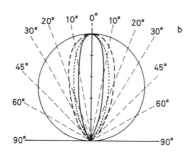

Fig.. 5.15a, b. Retarding field curves (**a**) and energy distributions of TSEE (**b**) from α-Al$_2$O$_3$ on SiO$_2$ normalized to the peak intensity N_{max}. $T_{max} = 170°$ C (—○—), 260° C (···Δ···), and 450° C (--□--) [5.40]

Fig. 5.16a, b. Normalized angular distributions of TSEE from α-Al$_2$O$_3$ on Si (**a**); in comparison (**b**) to a cosine distribution (circle); $T_{max} = 170°$ C (—○—), 260° C (···Δ···), and 450° C (--□--) [5.40]

A different interpretation of the width of energy distributions was given by *Fitting* et al. [5.101] who correlated the internal energy distribution of thermally liberated electrons inside the insulator with the measured external energy distribution of emitted electrons. These theoretical considerations show that special attention should be paid to the slope of the rise of external energy distributions. The liberated electrons inside the solid undergo reflection and refraction at the surface barrier χ. A small percentage overcomes the surface barrier. Thus, χ limits and deforms the energy distribution observed in a characteristic way. The mean energy of emitted electrons was found to increase slightly with increasing temperature. Experiments confirmed the theoretical prediction (Fig. 5.15).

Kortov [5.102] studied BeO and confirms overthermal energies. The energy even increases with decaying emission. The temperature dependence of the energy is not monotonic and is not in agreement with electron-statistical considerations in solids. In a separate paper *Kortov* and *Zolnikov* [5.39] calculated the energy and angular distribution of exoelectrons with the aid of a Monte Carlo method. Both a field-free and a field-assisted emission mechanism [5.103] is assumed. Calculated and measured curves do not contradict. Especially the calculated narrow angular distributions are of interest. They were also confirmed by measurements of *Fitting* et al. (Fig. 5.16) [5.40].

It seems that the combination of the computer-simulated emission processes with both energy- and angular-distribution measurements is most promising for future advancement in the understanding of the EE mechanism.

5.3.13 Effects of Nuclear Radiation

Various substances show a reproducible exoemission when excited by nuclear radiation. BeO was found to be an especially strong emitter. Alkali halides, alkali earth halides, sulphates, sulphides and a number of metal oxides have been investigated as well. BeO is excited to emit TSEE by α-, β- and γ-rays. Compared to γ irradiation, α irradiation produces an additional TSEE maximum, e.g., in LiF and $SrSO_4$. Fast neutrons excite the exoelectron emitter via recoil protons. Therefore, neutron detectors are always covered with recoil proton or α radiators [5.104] (e.g., polyethylene or LiF on a BeO disk). Both α and soft β irradiation yield high TSEE intensities, since the total energy is adsorbed near the surface of the TSEE emitter. Simultaneous measurement of TSL from a BeO disk of 1.5 mm thickness results in a weak TSL signal. On the other hand, γ irradiation causes very weak TSEE but a very intense TSL. Therefore, in first approximation, TSEE and TSL from the same emitter are capable of discriminating between α and γ, and soft β and γ doses, respectively.

A fast-neutron induced TSEE signal can be separated from the γ-ray TSEE by a subtraction procedure. A BeO disk is partially covered with a hydrogenous material (hydrocarbons, e.g., polyethylene) serving as recoil proton radiator. The remaining part of the surface is covered with nonhydrogenous material (fluorocarbons, e.g., teflon). The difference signal from both parts of the BeO disk is attributed to fast neutrons.

5.4 Theoretical Models

The attempts to explain exoemission from solid surfaces can be classified as follows:

I) those which are based on physical phenomena that occur in the bulk as well as on the surface ("volume concepts"), such as the thermionic emission

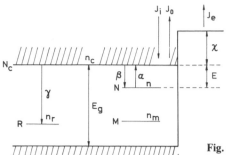

Fig. 5.17. Energy-level diagram forming the model for (5.20a–f) [5.13]

model, the field-assisted emission model and the Auger-emission or radiationless transition model, and

II) those which are based on surface-specific reactions ("surface concepts").

5.4.1 Thermionic Emission Model

The earliest attempts to interpret TSEE were based on the correlation of TSL and TSEE. Optical absorption measurements demonstrated that F centers in alkali halides can be responsible for luminescence and exoemission as well. Therefore, the reaction kinetics is described within the framework of a simplified energy band model.

After absorption of the required activation energy E, the system releases trapped electrons of the conduction band where they thermalize rapidly, assuming a Maxwellian energy distribution. These thermally released electrons may retrap, recombine with or without emitting radiation, or escape from the solid. Emission is possible if the electron affinity χ is sufficiently low and electrons with kinetic energy exceeding χ are available. The idealized model is shown in Fig. 5.17, where deep levels (M), thermally disconnected from the traps (N), are added to the conventional single trap model. Following *Kelly* [5.13], the complete set of equations describing the reaction kinetics is

$$\dot{n} + \dot{n}_c + \dot{n}_m = -\gamma n_c(n + n_c + n_m + N_e) - \dot{N}_e \qquad (5.20a)$$

$$\dot{n} = -\alpha n + \beta(N - n) \qquad (5.20b)$$

$$\dot{N}_e = J_e \frac{1}{ed_{\mathrm{eff}}}. \qquad (5.20c)$$

The probability for the transition $N \rightarrow N_c$ is $\alpha = v \exp(-E/kT)$, with $v =$ frequency factor. N, N_c, and M are the concentration of the trap levels, the density of states in the conduction band, and the concentration of deep trap levels, respectively; n, n_c, and n_m the concentration of electrons on these levels.

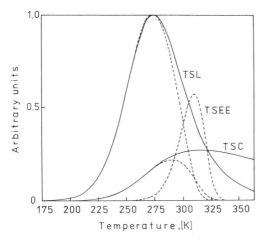

Fig. 5.18. TSL, TSC, and TSEE solutions of (5.20a–f) with $f^* = 1$ (solid lines) and $f^* = 0$ (dashed lines) for the following set of parameters: $\beta = \gamma = 10^{-14}$ cm³/K; $N_c = 10^{19}$ cm^{-3}; $M = 0$; $N = 10^{15}$ cm^{-3}; $d_{eff} = 5 \cdot 10^{-7}$ cm; $\chi = 2$ $E = 8000$ K; $m = m_0$; n_c $(T_0) = N$ [5.13]

The coefficients for retrapping and recombination are β and γ. The cross section for retrapping, S_n, is defined by $\beta = S_n \langle v \rangle$, where $\langle v \rangle$ is the mean thermal velocity of the electrons. If $n_m = - \gamma n_c n_m$ recombines radiationless, the luminescence intensity is proportional to $-(\dot{n} + \dot{n}_c)\eta$, where η is the luminescence efficiency. Conductivity and exoemission intensity are proportional to $-\dot{n}_c$. The emission-current density J_e results from

$$J_e = (1 - f^*)J_0, \tag{5.20d}$$

with $0 \leq f^* \leq 1$, and

$$J_i = f^* J_0. \tag{5.20e}$$

The latter statement takes into account that the emitter may charge by the emission process itself. A certain part (J_i) of the initially emitted current density J_0 returns. The thermionic emission current from the conduction band is

$$J_0 = n_c \left(\frac{kT}{2m}\right)^{1/2} e^{-\chi/kT}. \tag{5.20f}$$

The thermionic TSEE concept involves two steps, namely the liberation and the escape process. The emitted electrons originate within the small effective layer of thickness d_{eff} beneath the surface. A plot of computer calculated TSL, TSC, and TSEE curves is shown in Fig. 5.18. With $\chi = 0$, the peaks of TSC and TSEE coincide. $\chi > 0$ results in a shift of the TSEE peak to higher temperatures; $\chi < 0$ would shift the TSEE peak below the TSC peak. Comparison with experimental curves should, however, be carried out with caution and requires the knowledge of χ. Nevertheless, attempts to estimate a χ from TSEE measurements are reported in [5.102].

Equations (5.20a–f) can be reduced if the condition $N > 10^{15}\,\mathrm{cm}^{-3}$ is fulfilled. Then

$$\dot{n}_c = \alpha n - n_c \beta(N - n) - n_c \gamma(n + M) - \dot{N}_e, \qquad (5.21\text{a})$$

$$\dot{n} = -\alpha n + n_c \beta(N - n). \qquad (5.21\text{b})$$

The error is 1%, if

$$\frac{N_c}{N + M} \lesssim 10^4$$

is valid. In the case of negligible retrapping (5.21) reduce further:

$$\dot{n}_c = \alpha n - n_c \beta(n + M) - \dot{N}_e, \qquad (5.22\text{a})$$

$$\dot{n} = -\alpha n. \qquad (5.22\text{b})$$

In many papers observed kinetic parameters are evaluated by procedures based on

$$\dot{n}_c = \alpha n - \dot{N}_e \qquad (5.23\text{a})$$

$$\dot{n} = -\alpha n \qquad (5.23\text{b})$$

only. This assumption of direct emission from traps seems to be valid in the case of EE-active adsorption centers.

Most of the procedures used for determination of kinetic parameters from TSEE curves presume monoenergetic traps and first-order kinetics [5.105]. Only recent work [5.33, 35] considers the energy distribution of traps. The evaluation of true trap parameters is a complex problem. A prerequisite is the knowledge of the kinetic model which cannot be deduced from TSEE experiments alone.

Selective OSEE is a photothermal process. In the framework of this model an optical transition $N \to N_c$ takes place. The escape mechanism from the conduction band is supposed to be a thermionic process in accordance with (5.20f).

5.4.2 Field–Assisted Emission Model

This model has been envisaged because electric fields and charging occur in insulators after irradiation [5.3, 106] and upon heating of pyroelectrics. The special case of electron-beam excited EE in radiation-hardened insulators was discussed in [5.103]. Based on the investigation of electron penetration into solids [5.107, 108] and the drift behavior of excited secondary electrons in electric fields [5.109], the temporal and spatial buildup of the space charge $\varrho(x,t)$ and the action of the arising electric field $F(x,t)$ on the total current density $J(x,t)$ were calculated numerically by a self-consistent computer procedure [5.88]. The resulting field strength during electron bombardment is obtained from the continuity equation

$$-\frac{d}{dx}J=-\dot{\varrho}\,,\tag{5.24}$$

and the Poisson equation

$$\varepsilon_0\varepsilon_r\frac{d}{dx}F=\varrho\,,\tag{5.25}$$

where the total current density is

$$J(x)=J_{PE}+J_{SE}+J_h+J_{PF}+J_{FN}+J_G\,.\tag{5.26}$$

The subscript PE denotes primary electrons; SE secondary electrons; h holes; PF Pool-Frenkel, FN Fowler–Nordheim; G grid. Figure 5.19 shows the result of a 1 keV electron beam bombardment of a 1000 Å SiO_2 layer on silicon. The surface potential adjusted by the potential U_g of a grid in front of the sample has a considerable influence on $\varrho(x)$ and $F(x)$. The space charge is frozen in after electron bombardment ceases, as has been shown experimentally. The frozen-in state of the space charge inside the insulator is the basis of the field-assisted emission model of TSEE. The band bending, shown in Fig. 5.20, corresponds to the case $U_g=0\,V$. Electrons are liberated in the negative space charge region, and, under the influence of the internal field, escape from the solid if their energy exceeds χ. Sources of energy of the released electrons are the electric field and the thermal energy. The transport from the origin of the free electron to the surface is strongly field dependent. The surface potential barrier will affect the escape mechanism. The liberation comprises the detachment of the electron from the trap and the simultaneous relaxation of the lattice. The resulting relaxation energy may be transferred to the electron. The transport and escape is best described by a Monte Carlo process, as proposed in [5.39], which can explain intensity, energy, and angular distribution of the emitted electrons. The space-charge model is restricted to highly insulating materials. Appreciable electron, hole, and ion conductivity will prevent stable space charges.

5.4.3 Radiationless Transitions

Emission of electrons via an Auger transition has been observed during thermally stimulated release of self-trapped holes from V_k centers in NaCl:Tl

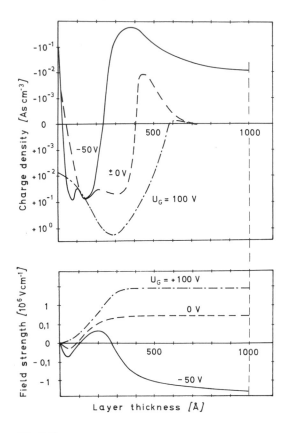

Fig. 5.19

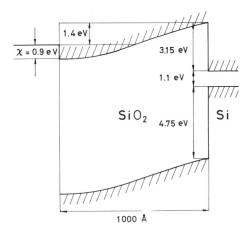

Fig. 5.20

Fig. 5.19. Calculated space charge $\varrho(x)$ and field strength $F(x)$ in the stationary final state during bombardment

Fig. 5.20. Energy-band model of the Si-SiO$_2$ system after bombardment with electrons of energy $E_0 = 1$ keV; $U_g = 0$ [5.88]

and KCl : Ag. The radiationless recombination of migrating holes and trapped electrons releases energy which is transferred to another trapped electron. If the condition $E_g \gg \chi$ is fulfilled, emission may take place from near surface regions.

5.4.4 Surface Concept

There is clear experimental evidence that adsorbed species can be the origin of EE. During the excitation process adsorbed atoms or molecules may either be ionized or dissociate and subsequently be ionized. In this case the emission process is a direct emission into the vacuum and electrons from the conduction band of the solid do not participate.

Most authors describe the kinetics by the simple set of equations (5.23) and by chemical reaction equations [5.10, 66, 69, 90], e.g.,[1]

excitation : $A_2 + e^- \rightarrow A_2^-$,

stimulation : $A_2^- + E_e \rightarrow A_2 + e^- \uparrow$.

A simplified potential curve model [5.110] for the case of adsorbed atoms or molecules is shown in Fig. 5.21. The excitation and stimulation process can be described from a uniform point of view [5.67, 89]. Thresholds, resonances for excitation, and anisotropic cross sections as well as the activation energy can be interpreted. The discrepancy between lattice temperature or "molecule temperature" and "electron temperature" of the released electrons is attributed to the relaxation of nuclei and the energy released during this process.

During excitation by electrons (e.g., Fig. 5.21b) Franck–Condon transitions take place from the maximum of the probability density distribution (PDD) of the oscillator in the ground state $A + B$ to the ionized state $A + B^-$. E_R is the resonance energy. If the hyperthermal vibrational energy dissipates into the solid via phonons, the system remains in the $A + B^-$ state, if not, the system dissociates.

Stimulation converts the system back to the ground state. The activation energy is needed to surmount the potential barrier E, the thickness of which determines the lifetime of the ionized state. The transition into the ground state is exothermal. The excess energy is transferred to the electron. Emission takes place if $E_\infty - E_{x_e} < E_{ph}^0$. The interpretation of the frequency factor $\nu = \kappa_t \cdot \bar{f}$ is given in terms of the theory of the transition state. Here $\bar{f}$ is the impact frequency of B^- against the potential wall and $(0 < \kappa_t \leq 1)$ is the coefficient for transmission. Following Zener [5.111], κ_t may vary over several orders of magnitude as a consequence of nonadiabatic transitions. This may explain the observed changes in ν. The maximal kinetic energy of the emitted electrons is $E_{ph}^0 - (E_\infty - E_{x_e})$.

---

1 A and B are the general symbols denoting atoms of different elements.

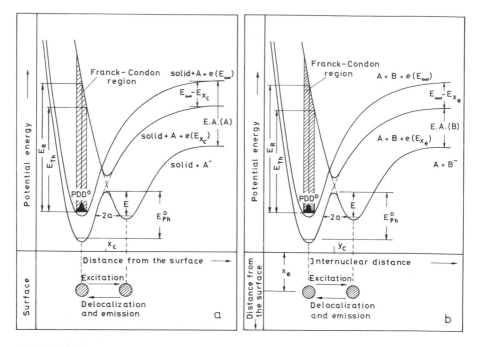

Fig. 5.21a and b. Simplified potential curve model for atoms (**a**) and diatomic molecules (**b**) adsorbed at the surface and residing within the range of the image force potential $x = E_\infty - E_c$ of the solid. The EE activity is localized in the adsorption bond (**a**) or inside the molecule (**b**) [5.89]

The potential curves in Fig. 5.21 correspond to calculated O_2 curves. The ground state of O_2 is of type $^3\Sigma_g^-$ and the state of O_2^- is the first excitation state ($^4\Sigma_u^-$). Of course, other configurations and transitions are possible. The potential curve model is also applicable to recombination processes of mobile species.

5.4.5 Energy-Band Model and Interpretation of Kinetic Parameters

The phenomenological treatment of the electron or hole kinetics in the framework of the energy-band model does not suffice for the interpretation of E_e and v_e. Localization and delocalization of electrons in the bulk and at the surface are generally accompanied by changes in the position of the nuclei (relaxation). The band scheme, a consequence of fixed nucleus coordinates, cannot account for this complex process. The solution of the complete Schrödinger equation has to be considered. The most simple case is the diatomic molecule, as demonstrated in Sect. 5.4.4. Band-structure calculations including traps and the thermal release of electrons including lattice relaxation seem to be adequate to solve the theoretical problem of TSEE from volume states.

5.5 Applications

Extended studies [5.112] of dosimetric properties of TSEE emitters, especially BeO, revealed that the dose response may range from as low as 10^{-6} rad to 10^2 rad. Ultrahigh sensitivity BeO:Li dosimeters are supposed to exhibit negligible fading at room temperature and a reproducibility of about $\pm 6\%$. A test under field conditions has not yet been made. A similar type of TSEE dosimeter tested under various conditions is BeO:Si. Figure 5.22 shows its gamma response. One attractive feature of this material is that the TSL and TSEE peaks are separated by 135 K. TSL and TSEE can be read out from the same dosimeter on separate occasions. A combined readout of TSL and TSEE has the advantage of discrimination between those types of radiation which cause bulk effects (TSL) and those causing surface effects (TSEE). TSL will be due mainly to γ radiation, whereas α, β and n radiation preferentially contribute on TSEE. In this way BeO:Si dosimeters discriminate between γ- and α-rays, γ- and n-rays, and γ- and weakly penetrating β-rays in mixed radiation fields. As shown in Fig. 5.23, the directional response of BeO:Si disk TSEE dosimeters is advantageous. Unfortunately tests under field conditions showed that their response depends strongly on humidity. Mechanical treatment was also found to influence the TSEE signal considerably. These shortcomings have not yet been overcome and have caused a certain pessimism with regard to application of TSEE dosimeters on a commercial scale in the near future. Nevertheless, the use of TSEE dosimeters under well-defined conditions seems to be attractive even today.

Direct and indirect imaging of exoelectron emitting surfaces has been brought about with the aid of microchannel plates (MCP) or scanning light beams. Direct observation of TSEE from LiF and $LiNbO_3$ [5.113] was possible using the principle of the thermionic emission microscope. The cathode has to be replaced by the TSEE emitter confronted by a MCP. Figure 5.24 shows a simplified scheme, omitting the immersion lens required for magnification. Results confirm that TSEE arises from certain spots at the surface. The domain structure of the ferroelectric $LiNbO_3$ could be made visible. The performance of an exoelectron microscope is characterized by three parameters. They are the resolving power of the electron optics, the resolution of the MCP and the brightness. The resolving power of the electron optics is

$$R_P = k^* E_0 / F_0, \tag{5.27}$$

k^* being a numerical factor somewhat smaller than unity. E_0 is the mean energy of exoelectrons and F_0 the field strength of the objective. With $E_0 = 0.3$ eV and $F_0 = 30$ kV cm^{-1}, δ is of the order of 1000 Å. With a total magnification of 2000 the image spots are about 20 μm in diameter, which can be resolved by the MCP. The TSEE microscope permits considerable reduction in sample temperature and the elimination of cesium activation usually applied in thermionic

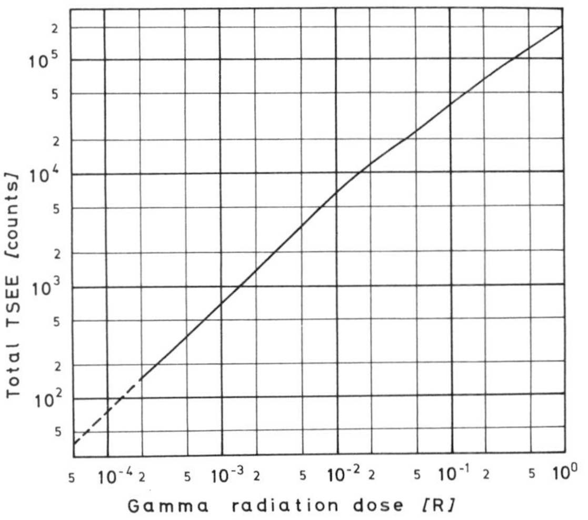

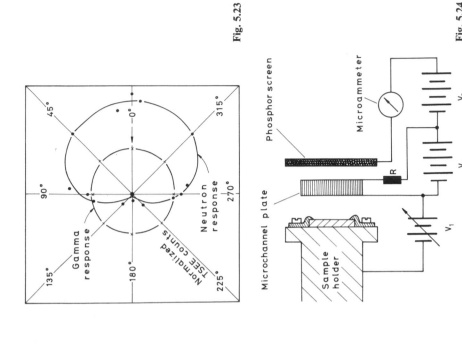

Fig. 5.23

Fig. 5.24

Fig. 5.22. Total number of TSEE counts in the main peak of BeO:Si emitters as a function of the gamma radiation dose [5.8]

Fig. 5.23. Directional response of ceramic BeO disk TSEE dosimeters to 14 MeV neutrons and gamma radiation, using thick polyethylene and teflon radiation covers [5.112]

Fig. 5.24. Schematic experimental arrangement for exo-electron proximity imaging [5.113]

microscopy, and further, the extension of its application to semiconductors and insulators.

OSEE microscopes were constructed by *Veerman* [5.114], and *Baxter* [5.115] and have been applied to the detection of fatigue damage in metals. The metal surface is scanned by a small spot (15 μm) of uv light. Simultaneously, OSEE is recorded using a scanning light spot. Fatigue deformation in aluminum and steel produces OSEE after less than 1 % of fatigue life and failure finally occurs in the region of maximum emission. The regions of maximum emission have been attributed to the development and propagation of fatigue cracks.

Other applications of imaging have been suggested. *Krylova* [5.10] points out the possibility to characterize the "energy relief" of the surface, which is responsible for different physico-chemical reactions, especially heterogeneous catalysis.

Kramer [5.116] proposes photography on the basis of EE from oxidized Al. The latent "image" is "inscribed" by high-energy radiation and read out by optical scanning and monitoring OSEE. *Schmidt* et al. [5.117] tried to develop a procedure enabling the determination of ion radiation-induced structure damage and ion implantation profiles within thin insulating layers without destroying them. TSEE is recorded as a function of energy of the exciting electrons. The validity of the space-charge model (Sect. 5.4) is presumed.

Robock [5.118] used TSEE in biochemical studies of toxic SiO_2, since silicosis is assumed to be caused by an electron transfer reaction between toxic SiO_2 dust particles and the cell membranes of the lungs.

In addition, first- and second-order phase changes, grinding and wear, and other mechanical processes have been characterized by EE [5.9].

5.6 Conclusion

Solutions to many problems concerned with exoemission and its mechanisms are not available yet. One problem is the creation of TSEE-active centers. Promising attempts in solving it are the investigations of the well-defined creation spectra of alkali halides using excitation by ultraviolet light. Analogous phenomena were observed in TSEE spectra, induced by electron bombardment, which depend in a reproducible manner on the energy of exciting electrons. The mechanism of creation of defects which serve as electron traps and are the origin of exoelectrons in materials other than alkali halides is not well understood. Frequently, a lack of retrapping is reported in spite of high frequency factors. This suggests that the capture process in EE-active centers requires energy. In terms of the potential curve model this energy would correspond to the energy required for a Franck–Condon transition. From this point of view, the discrimination between volume and surface states by varying the energy of the incident uv quanta or electron becomes difficult, at least when dealing with alkali halides, which are believed to show volume as well as surface emission.

The identification of the process by which exoelectrons are released from traps is a further problem. Activation energy and frequency factor cannot be interpreted in the framework of a simplified energy-band model. This energy scheme may serve as a basis of kinetic considerations, provided the model is adequate. It is likely that the TSEE activation energy is due to a thermally activated annealing of lattice defects generated by the excitation process in analogy to the mechanism proposed by the potential curve model of TSEE-active adsorbate reactions. The difference between the energy state of an "activated complex" (vacancy + interstitial + electron) and the ground state of the annealed center (recombined vacancy + interstitial) might supply the electrons with the commonly observed "overthermal" energies.

Another problem is concerned with the kind of process enabling electrons to overcome the surface potential barrier χ. In the framework of volume concepts the treatment of electron–phonon interaction by Monte Carlo methods including phonon creation and annihilation seems to be most adequate for exoemission since realistic energy and angular distributions may be obtained. This procedure may explain the role of χ as well as space charges.

A further problem is the generation of space charges at the surface and in the bulk of insulating materials. Especially excitation with electrons will cause complications, unless space charges are generated in a controlled way enabling volume emission or emission from surface regions. Energy *and* angular distribution measurements will be a helpful guide in deciding whether thermal, overthermal or field-assisted electrons are emitted, thus providing information on the emission process.

A still aggravating problem is a deficiency of rigorous theoretical treatments of exoelectron emission, both quantum mechanically and thermodynamically. Adequate kinetic considerations are difficult since a prerequisite for knowledge of the reaction kinetics is the knowledge of the involved kinetic parameters. As a consequence, accurate trap level spectroscopy with the aid of TSEE is very complex and difficult as in the case of TSC, TSL, etc. TSEE glow curves are usually recorded in the temperature range between liquid nitrogen temperature and several hundred degrees Celsius. Recent work in Poland extended this range down to 2.5 K, offering promising tests of TSEE concepts at low T.

Applications of EE phenomena have not yet become widespread on a commercial scale. But specialized applications seem to be of great practical interest as is evident from investigations concerned with dosimetry, material fatigue, toxicity in biology, heterogeneous catalysis in chemistry, material contamination and radiation damage, etc. Thus, further fundamental work is required; its progress is undoubtedly connected with common progress in the fields of experimental and theoretical physics related to EE phenomena.

List of Symbols and Notations

$2a$	potential wall thickness [Å]	E_{xc}	value of image force potential energy at x_c [eV]
A	frequency factor (desorption) [s^{-1}]; atom A	E_{∞}	value of image force potential energy at infinity [eV]
B	atom B	E_{ph}^0	energy separation between transition state and zero oscillator of a molecule [eV]
c_A	concentration of atoms A [cm^{-3}]		
c_B	concentration of atoms B [cm^{-3}]		
c_{AB}	concentration of compound AB [cm^{-3}]	E.A.(B)	electron affinity of atom B [eV]
C^*	relative concentration of excited TSEE-active centers; $0 \leq C^* \leq 1$; $C^*(T) = n_e(T)/n_e(T_0)$, without dimension	f^*	proportionality factor in $J_i = f^* J_0$, without dimension
		f_{exc}	external excitation function
		f_{st}	external stimulation function
d_{eff}	effective thickness of emission layer [Å]	F	electrical field strength [V cm^{-1}]
		h	hole
D	diffusion constant [cm^2 s^{-1}]		Plancks constant
e	elementary charge [As]	J	current density (total) [A cm^{-2}]
e^-	electron	J_{PE}	current density of primary electrons (PE) [A cm^{-2}]
e^0	exciton		
e^+	hole	J_{SE}	current density of secondary electrons (SE) [A cm^{-2}]
$e(E_{\infty})$	free electron		
$e(E_{xc})$	electron in the image force potential of the solid at x_c	J_h	current density of holes [A cm^{-2}]
$e(E_{xe})$	electron in the image force potential of the solid at x_e	J_{PF}	current density caused by Pool-Frenkel effect (vb of Si $\rightarrow cb$ of SiO$_2$) [A cm^{-2}]
E	activation energy [eV]		
E_r	activation energy for relaxation [eV]	J_{FN}	current density caused by Fowler-Nordheim tunneling [A cm^{-2}]
E_e	activation energy for TSEE [eV]		
E_{opt}	optical activation energy for EE [eV]	J_G	current density caused by the potential difference occuring between grid and insulating sample along electron bombardment resulting EE current density [A cm^{-2}], $J_e = J_0 - J_i$
E_{therm}	thermal activation energy for EE [eV]		
E_d	energy needed for defect creation [eV]		
E_p	primary energy of exciting electrons [eV]	J_e	
E_{exc}	excitation energy [eV]	J_0	EE current density "out" [A cm^{-2}]
$\bar{f}$	impact frequency of an atom in a molecule pushing against the potential walls [s^{-1}]	J_i	EE current density "in" [A cm^{-2}]
E_0	initial energy of electrons (electron gun) [eV]	k	Boltzmann's constant; proportionality factor in $\delta = kE_0/F_0$ [V eV^{-1}];
E_c	conduction band edge [eV]	K	reaction velocity [cm^3 s^{-1}]
E_v	valence band edge [eV]	m	electron mass
E_g	energy separation (gap) between E_c and E_v [eV]	M	deep trap thermally disconnected from N; concentration of these traps [cm^{-3}]
E_R	resonance energy for excitation of EE-centres [eV]	n	concentration of electrons in the traps N [cm^{-3}]; order of reaction, without dimension
E_{Th}	threshold energy for excitation of EE-centres [eV]		
E_{xe}	value of image force potential energy at x_e [eV]	n_c	concentration of electrons in N_c [cm^{-3}]

n_r — concentration of electrons in R [cm^{-3}]

n_m — concentration of electrons in M [cm^{-3}]

n_e — concentration of excited TSEE-active centres

n_A — number of species A, without dimension

$\dot{n}, \dot{n}_c, \dot{n}_m$ — temperature derivatives [K^{-1} cm^{-3}] [e.g. $n = dn/dT = (1/q)\cdot dn/dT$, with $q = 1$ [K s^{-1}]]

N — flat trap; concentration of these traps [cm^{-3}]

N_c — concentration of states at E_c [cm^{-3}]

N_e — rate of emitted exoelectrons [cm^{-3} s^{-1}]

N_{ec} — rate of counted exoelectrons [cm^{-3} s^{-1}]

p^*, p_i^* — perturbation

q — heating rate [K s^{-1}]

Q^* — thermal activation energy for diffusion [eV]; supplied amount of heat (energy) to activate chemical reactions

Q_{rec} — reaction heat

R — recombination centre; concentration of these centres; universal gas constant; resistance

t — time [s]

T — temperature

T_d — temperature at which defects are created [°C]

U_g — potential of a grid [V]

U_s — surface potential [V]

$\langle v \rangle$ — mean thermal velocity of electrons in the cb [cm s^{-1}]

x — space coordinate

x_e — spatial equilibrium separation between the adsorbed molecule and the solid [Å]

x_c — nuclear coordinate of the transition (transition state) [Å]

y — space coordinate

y_c — nuclear coordinate of the molecule transition (transition state) [Å]

α — probability per time for the transition $N \to N_c$ [s^{-1}]

β — coefficient for retrapping [cm^3 s^{-1}]

γ — coefficient for recombination [cm^3 s^{-1}]

δ — fraction of true secondary electrons (SE), without dimension; resolving power of an EE microscope [Å]

ε — relative strain [%]

ε_0 — permittivity of free space

ε_r — relative dielectric constant

η — fraction of scattered primary electrons (PE), without dimension; luminescence effectivity

ϑ — angle [deg]

θ — degree of coverage

$\varkappa_t$ — coefficient for transmission, without dimension

λ — wavelength

ν — frequency [s^{-1}]; frequency factor [s^{-1}]

ν_e — frequency factor for EE-active centres [s^{-1}]

$^{2S+1}\Pi_{g,u}^{-,+}$ — electronic state of a diatomic molecule ($\Lambda = 1$); superscript on the left: multiplicity $2S+1$; superscript on the right: − denotes change of sign of wave function on reflection of the electron coordinates in any plane through the nuclei of the molecule, + denotes that wave function does not change sign: subscript on the right: g–"gerade", u – "ungerade"

ϱ — resistivity [Ω cm]; space charge [As cm^{-3}]

$\dot{\varrho}$ — time derivative of space charge: $d\varrho/dt$

σ — total of δ and η: $\sigma = \delta + \eta$, without dimension

σ_β — cross section for retrapping [cm^2]

$^{2S+1}\Sigma_{g,u}^{-,+}$ — electronic state of a diatomic molecule ($\Lambda = 0$)

τ_r, τ — relaxation time [s]

τ_e — relaxation time of TSEE-active perturbations [s]

τ_L — relaxation time of TSL-active perturbations [s]

τ_C — relaxation time of TSC-active perturbations [s]

τ_{exc} — life time of excitons e^0 [s]

τ_{vib} — a period of the vibrations of a lattice ion [s]

φ — Fermi level [eV]; angle

ϕ — work function

χ — electron affinity [eV]

References

5.1   J.C.McLennan: Philos. Mag. **3**, (6) 195 (1902)
5.2   W.B.Lewis, W.E.Burcham: Proc. Cambridge Philos. Soc. **32**, 503 (1936)
5.3   M.Tanaka: Proc. Phys.-Math. Soc. Jpn. **22**, 899 (1940)
5.4   J.Kramer: Z. Phys. **125**, 739 (1949)
5.5   J.Kramer: *Der metallische Zustand*, (Vanderhoeck und Ruprecht, Göttingen 1950)
5.6   K.Lintner, E.Schmid: Acta Phys. Austr. **10**, 313–480 (1957)
5.7   A.Scharmann: "Elektronennachemission", in *Festkörperprobleme*, Vol. **6** (Akad.-Verl., Berlin 1967) pp. 106–126
5.8   K.Becker: CRC Crit. Rev. Sol. State Sci. **3**, 74 (1972)
5.9   J.A.Ramsey: "Exoelectric Emission", in *Progress in Surface and Membrane Science*, Vol. **11** (Academic Press, New York 1976) pp. 117–180
5.10   I.V.Krylova, Usp. Biol. Khimii **45**, 2138 (1976) [English transl.: Russian Chem. Rev. **45**, 1101 (1976)]
5.11   R.Seidl: Czech. J. Phys. B**10**, 931 (1960)
5.12   M.Kh.Shorshorov, D.A.Zhebynev, V.P.Alekhin: Proc. 5th Intern. Symp. on EE and Dosimetry, ed. by A. Bohun (Inst. Solid State Physics, Acad. Sci., Prague 1976) and A.Scharmann (1. Phys. Inst. Justus-Liebig-Univ., Gießen, 1976) pp. 229–246
5.13   P.Kelly: Phys. Rev. B**5**, 749 (1972)
5.14   T.F.Gesell, E.T.Arakawa, T.A.Callcott: Surf. Sci. **20**, 174 (1970)
5.15   E.Linke, C.Grabo: In Ref. 5.12, p. 255
5.16   A.Gieroszynski, B.Sujak: Acta Phys. Polon. **28**, 311 (1965)
5.17   D.R.Arnott, J.A.Ramsey: Surf. Sci. **28**, 1 (1971)
5.18   V.S.Kortov, Yu.D.Semko: Phys. Status Solidi (a) **11**, K35 (1972)
5.19   J.Drenckhan, H.Gross, H.Glaefeke: Phys. Status Solidi (a) **2**, K79 (1970)
5.20   V.Bichevin, H.Käämbre, Ch. Lushchik: Phys. Status Solidi (a) **5**, 525 (1971)
5.21   H.Gross, H.Glaefeke: Phys. Status Solidi (a) **1**, K61 (1970)
5.22   G.Holzapfel, R. Nink: PTB Mitt. **4**, 207 (1973)
5.23   R.Nink, G.Holzapfel: J. Phys. (Paris) **34**, C-9-491 (1973)
5.24   M.Pirog, B.Stepniowski, B.Sujak: Acta Phys. Polon. **26**, 3 (1964)
5.25   K.Becker: Proc. 4th Intern. Symp. on EE and Dosimetry, ed. by A.Bohun (Inst. Solid State Physics, Acad. Sci., Prague, 1974) pp. 218–235
5.26   A.Bohun: PTB Mitt. **80**, 318 (1970)
5.27   M.Balarin, A.Zetzsche: Phys. Status Solidi **2**, 1670 (1962)
5.28   G.Holzapfel: Z. Angew. Phys. **29**, 107 (1970)
5.29   G.Holzapfel, M.Krystek, L.Wolber: Le Vide, les Couches Minces **30**A, 107 (1975)
5.30   H.Gobrecht, H.Nelkowski, D.Hofmann, J.Pachaly: Z. Angew. Phys. **26**, 209 (1969)
5.31   H.Gobrecht, D.Hofmann: J. Phys. Chem. Sol. **27**, 509 (1966)
5.32   P. Bräunlich: J. Appl. Phys. **38**, 2516 (1967)
5.33   J.Becherer, G.Rudlof, H.Glaefeke: Proc. 8th Conf. "Physik der Halbleiteroberfläche" ed. by H.Flietner (ZIE, Acad. Sci. GDR, Berlin 1977) pp. 157–163
5.34   J.A.Tale: v *Kibernetisaziya Nauchnogo Eksperimenta* (Uchenye Zapiski Latviiskogo Gosudarstvennogo Universiteta im. Petra Stuchki, Tom 170, Vijpusk 4, Riga 1972)
5.35   G.Holzapfel, M.Krystek: Phys. Status Solidi (a) **37**, 303 (1976)
5.36   R.Chen: J. Mater. Sci. **11**, 1521 (1976)
5.37   U.Brunsmann, A.Scharmann: Phys. Status Solidi (a) **43**, 519 (1977)
5.38   V.S.Kortov, R.I.Mints, I.E.Myasnikov, Yu.A.Shevchenko: Phys. Status Solidi (a) **2**, 55 (1970)
5.39   V.S.Kortov, P.P.Zolnikov: Phys. Status Solidi (a) **31**, 331 (1975)
5.40   H.-J.Fitting, H.Glaefeke, W.Wild, J.Lange: Phys. Status Solidi (a) **42**, K75 (1977)
5.41   H.-J.Fitting: Exp. Techn. Phys. **21**, 495 (1973)
5.42   U.Brunsmann, A.Scharmann: Phys. Status Solidi (a) **26**, K123 (1974)
5.43   L.I.Samuelsson: Phys. Status Solidi (a) **32**, K155 (1975)

5.44   V. Bichevin, H. Käämbre: Proc. 4th Czech. Conf. on Electronics and Vacuum Physics, Prague 1968, Text of contributed papers, p. 384

5.45   M. Kawanishi, R. Kikuchi, J. Okuma, T. Ikela: In Ref. 5.25, pp. 142–144

5.46   L. H. Ford, G. Holzapfel, W. Kaul: Z. Angew. Phys. **30**, 259 (1970)

5.47   W. T. Pimpley, E. E. Francis: J. Appl. Phys. **32**, 9 (1961)

5.48   C. Simoi, I. Hrianca, P. Gracium: Phys. Status Solidi **29**, 761 (1961)

5.49   A. Bohun: In Ref. 5.26, p. 320

5.50   R. I. Mints, V. P. Melekin, V. M. Segal, I. Yu. Icolev: Sov. Phys. Solid State **14**, 2153 (1973)

5.51   V. S. Kortov, R. I. Mints, V. G. Teplov: Phys. Status Solidi (a) **7**, K89 (1971)

5.52   R. F. Tinder: J. Appl. Phys. **39**, 335 (1968)

5.53   M. B. Partenskii: Phys. Metals Metallogr. **32**, 59 (1972)

5.54   L. A. Andreev, Ya. Palige: Sov. Phys.-Dokl. **8**, 1003 (1964)

5.55   T. J. Lewis: Proc. Phys. Soc. **67**B, 187 (1954)

5.56   G. E. Rhead: Surf. Sci. **68**, 20 (1977)

5.57   O. Buck, W. J. Pardee, F. J. Szalkowski, D. O. Thompson: Appl. Phys. **12**, 301 (1977) W. J. Pardee, O. Buck: Appl. Phys. **14**, 367 (1977)

5.58   D. Born, E. Linke: In Ref. 5.12, p. 265

5.59   I. V. Krylova, V. I. Svitov, N. I. Konyushkina: Zh. Fiz. Khimii **50**, 933 (1976) [English transl.: J. Phys. Chem. **50**, 555 (1976)

5.60   A. Bohun: Czech. J. Phys. **11**, 819 (1961)

5.61   Yu. P. Sitonite, F. S. Zimin, I. V. Krylova: Russ. J. Phys. Chem. **44**, 1023 (1970)

5.62   N. Jakowski, H. Glaefeke, W. Wild: Wiss. Z. Univ. Rostock, Mathem.-Naturwiss. Reihe **24**, 659 (1975)

5.63   B. Rosenblum, P. Bräunlich, C. Himmel: J. Appl. Phys. **48**, 5262 (1977)

5.64   H. Glaefeke, N. Jakowski, W. Wild, H.-J. Fitting: In Ref. 5.12, pp. 35–39

5.65   I. V. Krylova: Phys. Status Solidi (a) **7**, 359 (1971)

5.66   M. Euler, W. Kriegseis, A. Scharmann: Phys. Status Solidi (a) **15**, 431 (1971)

5.67   N. Jakowski, H. Glaefeke, W. Wild, H.-J. Fitting: Wiss. Z. Univ. Rostock, Mathem.-Naturwiss. Reihe **25**, 587 (1976)

5.68   J. P. Hiernaut, R. P. Forier, J. Van Cakenberghe: Vacuum **22**, 471 (1972)

5.69   W. Kriegseis, A. Scharmann: Phys. Status Solidi (a) **29**, 407 (1975)

5.70   E. Huster: Naturwissenschaften **64**, 448 (1977)

5.71   B. Sujak, T. Gorecki: Wiad. Chem. **6**, 361 (1973)

5.72   B. Sujak, L. Biernacki, T. Gorecki: Acta Phys. Polon. **35**, 679 (1969)

5.73   W. Hanle, A. Scharmann, G. Seibert: Z. Phys. **171**, 497 (1963)

5.74   B. Rosenblum, P. Bräunlich, J. P. Carrico: In Ref. 5.25, pp. 100–101

5.75   W. A. Syslo: In Ref. 5.12, pp. 207–211

5.76   U. Brunsmann, A. Scharmann, U. Wiessler: In Ref. 5.12, pp. 283–287

5.77   B. Rosenblum, J. P. Carrico, P. Bräunlich, L. Himmel: J. Phys. E **10**, 1056 (1977)

5.78   W. D. von Voss, F. R. Brotzen: J. Appl. Phys. **30**, 1639 (1959)

5.79   R. N. Claytor, F. R. Brotzen: J. Appl. Phys. **36**, 3549 (1965)

5.80   A. Maiste, B. Sorkin, M. Elango, H. Käämbre: Phys. Status Solidi (a) **20**, K83 (1973)

5.81   A. A. Alybakov, V. A. Gubanova, S. Halmursajev: In Ref. 5.12, pp. 153–156

5.82   G. Holzapfel, M. Krystek: J. Phys. (Paris) **37**, C-7-238 (1976)

5.83   A. Tomita, N. Hirai, K. Tutsumi: In Ref. 5.12, pp. 157–160

5.84   W. Maenhout-Van der Vorst: In Ref. 5.12, pp. 175–178

5.85   V. Bichevin, H. Käämbre: Phys. Status Solidi (a) **4**, K235 (1971)

5.86   P. D. Townsend: J. Phys. C **9**, 1871 (1976)

5.87   B. Carriere, B. Lang: Surf. Sci. **64**, 209 (1977)

5.88   H.-J. Fitting, H. Glaefeke, W. Wild, M. Franke, W. Müller: Exp. Techn. Phys. **27** (1), 13–24 (1979)

5.89   N. Jakowski, H. Glaefeke: Thin Solid Films **36**, 195 (1976)

5.90   M. Euler, A. Scharmann: Z. Phys. B **25**, 313 (1976)

5.91   M. Euler, A. Scharmann: Phys. Status Solidi (a) **34**, 297 (1976)

5.92   M. Euler, A. Scharmann: In Ref. 5.12, pp. 184–189

5.93   A. Bohun: Phys. Status Solidi **3**, 779 (1963)
5.94   W. Schlenk: Phys. Status Solidi (a) **33**, 217 (1976)
5.95   H. Glaefeke, N. Jakowski, W. Wild, H.-J. Fitting: Wiss. Z. Univ. Rostock, Mathem.-Naturwiss. Reihe **25**, 535 (1976)
5.96   E. Linke, J. Wollbrandt: In Ref. 5.25, pp. 75–78
5.97   U. Brunsmann, A. Scharmann: Phys. Status Solidi (a) **15**, 525 (1973)
5.98   U. Brunsmann, R. Huzimura, A. Scharmann: Phys. Status Solidi (a) **26**, K149 (1974)
5.99   Y. Hayakawa, N. Oda: Phys. Status Solidi (a) **29**, K117 (1975)
5.100 W. Schlenk: Phys. Status Solidi (a) **29**, K151 (1975)
5.101 H.-J.-Fitting, H. Glaefeke, W. Wild: Surf. Sci. **75**, 267 (1978)
5.102 V. S. Kortov: Phys. Status Solidi (a) **19**, 59 (1973)
5.103 J. Drenckhan, H. Gross, H. Glaefeke: Phys. Status Solidi (a) **2**, K51 (1970)
5.104 K. Becker, M. Abd-El Razek: Nucl. Instrum. Methods **113**, 611 (1973)
5.105 M. Krystek: In Ref. 5.12, p. 179
5.106 B. Sujak: Acta Phys. Polon. **20**, 969 (1961)
5.107 H.-J. Fitting, H. Glaefeke, W. Wild, G. Neumann: J. Phys. D**9**, 2499 (1976)
5.108 H.-J. Fitting, H. Glaefeke, W. Wild: Phys. Status Solidi (a) **43**, 185 (1977)
5.109 H.-J. Fitting, H. Glaefeke, W. Wild, R. Ulbricht: Exp. Techn. Phys. **24**, 447 (1976)
5.110 N. Jakowski, H. Glaefeke, W. Wild: Proc. 4th Conf. "Physik der Halbleiteroberfläche" ed. by H. Flietner (ZIE, Acad. Sci. GDR Berlin 1973) pp. 109–116
5.111 C. Zener: Proc. R. Soc. London A **137**, 696 (1932)
5.112 R. B. Gammage: In Ref. 5.12, pp. 107–121
5.113 P. Bräunlich: In Ref. 5.25, pp. 30–54
5.114 C. Chr. Veerman: Mater. Sci. Eng. **4**, 329 (1669)
5.115 W. J. Baxter: J. Appl. Phys. **44**, 608 (1973)
5.116 J. Kramer: In Ref. 5.26, p. 343
5.117 M. Schmidt, H. Glaefeke, J. Drenckhan, W. Wild: Radiat. Eff. **17**, 185 (1973)
5.118 U. Robock, U. Teichert: In Ref. 5.26, p. 353

6. Application of Thermally Stimulated Luminescence

L. A. DeWerd

With 7 Figures

Luminescence has been studied for many decades, but the development of applications for thermally stimulated luminescence (TSL) has spanned only the past 20 years. The major applications have been principally in the areas of radiation dosimetry and archaeological dating [6.1]. The most widespread application is in the detection of ionizing radiation, for example, in safety monitoring of X-ray machines in health care facilities and in monitoring around nuclear power plants. In addition, since natural background radiation always has been present, the same phenomena can be used to establish the ages of certain archaeological artifacts, especially pottery, and the ages of some geological formations. Thermoluminescence (TL)[1] techniques have also been applied to the detection and measurement of ultraviolet radiation [6.2, 3]. The purpose of this chapter is to explore some of the applications of thermally stimulated luminescence phenomena in conjunction with appropriate background material.

The applications of luminescence have increased greatly as reflected by a large number of publications. Five international conferences [6.4–8] on luminescence dosimetry and related applications have been held since 1965. In the following sections, instead of an exhaustive list of references, some reviews or most pertinent references are cited. These can be used to expand to other references that are more detailed for specific applications.

6.1 General Applications

Generally, luminescent phenomena can be divided into two general groupings through the time relation of the stimulation and resulting light emission; the light emission may occur only during stimulation – fluorescence – or it may continue for some time after the excitation has ceased – phosphorescence. Each of the various classes of luminescence has found an application in our modern world; some applications have affected our lives significantly.

---

1 Thermally stimulated luminescence (TSL) and thermoluminescence (TL) designate the same process. Those involved in the application of this phenomenon commonly use the term thermoluminescence (TL). Although either terminology could be used, this chapter will use the term commonly accepted in the application of these phenomena, i.e., thermoluminescence (TL).

6.1.1 Applications of Fluorescence and Phosphorescence

The most familiar application of fluorescence is the fluorescent lamp which produces light without a significant rise in temperature. Electrodes at each end of the glass tube of the lamp vaporize a drop of mercury by electrical discharge. This excitation results in the emission of ultraviolet light via transitions to a lower energy level, which stimulates the coating material on the inside of the glass tube. This coating in turn emits visible light. The emitted light results from the absorption of energy and not necessarily from the temperature of the material. Various materials or phosphors can be used for the lamp coatings resulting in light with selected colors, e.g., cadmium borate for pink light, zinc silicate for green, calcium tungstate for blue. Mixtures are used to produce white light. The fluorescent light is an example of "prompt luminescence", where an observer would report the stimulus and emission as starting and stopping together. More fundamental considerations indicate that the stimulus and emission are separated by less than 10^{-8} s with the process explained by a configuration coordinate diagram.

Another common application of fluorescence is the television picture tube. In this case, the phosphor is excited with energetic electrons aimed and controlled electronically in the tube. Color televisions use many small dots of three different phosphors, which may be excited individually to produce red, green or blue luminescence. Our eyes and brain work together to interpret combinations of these three as the variations in colors composing normal scenes. The amount of brightness or intensity is controlled by varying the excitation.

A comparison of these two examples shows that diverse sources of excitation, from photons to electrons, can be used; the necessary ingredient is the absorption of energy by the luminescing material. This energy is then partially converted and emitted optically in wavelengths which are characteristic of the composition of the phosphor and not of the exciting source.

Another recent application of fluorescence is growing in the area of forensics, i.e., crime detection [6.9]. This application is based upon the premise that a transfer of material through contact of two objects has occurred at the scene of a crime. For example, the tires of the "getaway car" or that of the hit-and-run driver left material on the pavement. This happens by driving on the parking lot and does not require the "squealing of tires." Recent tire prints, caused by the transfer of tire rubber extender oil left on hard surfaces, can be observed by the fluorescence resulting from ultraviolet light. These fluorescent tire prints provide evidence of tire patterns, weight distribution of the wheels and chemical composition of the tires. Problems with this technique are the possibility of the many overlapping tire tracks in parking lots and the presence of bitumen or tar which are strongly fluorescent and inhibit print formation.

Probably the most widespread application of phosphorescence is in luminescent watch dials or "glow in the dark" toys. Again a phosphor is excited by near ultraviolet and/or visible light. However, in this case, the emitted light continues after the exciting source is removed. Thus, the watch dial continues to glow in the dark while the phosphorescence decays.

6.1.2 Thermally Stimulated Luminescent Phenomena

While fluorescence and phosphorescence are of general interest and have many applications, the luminescent materials of most interest in this chapter are those in which the luminescent capability is stored after the primary excitation. This stored luminescence is not released until a separate, secondary stimulation is administered at a later time. Two phenomena having this characteristic are thermoluminescence (TL) and photoluminescence (PL). The primary excitation of interest is usually X-rays or other ionizing radiation, including ultraviolet light. For TL, this stored luminescent information is released by the subsequent application of heat. TL can be considered to be a "frozen-in" phosphorescence; the heat serves only as the required secondary stimulation to release the stored luminescence. Similarly, for PL, the secondary stimulation is provided by light, usually ultraviolet. In both cases, no emission would occur if the primary radiation had not occurred. If the primary stimulation is by ionizing radiation, the term radiophotoluminescence (RPL) is used. The equivalent term for TL would be radiothermoluminescence (RTL); however, the term TL is in general use for this process. The luminescent process that has found the widest application is TL.

Phenomena physically related to the thermally stimulated process have been explored for applications similar to those for TL. These phenomena are related to the thermoluminescence processes in that the primary absorbed energy is stored as mentioned above. The difference arises in the manner in which the stored signal subsequently is obtained after heating. The two phenomena that are of the greatest interest are thermally stimulated exoelectron emission (TSEE) and thermally stimulated currents (TSC). In both of these cases the stored signal is obtained by means of secondary heat stimulation; however, the signal is obtained by electrical current measurement rather than light detection. For TSEE, the thermally stimulated signal results in electrons emitted from the surface of the solid being collected. *Becker* [6.10, 11] has given a review of TSEE and its applications. Many papers have been written in attempts to relate TL and TSEE signals quantitatively, but with mixed success [6.12, 13]. The TSEE phenomenon is greatly dependent on the surface of the solid as well as the many different impurities that can be within the surface of the solid. TSC, on the other hand, is a property of the bulk of the material and thus does not suffer from this problem. For TSC, the thermally stimulated signal results in a current flowing in the material. The magnitude of the current is proportional to the stored signal. Recently efforts have been reported [6.14–16] to demonstrate the use of TSC for radiation dosimetry.

6.2 Preparation and Properties
of Thermally Stimulated Luminescent Materials

The luminescent properties of any solid depend on its origin, detailed microscopic structure, and treatment. Materials exhibiting luminescence include rocks
and minerals, artificial single crystals and polycrystalline solids; these may be
used in bulk or crushed into powder. Whether single crystals, polycrystals or
powder, a phosphor gives a characteristic luminescence. Glasses, which do not
have the long-range orderly structure, also luminesce; most RPL dosimeters
are polished glass blocks.

All of the processes used for radiation dosimetry (TL, TSC, TSEE, etc.)
depend strongly on the type and concentration of impurities in the material.
Different impurities are involved in the storage of the absorbed energy and the
emission process. This has resulted in the proposal of models to explain these
processes. This section is meant only to be a brief overview of the salient
features involved in the particular models and defects for the thermoluminescent processes of interest for dosimetry.

Artifically grown single crystals give more uniform luminescent properties
than do natural materials, since their impurity levels can be closely controlled,
and this is usually essential for reproducible results. Afterwards, grinding can
be performed if a powder is desired. Several common methods exist for single-
crystal growth; however, the Czochralski growth technique is the most
frequently used for crystal growth, as it provides good control of purity with
relative simplicity. However, since single crystals often do not have uniform
impurity distributions, the crystals are ground into powders which can be mixed
to average the impurity distributions to give a more uniform luminescence. This
powder is then pressed through a die at high temperature and high pressure to
make "extruded ribbons." This is commonly done by the manufacturer of the
widely used phosphor, LiF, for radiation dosimetry.

6.2.1 Simple Models

A simple model can be used to explain the luminescent process. Figure 6.1
shows a series of events which result in TL. Figure 6.1a shows how the primary
stimulation – in this case ionizing radiation – excites an electron out of the
valence band to the conduction band where it wanders until it finds a localized
defect where it is trapped or becomes immobile. A similar process occurs for the
hole. Carriers thus metastably trapped in defects need a greater amount of
energy to escape than is normally available thermally. This results in the stored
energy situation. Figure 6.1b illustrates what results when a secondary
stimulation is applied. For TL, the thermal stimulation excites a carrier; in Fig.
6.1b, the electron is shown. The hole could as well be the entity excited to move
to recombine with the electron. Upon recombination, the resulting energy is

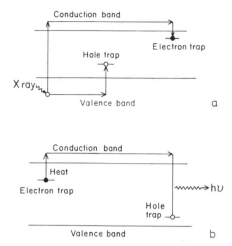

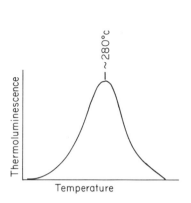

Fig. 6.1a and b. Simple schematic band model for TL. **(a)** Carriers (electrons, closed circle; hole, open circle), are excited to the appropriate band by X-radiation and then trapped; **(b)** heat is applied to excite the electron into the conduction band. The electron then recombines with a hole giving off TL output (*hv*)

Fig. 6.2. TL glow peak for CaF$_2$:Mn

given off in the form of light for TL. The light emission is recorded as a function of temperature and the resulting plot is called a glow curve. In most applications, the temperature is programmed to rise linearly with time during readout so that time scale and temperature scale are approximately equivalent. For example, a typical glow curve for calcium fluoride doped with manganese (CaF$_2$:Mn) is shown in Fig. 6.2. Initially, as the temperature rises, the thermal release rate of the trapped electrons increases rapidly to give the increasing light emission. At the same time, however, this release is depleting the number of trapped electrons. The maximum (or peaking) of the glow curve occurs when the depletion effect begins to dominate; as the temperature rises further, the emission drops rapidly to zero as the trapped electrons become completely exhausted. The temperature at which the glow peak occurs is characteristic of the particular type of trapping center.

A simplified model for RPL uses the configuration coordinate diagram.[2] The RPL trapped charge is detected by illumination, usually with ultraviolet light, resulting in fluorescence at lower energy than the incident exciting light as dictated by the familiar Stokes shift. Note that in the RPL process, the electron does not leave the localized trap. Thus, in contrast to TL, TSC and TSEE, the RPL may be read again. The detection of RPL is facilitated by the Stokes shift, since the exciting light can be excluded from the detection system.

2   For a detailed discussion see Chapt. 2

6.2.2 Importance of Impurities and Complexity of the Thermoluminescence Model

Most thermoluminescent materials are ionically bonded materials. The most important TL dosimeter material, LiF, is an alkali halide having the sodium chloride structure. The defect properties and more detailed model involved in TL will thus be described via the alkali halide (LiF) lattice. The appropriate model for TL in LiF is known to be more complicated than the one described above but a complete understanding is still lacking. For example, impurities play a major role in the TL process.

The history of the development of LiF dosimetry phosphor can be given as an example of the importance of impurities. When *Daniels* [6.17] began work in 1947 on the application of TL to measurement of ionizing radiation, he used the "pure" LiF then available. When *Cameron* reactivated TL research in 1960, the LiF phosphor had been purified to the extent that it did not produce as good a signal. Cooperative research in the addition of impurities to the LiF melt resulted in a patent [6.18] and a phosphor designated LiF (TLD 100). The amount and type of impurities necessary for high TL sensitivity in LiF have been shown to contain concentrations of 100 to 200 ppm Mg and 10 to 20 ppm of Ti [6.19, 20]. These dopants are grown into the LiF crystal structure from the melt, a process that has some inherent difficulties, since other impurities may be unknowingly added in small amounts, as was Mg before 1960. As an example, research in recent years has indicated the importance of the hydroxide ion in the TL process [6.21–23].

One controversial question for the TL model in LiF (TLD100) has been whether the TL traps are electron or hole traps. LiF (TLD100) displays five glow peaks, shown in Fig. 6.3, indicating that it has five trapping centers. *Claffy* [6.24] found that optical absorption bands, thought to be involved with TL in LiF, bleach as the F-band bleaches. Thus she proposed that the TL traps are hole traps (cf [6.25]). However, the strongest evidence to date suggests electron traps are responsible for TL in LiF; the work reported in [6.26] used F-band light in LiF (2500 Å) to bleach the F center and repopulate all of the TL peaks.

The optical absorption bands and defect complexes have had much experimental work in attempts to correlate them with TL glow peaks in LiF (TLD100). For example, using dielectric loss experiments, *Grant* and *Cameron* [6.27] have associated peak 2 of the LiF (TLD100) glow curve (Fig. 6.3) with an optical absorption band at 3800 Å and postulated an impurity–vacancy dipole defect model for this peak. This work has been verified and extended to show that peak 5 consists of impurity–vacancy dipole complexes [6.24, 28, 29]. Thus, the suggested defect model for the traps in LiF would be Z-type centers. The optical absorption band associated with peak 5 is generally accepted to occur at 3100 Å [6.26, 30] with a possible suggestion of another band occurring at 3400 Å [6.24]. The optical absorption band associated with the recombination center lies in the 2000 Å region [6.23, 31].

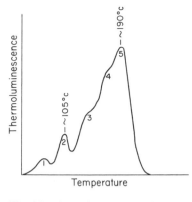

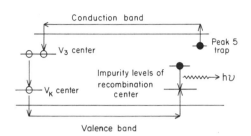

Fig. 6.3. Thermoluminescent glow curve for LiF(TLD100) with glow peaks numbered in conventional fashion. Each glow peak results from a trapping level

Fig. 6.4. Schematic band model for glow peak 5 in LiF(TLD100) showing the greater complexity involved in TL

A comprehensive model incorporating the above facts has been suggested [6.26]. This is not firmly established as the final model, but it serves to illustrate the complexity involved. Figure 6.4 is a sketch of the basis of the model for peak 5 of the LiF (TLD100) glow curve. The electron is released from a peak 5 trap to combine with a hole in a V_3 center which then becomes a V_K center. Since the V_K center is thermally unstable at temperatures above 77 K, the hole is released and wanders in the valence band until it can recombine with an electron at a Ti recombination center. The luminescence is then characteristic of this Ti center. Note that this model has both holes and electrons as carriers but that the TL traps are electron traps.

6.3 Application of Thermally Stimulated Luminescence to Radiation Dosimetry

One of the major applications of thermoluminescence is radiation dosimetry, which requires the sensitive detection and accurate measurement of ionizing radiation. Many specific applications appear in the literature using thermoluminescent dosimetry (TLD); it is not the purpose here to review each specific application but instead to present general procedures with specific applications used for illustrative purposes. A description of the criteria to be considered for selection of a luminescent dosimetry system and the possible difficulties encountered during use are of value for future selection of potential materials. The areas of application of luminescent dosimetry fall into personnel dosimetry and environmental monitoring, radiation therapy dosimetry and diagnostic radiology dosimetry. Applications of luminescence dosimetry in these areas can be found in a number of books and manuscripts [6.1, 11, 32] as well as the proceedings of the five international symposia [6.4–8]. A basic text for further information on the field of radiation physics is given by *Johns* and *Cunningham* [6.33]. A review of the field of radiation dosimetry is given in a series edited by *Attix* et al. [6.34].

A large amount of dosimetry has been and still is based upon the darkening of photographic film by irradiation. When dosimetry systems have been compared, especially in the past, film is the standard of comparison. The first luminescent phenomenon used for dosimetry was the RPL process in glasses developed in the 1950s by the US Navy. Unfortunately, this process worked only for large exposures. A review of RPL is given by *Piesch* [6.35]. Recently the use of a similar dosimeter system by the bleaching of F centers created by radiation has also been tried. Generally these processes are not sensitive enough except for accident dosimeters. The most widely used luminescent dosimeter system today is TLD.

Since ionizing radiation is the basis for dosimetric applications of luminescence as well as the dating applications, it is appropriate to include some background material on radiation and its units. The units for measurement will be used in the discussion on applications.

6.3.1 Units and Hazards of Ionizing Radiation

X-rays were first discovered in 1895 by W. C. Roentgen while experimenting with his Crookes tube; in 1901 he was awarded the first Nobel prize in physics for this discovery. Roentgen coined the name X-rays, but many of his contemporaries and some people today refer to them as Roentgen rays. Actually Crookes himself unknowingly observed the effects of X-rays on some photographic plates kept in his laboratory; periodically he noticed these plates would be fogged even while enclosed in their protective containers. Crookes solved the problem by keeping the plates outside the laboratory, thus leaving the discovery of X-rays to Roentgen. Since that time many types of ionizing radiation have been discovered.

The measurement of radiation has led to a system of units to quantify the measurements. The most familiar unit is based on the amount of ionization produced in air by X-rays and gamma rays and is called the Roentgen in honor of the discoverer of X-rays. The Roentgen (R) is defined as the amount of gamma or X-ray radiation which produces ions in air carrying one electrostatic unit of charge of either sign per 0.001293 grams of air (equal to 1 cubic centimeter of air at standard temperature and pressure). Notice that the definition only applies to X or gamma rays and only to the radiation's effect in air. The phrase "a person was exposed to 0.1 R" means that the person entered a radiation field for a time such that the same radiation would cause ionization in one cubic centimeter of air at standard temperature and pressure of 0.1 esu. The word exposure is used for radiation intensities measured in Roentgens.

The definition of the Roentgen limits its use to X and gamma radiation in air. A unit for the measurement of other radiations, e.g., α and β particles, as well as the ionizing effect in materials other than air, led to the need of another unit of absorbed dose, the rad. The rad is defined as the amount of radiation which deposits 100 ergs of energy in each gram of material. The new SI unit for the rad is called the Gray (1 Gray = 10^2 rads). The rad is often used in reference to soft tissue or to water. The amount of X or gamma radiation that would cause a certain ionization in air would alternately result in a certain energy deposited in some other material. Hence, the rad and the Roentgen can be related to one another, once the material is specified, by comparing the absorption of the radiation in the material to the absorption in air. Conceptually, these units are quite separate, since the rad measures energy absorbed or dose, while the Roentgen measures only ionization in air.

Since biological effects depend on microscopic details of how the dose is absorbed, a third unit, the rem has been introduced. The rem is intended to gauge the biological harm or alteration that a radiation will produce. The rem is defined as the product of the absorbed dose (rads) and other modifying factors which quantify the relative biological effectiveness. Neutrons, for example, do more harm biologically than an equal absorbed dose of X or gamma rays, and thus the neutrons would give a higher dose in rem.

It is well known that ionizing radiation can be harmful. The early investigators of X-rays noticed that high exposures caused skin reddening to the point of burns. However, the pioneers studying radiation did not completely understand its inherent dangers and did not take adequate protective measures. One of Thomas Edison's assistants probably became the first person to die from X-ray induced cancer. In 1936, a monument with 110 names was dedicated "to the roentgenologists and radiologists of all nations who have given their lives in the struggle against the diseases of mankind." Since that time, high exposures of radiation have been shown to increase the probability of contracting leukemia and other forms of cancer. In general, the protective measures involve distance and shielding, such as lead-lined covers. The hazard that radiation presents is dependent on a number of factors. For example, harm from radiation depends on the exposed area and location on the body; certain organs are more radiation sensitive than others. Also, limited areas of the body can withstand higher amounts of radiation than if the whole body was exposed. For example, when radiotherapy is used for cancer treatment, a small tumorous region may be exposed to a few thousand rads in stages over a few weeks without killing the patient, but with the hoped-for effect of killing the cancerous growth.

The human population receives radiation exposure every day from natural background. This varies depending on geographical location but on the average about 0.4 mR is received each day or about 125 mR per year. However, the greatest source of man-made radiation exposure to the population is from medical diagnostic X-rays.

The exposure from a diagnostic X-ray examination varies with the type of examination performed as well as the care taken to determine the lowest exposure for the best image. Some typical exposures might be 60 mR for a chest radiograph and 300 mR for a dental radiograph. The skin reddening, used by early investigators as a radiation monitor, occurs at less than 300 R for low-energy X-rays, but not until about 700 R for high-energy X-rays. This points out the necessity for measuring radiation and the importance of radiation units, other than reddening of human tissue.

6.3.2 Thermoluminescent Dosimetry Characteristics

General characteristics that are important in the selection of any dosimeter system are stability, sensitivity and energy dependence. A dosimeter's overall stability determines how long after exposure the readout may occur. The major physical factor determining the stability of the thermoluminescent material for dosimetry is the temperature at which the peak of the glow curve occurs. Most commercially available TL and RPL dosimeters have high glow peak temperatures – deep electron traps – so that they are stable for many months or years. In LiF (TLD100) for example, the main dosimetry peak decreases only about one percent per year [6.36]. If a dosimeter is to be used in an unusual temperature or humidity environment, its stability must be checked under these conditions. Film is notoriously unstable in warm, moist climates; for example, *Becker* et al. [6.37] reported the fading of the response of film in tropical countries. Some solid dosimeters are hygroscopic, in which case moist conditions are to be avoided, while others are unstable when exposed to light.

At low radiation exposures both TL and RPL dosimeters normally exhibit a linear response. The TL response of LiF (TLD100) and CaF_2:Mn are shown as functions of exposure in Fig. 6.5, the darkening of a film example is shown in Fig. 6.6. The linear responses are especially convenient; notice that film does not have this advantage at some exposures. LiF (TLD100) has a supralinear response for exposures greater than approximately 10^3 R. Once the response is known to be linear, a single control reading establishes the relation between the signal magnitude and exposure; that is the reading *calibrates* the system. All subsequent readings are then expressed readily in terms of exposure. In practice several control dosimeters may be read interspersed with the unknown readings to be sure that no change in calibration has occurred. Figure 6.5 also shows that the linear response of luminescent materials continues to high exposures, up to at least 500 R, which is high enough to cover most exposures of interest. Film's opacity stops growing sooner, after about 10 R, and a second high exposure film must be included to cover the range up to 1000 R.

From Fig. 6.5 it is clear that CaF_2:Mn shows considerably more TL than does LiF. The measure of this efficiency is called *sensitivity*. The lowest detectable exposure will usually depend on the sensitivity of a material, although other considerations such as the nature and properties of the

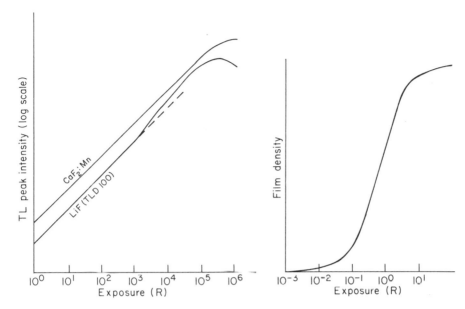

Fig. 6.5. Thermoluminescent response with exposure for CaF_2:Mn and LiF(TLD100). Note supralinearity of LiF(TLD100)

Fig. 6.6. Optical density of photographic film with exposure. Note the smaller range of exposure for film than for TL phosphors (Fig. 6.5)

detecting apparatus are also important. For example, $Li_2B_4O_7$ emits yellow TL light while LiF emits blue light. If the optical system of the TL reader is optimized for sensing blue light, then it may not see the yellow light very well at all. Hence, for the same mass, $Li_2B_4O_7$ might emit as much light as LiF, but the reader would register a signal much lower than for LiF. Questions have been raised in the past whether the TL emission was the same for both low and high exposures; research has shown that there appears to be no difference [6.38]. The sensitivity of RPL dosimeters is also subject to similar matching considerations, and both the exciting light and the detector's response must be considered.

The lowest detectable exposure is not determined by the sensitivity alone. TL readers detect incandescence from both the sample and heating pan (called black body signal). Readers have been designed [6.39, 40] using hot nitrogen gas to alleviate part of this problem. If the TL signal is much smaller than the black body signal, the TL will be undetectable and the black body radiation will limit the lowest detectable exposure. Generally, the black body signal is small below 250 °C; so the most useful dosimeters have their TL peak near 200 °C, which is sufficiently high for adequate stability yet low enough to be relatively free from black body signals during heating. For some phosphors emitting light in the blue region, the black body signal can be reduced by the

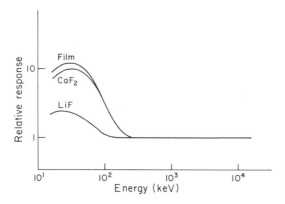

Fig. 6.7. TL response relative to TL response from [137]Cs (662 keV) radiation

use of optical filters, which preferentially absorb the near infrared region produced by the black body signal.

Another interference influencing the lowest detectable exposure is a signal in the sample which is not due to irradiation. Unirradiated samples may give some signal during readout, and it is reasonable to assume that an irradiated sample gives a signal due to irradiation (radiation-induced) plus the signal observed in unirradiated material (nonradiation-induced, or *spurious*). The spurious signals appear to result from surface phenomena. They are generally higher in powder samples, most likely caused by mechanical surface effects. As in the case of interference from black body radiation, spurious luminescence can limit the lowest detectable exposure when it is larger than the radiation-induced signal. The use of single-piece dosimeters greatly reduces spurious TL compared with that in a powder, and readout in an inert atmosphere such as nitrogen further reduces the spurious signal to the point where it is not a problem [6.41]. In contrast, RPL is limited by a high spurious signal. Even after careful washing to eliminate spurious signals from dirt on the surface, a bulk signal, called the predose, is present. Tedious washing procedures and predose effects are major disadvantages of RPL compared to TL. Similarly, the opacity of unexposed film ("fog") is relatively large and variable, making film difficult to apply at low exposures.

Handling characteristics of the TL materials can be very important. Studies on the effects of mechanical deformation on TL [6.42, 43] have shown that the crystal structure must be deformed to induce changes. Essentially one would have to "stand on the crystal." However, handling techniques are important [6.44] and care must be exercised.

When a specific dosimetry problem is at hand, stability, sensitivity and freedom from spurious signals are not the only considerations. Another is tissue equivalence, also called energy dependence. The number of photons absorbed in a given material varies with energy. The most convenient dosimeter is one that absorbs radiation in a manner similar to tissue – if it is being used for dosimetry in human applications. Figure 6.7 shows the relative response of several dosimeter materials for equivalent exposures to radiations of different energies, as compared with ^{137}Cs gamma rays (662 keV).

As opposed to soft tissue, bone absorbs low-energy radiation much more readily than high-energy radiation. This is why low energies are used to make X-ray images of bones. High-energy radiation is used in radiotherapy because it has the penetration necessary to reach tumors deep in the body. At these energies, bone absorbs radiation similarly to soft tissue and all of the body will absorb more equally. The TL response of CaF_2 varies with energy like the dose in bone, and can be used for specialized applications to measure the absorbed dose of the bone. If low-energy radiation is being measured, the energy response of the phosphor must be considered so that the exposure is not over estimated.

While the considerations of stability, sensitivity, spurious signals and energy dependence are of general importance, additional complications must also be considered in certain phosphors. First, at high exposures, usually above 500 R or so, many dosimeters lose their linear response and become supralinear, as shown in Fig. 6.5. Moreover, the degree of supralinearity may itself be energy and radiation type dependent [6.45, 46]. Above 10^5 R, most dosimeters' response ceases to grow with increasing exposure; we say that the response saturates (see Fig. 6.4). Hence at these very high exposures, special dosimetry techniques must be used. A history of previous high exposures may render a dosimeter unuseable at low exposures, for several reasons. In some cases the high exposure changes the sample's subsequent sensitivity, usually reducing it. In this case we say that the dosimeter suffered radiation damage. Sometimes the radiation and thermal histories together can be important. Samples receiving a high dose and a treatment at a low temperature (300 °C) often exhibit a higher response than normal or the sample is "sensitized" [6.1, 47]. Usually it is best if some knowledge of the previous radiation and thermal history of the dosimeters is kept for deciding upon high and low exposure applications.

For many materials a high temperature annealing (400–500 °C) is sufficient to empty all of the filled traps and to ready the material for reuse. Often the reading cycle is sufficient for this purpose. In some cases the dosimeter's response depends critically on its recent thermal history. The most notable example is LiF (TLD100), which must be heated *and* cooled in a certain reproducible way if consistent results are to be obtained. Despite this inconvenience, LiF receives wide use because of its low energy dependence and relatively high sensitivity.

The heating and cooling characteristics of a phosphor are important, since this annealing process can alter the glow curve structure of a phosphor. As an example, we will discuss the characteristics of the widely used phosphor LiF (TLD100). The standard procedure consists of a pre-irradiation annealing at 400 °C to "reset" the material by eliminating the effects of previous treatments the crystal may have received. If the previous history of exposures has been low (less than ~100 R), the readout cycle is usually sufficient and this 400 °C annealing need not be done before each irradiation. It is important that a uniform cooling procedure be used after any annealing, e. g., the phosphor is placed on an aluminum block at room temperature. This 400 °C annealing procedure for LiF will result in a glow curve consisting of five glow peaks after the crystal is irradiated.

The intensity of these glow peaks decreases with time after irradiation because of the thermal energy at room temperature [6.1]. A half-life, similar to the half-life of radioactive materials, can be assigned to each peak. The high temperature peaks in LiF are very stable, e.g., a half-life of approximately 80 years for peak 5, but the half-life of the low temperature peaks is fairly short, e.g., about 10 hours for peak 2. This could be a major problem in radiation dosimetry if the signal from all of the glow peaks is integrated, since the low-temperature peaks fade with time and would give a different result depending on when the signal was read out. It was discovered by accident that an additional pre-irradiation annealing at 80 °C for 24 hours would eliminate the low-temperature peaks without greatly affecting the high-temperature peaks. A similar result can be obtained if the phosphor is annealed at 100 °C for 10 or 15 minutes after irradiation (post-irradiation annealing) instead of the 80 °C annealing [6.36]. Experiments mentioned previously have identified the low-temperature peaks with a defect composed of an impurity–vacancy dipole. The 80 °C annealing alters this defect equilibrium in the crystal so that the dipoles are no longer present when the sample is subsequently irradiated. The 100 °C annealing empties the low-temperature traps before readout so that no signal remains in them [6.48]. The cooling rates and annealing temperature can affect the concentration of the trap defects [6.29, 48]. Thus, there is an optimum annealing procedure for a given application.

These characteristics must be considered for any luminescent dosimeter before application to any area for dosimetry. In the next sections particular details for each area of application of radiation dosimetry will be discussed.

6.3.3 Personnel Dosimetry and Environmental Monitoring

Personnel dosimetry is concerned with monitoring the amount of radiation received by radiation workers over a long time period. Thus, these measurements have several special requirements. Background exposures amount to about 10 mR per month; the dosimeters are read every one to three months, and consequently, the dosimetry system must be sensitive enough to detect 10 mR or less. In addition, the dosimeter should be able to measure the higher exposures which might be encountered in accidents. Accidental exposures near X-ray machines, radioisotopes, accelerators or reactors can range from a few R to thousands of R. Obviously, the physical change used to detect the radiation must persist intact for the several months that may separate issue and reading; that is, the dosimeter's response must be stable. Further, different energies of radiation are likely to be encountered in personnel dosimetry, and since people are being monitored, the dosimeter should give a response which reflects the actual dose to tissue, regardless of the energy of the irradiation. This is called tissue-equivalent response.

TL dosimeters satisfy these requirements. The use of TLD for personnel monitoring has progressed slowly but a number of commercial companies now offer the service and its use is increasing [6.49]. The composition of the badge varies according to the expected type of radiation the personnel will encounter; an example of a TLD personnel badge is given by *Kocher* et al. [6.50]. A typical badge might consist of two or three TLD chips in a plastic holder with identification. Different metal absorbers can be placed in front of individual chips to distinguish between radiation type and energies, or for other desired information. The main competitor for TL is photographic film which is still largely used for personnel monitoring. Some studies [6.37, 51] have compared TLD and film, showing the advantages of the use of TLD for personnel monitoring.

Environmental monitoring has much the same requirements as personnel monitoring except that there may be greater extremes of conditions for the dosimeter. A large amount of environmental monitoring has been done around reactors, although the use of TLD for other types of environmental monitoring is of interest. A recent application in environmental monitoring has been to measure the thermal conditions [6.52]. In this case, a large known exposure is given, and the decay in the peak height caused by thermal fading then yields a measure of the thermal conditions.

6.3.4 Radiation Therapy Dosimetry

The small size and large range of TLDs are their principal advantages for clinical applications. In the application of TLD to radiotherapy, sensitivity and stability, are not as important as in personnel monitoring, since doses are high (100–300 rads per application), the exposures are short, and the reading is made soon after exposure.

Dose planning for external beam radiotherapy involves knowing the relative values of the absorbed dose to the tumor and the surrounding healthy tissue. The tumor dose is often maximized with respect to the radiation tolerance of the surrounding tissues. Dose planning is based on calculated isodose curves. Irregularities in tissue homogeneities can perturb the absorbed dose distribution. Thus, dose measurements in the body (*in vivo*) and in anatomical phantoms are necessary to accurately determine the dose distribution or isodose curves in realistic conditions. The small physical size of TLDs is especially advantageous here since they may be used in body cavities, or to monitor particular small regions, such as the cornea of the eye. TLDs have been used for a number of applications in therapy. The radiation dose to critical organs within or outside the radiation field can be checked. TLDs can be used to discover systematic errors in radiation technique or dose planning. They also can be used for verification of dose for new or old treatment schemes. TLDs have become part of the standard equipment of a radiation therapy clinical setting.

6.3.5 Diagnostic Radiology Dosimetry

Diagnostic radiology employs many radiation techniques to diagnose diseases of the body. The most standard technique is the radiograph or "X-ray" and this is the major source of radiation exposure for the general public. Surprisingly, until recently, very little attention has been given to the measurement of radiation exposure to the patient. Of course, this is very difficult to do even with proper ionization chambers, since there are many X-ray sets in small clinics throughout the country. This is where TLD can play a major role; because of its small size, TLDs can be mailed to clinics, etc., exposed on a patient and returned to a facility for measurement [6.53]. This type of mail monitoring of diagnostic X-ray machines is now being done by the Medical Physics Laboratories of the University of Wisconsin as a public service. The TLDs are being used to measure the radiation exposure for various diagnostic examinations, such as mammography, chest, dental, etc. Because of variation in TLD chips, groups with sensitivities within 2% are used to improve accuracy. Experiments have shown that the exposure measurement via this method is accurate to within 10%.

The applications in diagnostic radiology are continually increasing; many similar to those mentioned for therapy are possible. For example, the exposure to an internal organ can be measured during a diagnostic examination; these measurements then can be used to verify calculation procedures. In computer tomography (CT) applications, the spread of the X-ray beam perpendicular to the CT slice can be measured as well as the exposure. Thus the actual width of the slice can be measured. The small size of the TLDs allow their use for radiation measurement in fluoroscopic procedures, since the TLD will not interfere with the image. Also, TLDs have found use to measure some characteristics of X-ray machines; the half value layer of diagnostic X-ray machines can be determined via TLDs [6.53].

The requirement necessary for the application of TLD in diagnostic area is the ability to detect low exposures (e.g., 30 mR to 60 mR for a chest radiograph). Also great care must be exercised to take into account the energy response of the TLDs, since diagnostic X-rays are in the low-energy regime. For example, if the TLDs are calibrated at a higher energy, such as ^{60}Co (1.2 MeV) and then used for the diagnostic region, there can be a factor 1.3 to 6 overresponse depending on the type of phosphor.

6.3.6 Dosimetry of Particulate Radiation

For radiations other than X and gamma radiation, special efforts are sometimes required for dosimetry. Generally the dosimetry of energetic beta particles and accelerator-produced electrons are not very different from that for X and gamma rays. However, low-energy beta particles are absorbed easily and very little material should intervene between the source and the dosimeter.

Usually beta dosimetry is done with thin dosimeters that are packaged so that one side is uncovered. Alpha radiation or protons are more readily absorbed than beta particles so even more care must be observed, and the application to these radiations is not widespread. Dosimetry of thermal neutrons (energies up to a few eV) has been accomplished by using ^{6}Li as the isotope in the LiF lattice. The ^{6}Li isotope has a high cross section for thermal neutrons to result in an alpha particle and tritium. When this is used in conjunction with a phosphor rich in the ^{7}Li isotope, the TL response of both phosphors can be compared to separate out any neutron–gamma ray mixed fields. This is generally a complicated procedure.

Fast neutron dosimetry (energies of several keV or more) is one of the principal challenges facing present-day luminescent dosimetry research. Most schemes rely on detecting indirect results of the neutron's passage, such as activated nuclei, fissioned nuclei or energy from recoil protons. These processes can result in a luminescence detection or a visualization of the tracks resulting from the neutron passage. Systems for personnel dosimetry are not yet on hand.

Recent work [6.54] has indicated a separate glow peak characteristic of an impurity diffused into the surface. The glow peak is only present as a result of ionizing radiation interacting on the surface (or whatever the diffusion depth of the impurity). This diffused phosphor readily lends itself to separation of the ever-present mixed gamma–neutron field. For example, a hydrogencous material could be used to generate recoil protons which are shallowly penetrating and interact with the surface layers. The gamma radiation would interact with the crystal bulk. By observing the height of the two appropriate glow peaks, the neutron and gamma exposures could be individually determined with the use of one dosimeter. A similar application is possible for mixed beta–gamma radiation.

6.4 Dating Methods Using Thermally Stimulated Luminescence

Humans have wandered on the earth for eons. While some have left written records, others leave only clay pots or some other artifacts. Often the archaeologist has only such possessions for clues to the past. Fortunately, surprising amounts of information can be extracted from artifacts, often including their ages. Since knowing an artifact's age is of fundamental interest, dating techniques themselves are of considerable importance. The most common luminescent technique for determining ages is TL, and the use of TL dating methods in the fields of archaeology and geology will be explored here. Since there is a greater application of TL dating methods to archaeology, most of the discussion will be devoted to this area.

The natural sciences have had a large impact on archaeology [6.55, 56], beginning with radiocarbon dating [6.57]. The basis for comparing any new dating method is to check if it is consistent with known dating methods. Thus, it is reasonable to give some background on other dating methods.

6.4.1 Radiocarbon and Tree Ring Dating

Radiocarbon dating has provided the basis for a chronological framework. Any radiometric dating method, such as radiocarbon dating, is based on the use of the decay rate of radioactive isotopes.

Carbon 14 is a radioactive isotope, present in the atmosphere, produced when cosmic rays create neutrons, which convert Nitrogen 14 into radiocarbon. This radioactive carbon isotope, ^{14}C, has the same chemical behavior as normal stable carbon, ^{12}C, and thus forms carbon dioxide which is distributed worldwide. Dating via this method is based upon the decay of radioactive ^{14}C and the $^{14}C/^{12}C$ ratio.

The basic assumption of radiocarbon dating is that the ratio $^{14}C/^{12}C$ in the atmosphere has remained constant over the years. This postulate has been tested by comparing radiocarbon dating with tree rings and found to be in error [6.58]. In a living tree only the outer ring is growing and is the only one in equilibrium with carbon in the atmosphere. Therefore as far as radiocarbon dating is concerned, the inner tree rings can be considered to be dead. The true calendar ages for the tree and its rings are determined from dendrochronology, which bases its time scale on the regular formation of rings (usually one ring per year). By matching distinctive groups of rings (a pattern of wide and narrow rings), a series of trees including those that have died in the past, can be dated. This has been especially profitable in the bristlecone pines located in inland California; these trees are long-lived and their wood withstands erosion by the elements for millennia after death. By a comparison of data from long-dead specimens to the 4600 year record of living trees, a 7100 year tree ring chronology has been developed for the bristlecone pine.

The comparison of a tree ring's known age with its radiocarbon content has shown appreciable variation in the $^{14}C/^{12}C$ ratio over the years [6.58]. Throughout the past 2000 years, the production of ^{14}C in the atmosphere has varied only slightly, but earlier than 2000 years ago, the carbon 14 dates are consistently more recent than the actual tree ring ages. Although an explanation for this variation is not yet fully understood, the tree ring comparison allows the radiocarbon dates to be corrected. The thermoluminescent dating results explained below agree with the tree ring dating.

6.4.2 Dating by Thermoluminescence

TL dating has as its basis the accumulation of luminescence from natural radiation over a period of time. Natural TL phosphors, e.g., rocks or quartz in clay found all over the earth, are usually saturated and no more TL can be induced. Thus, they are not very useful for dating. Some natural phosphors, however, have been heated in one way or another, e.g., firing of clay pots to complete their fabrication after shaping. This thermal treatment will have emptied the TL traps, so that the TL is no longer saturated; these materials can be used to date the occurrence of the event that last emptied their TL traps. For example, for archaeological pottery samples, TL is induced in the pottery from the date of firing. Usually the time between manufacture and burial is assumed to be short enough so that they may be considered to be the same historical time. Funeral urns, especially constructed for human burial, are an example where this supposition is clearly valid. The bones enclosed in the urns can be dated by the radiocarbon method, providing a very good comparison with TL dating of the urn. One basic assumption behind the dating process is that the TL used is stable during the years of burial time. To establish a definite age, one must determine how much TL is induced per year. This is done indirectly by determining the local dose rate (rads per year), and the phosphor's sensitivity to radiation (TL per rad) to give the TL induction rate (TL per year.) Knowing the total TL, the age is obtained directly from:

$$Age = \frac{Accumulated\ TL}{(TL\ per\ rad) \times (rads\ per\ year)}$$

In principle this technique is simple, but in practice TL dating is considerably more difficult. TL as applied in archaeological dating is reviewed by *Aitken* and *Fleming* [6.59].

Ancient artworks and artifacts, rare keys to the past, are often coveted possessions valued at high sums which tempt thieves and attract clever forgers proportionately. In several cases, thermoluminescent measurements have played a decisive role in the determination of the authenticity of artifacts [6.60, 61]. Clay constructions may be imitated, but it is virtually impossible for a forger to duplicate the effects of the long, slow irradiation that a genuine article has received while buried. Hence, the techniques of TL dating can be applied to detect forgeries and to authenticate ancient pieces, provided they contain some thermoluminescent material. An object suspected of being a fake has usually been removed from its burial spot sometime earlier, and it is not generally possible to measure the environmental radiation at the burial site. Thus, the precise age for such artifacts cannot be determined. However, since the problem is one of whether the object is ancient or made recently, the precise date is not important and rough ages are acceptable. The TL from a small amount of material removed from both can be compared. A forgery will have practically no TL in comparison to the TL of the real ancient pot. This technique can be used for cases of forgery that cannot be easily detected by an expert. An example of this was reported in Time magazine [6.62]. A Turkish farmer guided an archaeologist to an archaeological find known as the Hacilar deposit. Soon after this 1956 find, many artifacts of this same style appeared on the market. Although many of these were suspected of being forgeries, the artifacts could not be distinguished on the basis of style alone: The forger was a clever craftsman. TL investigations were used to prove that many were not authentic. Following this, the "forgery deposit" together with some authentic artifacts was found in a shed behind the farmer's house. The farmer, by then a rich man, was arrested. Other examples of authenticity verification are given in the literature [6.61, 63, 64].

The principles of geological TL dating are the same as those for archaeological TL dating. The major difference is that the most crucial phosphor in the archaeological case is quartz. Geological dating is limited by the time of formation of the rock or the last thermal event of the rock. In the absence of a thermal event in the last hundredthousand years or so, most materials would have saturated TL, and would not be good for TL dating.

Two geological applications of TL dating will be described as examples. The first is the determination of the age of a lava flow and/or an idea of the thermal profile around the flow. At the time of the volcanic eruption, the lava is hot enough so that it releases all of the TL from rocks that it touches. When the lava cools, the natural radioactivity present in the environment induces TL in these rocks over a period of many years. The subsequent TL intensity can then be recorded, and by measuring the local radiation rate, one may determine the age of the lava flow. The greater the TL output, the older is the eruption of interest. TL has also been used in the study of lunar materials [6.65–67]. The TL in surface samples is thought to result from the natural radioactivity of lunar soil, and from cosmic rays bombarding the lunar surface, just as in the case of minerals on the earth. However, on the moon, the sun's heating is very important, since it causes higher temperatures than on earth, and hence, changes the TL response of surface minerals. The TL generally increases with distance down to about 10 cm below the lunar surface due to reduced penetration of the sun's heat. [6.65] Rocks adjoining a lava flow will show a similar increase in TL as one moves away from the flow, since the rocks farther away will have been heated less. Geological TL dating applications have not been as fruitful as the archaeological TL dating efforts. Many geological applications are outlined in the book edited by *McDougall* [6.68].

A number of methods have been proposed to accomplish TL dating of archaeological artifacts; among these are the quartz inclusion technique [6.69], the radioactive inclusion (zircon) technique [6.70], the fine grain technique [6.71], the subtraction technique [6.72] and the predose technique [6.73]. Each technique has some advantages and disadvantages compared to the other. The reader is referred to these individual references for further information on these techniques. However, to illustrate the complexity of dating techniques, a brief description of the quartz inclusion technique will be given. In this method, the pottery is dated using the TL in small quartz inclusions found in the clay.

In order to measure the TL, the quartz is first separated from the fired clay in which it is contained. A piece of pot fragment (sherd) is ground to remove the surface layer (1–2 mm), then the remainder is carefully crushed, for example in a vise. Crushing the sherd is preferred over strict grinding because it is important to obtain the quartz grains intact without fracturing them, and having as little stress-induced TL as possible. Crushing favors this situation since the clay tends to crumble from around the grains before they suffer extensive breakage. Some poorly constructed South American Indian artifacts disintegrate after a few days in an acid bath, and physical crushing is not necessary. Grains between 100 and 300 microns are selected after crushing and the quartz grains are separated from the rest of the clay matrix. The quartz grains are washed to increase the detectable TL by reducing the light absorption from dirty surfaces.

Next, the extracted quartz is divided into two parts. In one portion the TL is read directly to give the total TL induced during burial. The second portion is irradiated in the laboratory to a known dose, then its TL is read. This second reading gives the TL for the burial dose *plus* the laboratory dose. By subtracting the two readings the laboratory induced TL is determined, and the TL sensitivity (i.e., TL/rad) of the quartz is then known. This additive irradiation technique allows measurement of the TL sensitivity without any prior thermal treatment in the laboratory. Obviously, using the difference between burial TL and burial-plus-laboratory TL implicitly assumes that the TL grows strictly linearly with increasing dose. Strict linearity may not, in fact, be observed, as noted for other phosphors previously. Sometimes the additive technique is not used, and as an alternative, the TL sensitivity is measured in the first sample by a laboratory irradiation after reading the burial TL. This method runs the risk that the thermal treatment (the first readout) may alter the TL sensitivity, making the determination false, but has the advantage that two samples are not necessary.

To determine the absolute age of a sherd, and not just its age relative to another ceramic piece, the natural dose rate at the artifact's burial site must be determined. This part of the procedure is complicated by the fact that the quartz in the buried earthenware was subjected to different kinds of radiation, all with different characteristics. The three important radiations – gamma, beta, and alpha – have radically different ranges in the pot or soil, and as a result they must all be treated differently when determining the natural dose rate.

Gamma rays are the most penetrating of the three radiations, with estimates of their range being many tens of centimeters in soil and pot. Hence, gamma rays from the soil, from cosmic rays and from the pot itself must all be considered as sources irradiating the quartz. In the case of unhurried work, the gamma exposures from the soil and cosmic rays can be measured directly by burying TL dosimeters at the excavation site. A year's wait is best since the dose may vary due to seasonal variations of water content in the soil. Typical values for a burial site range from 50 to 150 mrads per year.

Beta rays, energetic electrons, have a range which is much shorter and can be estimated as a maximum of 1 to 2 mm in soil or earthenware. Hence, if the outer layer of the pot is removed, as in the technique described above, the quartz selected from the remainder will have received beta irradiation only from sources within the sherd, and none from the soil. Removing the outer layer also guarantees that the quartz will not have suffered TL alterations due to light. Although direct measurement of the beta dose from within the sherd is difficult, it is possible to do in the properly equipped laboratory.

Alpha rays have the shortest range of all, about 20 to 50 microns in the pottery. As a result, quartz in the ceramic also receives alpha radiation only from sources within the clay matrix, and like betas, sources in the soil are unimportant. Moreover, for large grains of quartz (over 100 μm) the alpha particles from the surrounding clay cannot irradiate the entire volume because their range is less than a full radius. On the other hand, the betas and gammas are effective over the entire volume. Thus, the TL from betas and gammas predominates for large grains and the alpha-induced TL can be ignored.

Although direct measurement of the beta and gamma radiation is often the most suitable, it is not the only method for determining the natural dose rate. The most widely used alternative is based on the detection of radioactive elements in the soil and pot, and the determination of their concentrations. The radioactive elements that are considered are the ^{40}K and the uranium and thorium series. The advantage of this manner of measurement of the dose rate is that, given the sherd and a soil sample, the dating can be completed in about a week. This date can be refined later using data from a buried dosimeter. Presently the precision in dating is about 10 %.

The underlying assumption for both archaeological and geological TL dating is that the TL is stable over the times being considered, or that the rate of emptying is constant and well known. Recent work has indicated that this essential feature may not be realized for some materials [6.74]. For as yet unknown reasons, TL is lost in these materials at appreciable rates even though it occurs at high temperatures where one would expect the much slower emptying characteristic of deep, stable traps. Quartz does not appear to exhibit this. Several common geological materials do have anomalous fading [6.74] however, and TL dating in this field must be viewed with renewed caution, as must pottery dating using phosphors other than quartz. Other problems involve the loss of radon as it drifts out of the pot or soil. This alters the age. Estimates can be made of the amount escaping but the absolute conditions are not known. Also, variation in water content of the soil can affect the results since the water may absorb some of the incident radiation. Again estimates of this effect can be made. Even with these problems, the TL dating method is quite accurate. For example, at one point, the TL dates and radiocarbon dates disagreed, and it was assumed the TL dates were in error. However, by comparison with the tree ring data, the radiocarbon dates were found to be in error whereas the TL dates agreed with the tree ring data.

6.5 Summary

Thermally stimulated luminescent processes have found applications over a wide range; notably important among these are radiation dosimetry in the fields of personnel and environmental monitoring, diagnostic radiology and radiotherapy as well as in the fields of archaeology and geology for dating. Specific examples have been given in this chapter but the limits of application seem to be left to the imagination. The thermoluminescent phenomenon has found the greatest practical application; but other luminescent processes are beginning to be used in areas where they have particular advantages.

Clearly, among the most important characteristics for thermally stimulated luminescent applications are ease of use, small size, stability, sensitivity, and wavelength of light emission. Depending upon the application, one or more of these characteristics become exceedingly important. Of course the basis of the characteristics lies in the physical phenomenon involved in the solid-state physics. As can be noted from Sect. 6.2.2, there is a large amount of complexity involved in models for the thermally stimulated luminescent phenomena and the degree of relationship between defect processes and luminescent properties. Many of these luminescent materials can be applied to specific problems without a detailed knowledge of the defect properties of the solid. However, for advancements to other future applications and modification for better or more suitable characteristics of the thermally stimulated luminescent materials, a knowledge of the luminescent process, the associated defects and impurity concentrations as well as their interrelationship is of utmost importance.

Acknowledgement. I would like to extend my appreciation to Professor P. R. Moran for his suggestions and discussions.

References

6.1 J.R.Cameron, N.Suntharalingam, G.N.Kenney: *Thermoluminescent Dosimetry* (The University of Wisconsin Press, Madison 1968)
6.2 A.Dhar, L.A.DeWerd, T.G.Stoebe: Med. Phys. **3**, 415 (1976)
6.3 E.C.McCullough, G.D.Fullerton, J.R.Cameron: J. Appl. Phys. **43**, 77 (1972)
6.4 F.H.Attix (ed): Proc. Intern. Conf. on Luminescence Dosimetry, Stanford, 1965, Conf. 650637 (available from N.T.I.S.)
6.5 J.A.Auxier, K.Becker, E.M.Robinson (eds.): Proc. 2nd Intern. Conf. on Luminescence Dosimetry, Gatlinburg, 1968, Conf. 680920 (available from N.T.I.S.)
6.6 V.Mejdahl (ed.): Proc. 3rd Intern. Conf. on Luminescence Dosimetry, Risø, Denmark, 1971, Risø Rpt. 249 (Danish Atomic Energy Commission)
6.7 T.Niewiadomski (ed.): Proc. 4th Intern. Conf. on Luminescence Dosimetry, Krakow, Poland, 1974, Vol. 1–3 (available from Institute of Nuclear Physics Library, Krakow, Poland)
6.8 A.Scharmann (ed.): Proc. 5th Intern. Conf. on Luminescence Dosimetry, Sao Paulo, Brazil, 1977 (Available from I. Physikalisches Institut, Universität Giessen, Giessen, Germany)
6.9 J.B.F.Lloyd: Ind. Res. **19**, 29 (1977)
6.10 K.Becker: Crit. Rev. Solid State Sci. **3**, 39 (1972)
6.11 K.Becker: *Solid State Dosimetry* (CRC Press, Cleveland, 1973)
     K.Becker, A.Scharmann: *Einführung in die Festkörper Dosimetry* (Thiemig, München 1975)
6.12 A.Bohun: Proc. 3rd Intern. Symp. on Exoelectrons, Braunschweig, Germany, 1970 (available from Physikalisch-Technische Bundesanstalt)
6.13 J.Zimmerman: J. Phys. C: **4**, 3277 (1971)
6.14 G.D.Fullerton, P.R.Moran: Med. Phys. **1**, 161 (1974)
6.15 P.R.Moran, E.B.Podgorsak, E.B.Fuller, G.D.Fullerton: Med. Phys. **1**, 155 (1974)
6.16 G.D.Fullerton, P.R.Moran, J.R.Cameron: Proc. I.A.E.A. Symposium on Biomedical Dosimetry, Vienna, 1975 (available from I.A.E.A.)

6.17 F. Daniels: Proc. Intern. Conf. on Luminescence Dosimetry, Stanford, 1965, ed. by F. H. Attix, Conf. 650637 (1967) p. 34

6.18 C. F. Swinehart: "Thermoluminescent Doubly Doped LiF Phosphor", U.S. Patent 3320180 (1967)

6.19 M. J. Rossiter, D. B. Rees-Evans, S. C. Ellis: J. Phys. D: **3**, 1816 (1970)

6.20 D. E. Jones, A. K. Burt: Proc. Intern. Conf. on Luminescence Dosimetry, Stanford, 1965, ed. by F. H. Attix, Conf. 650637 (1967) p. 103

6.21 L. A. DeWerd, T. G. Stoebe: Proc. 3rd Intern. Conf. on Luminescence Dosimetry, Risø, Denmark, ed. by V. Mejdahl, Risø Rpt. 249 (1971) p. 78

6.22 H. Vora, L. A. DeWerd, T. G. Stoebe: Proc. 4th Intern. Conf. on Luminescence Dosimetry, Krakow, Poland, ed. by T. Niewiadomski, (1974) p. 143

6.23 H. Vora, J. H. Jones, T. G. Stoebe: J. Appl. Phys. **46**, 71 (1975)

6.24 E. W. Claffy: Proc. Intern. Conf. on Luminescence Dosimetry, Stanford, 1965, ed. by F. H. Attix, Conf. 650637 (1967) p. 74

6.25 C. C. Klick, E. W. Claffy, S. G. Gorbics, F. H. Attix, J. H. Schulman, J. G. Allard: J. Appl. Phys. **38**, 3867 (1967)

6.26 M. R. Mayhugh: J. Appl. Phys. **41**, 4776 (1970)

6.27 R. M. Grant, J. R. Cameron: J. Appl. Phys. **37**, 3791 (1966)

6.28 R. W. Christy, N. M. Johnson, R. R. Wilbarg: J. Appl. Phys. **38**, 2099 (1967)

6.29 A. Dhar: "Effects of Lattice Defects on Thermoluminescence in Lithium Fluoride (TLD-100) Single Crystals"; MS Thesis, University of Washington, Seattle (1971) (unpublished)

6.30 M. R. Mayhugh, R. W. Christy, N. M. Johnson: J. Appl. Phys. **41**, 2968 (1970)

6.31 D. W. Zimmerman, D. E. Jones: Appl. Phys. Lett. **10**, 82 (1967)

6.32 L. A. DeWerd, T. G. Stoebe: Am. Sci. **60**, 303 (1972)

6.33 H. E. Johns, J. R. Cunningham: *The Physics of Radiology*, 3rd ed. (Charles C. Thomas, Springfield 1974)

6.34 F. H. Attix, W. C. Roesch, E. Tochilin (eds.): *Radiation Dosimetry* (Academic Press, New York 1969)

6.35 E. Piesch: *Topics in Radiation Dosimetry*, Suppl. 1, ed. by F. H. Attix (Academic Press, New York 1972) Chap. 8, p. 462

6.36 D. W. Zimmerman, C. R. Rhyner, J. R. Cameron: Health Phys. **12**, 525 (1966)

6.37 K. Becker, R. Lu, P. Weng: Proc. 3rd Intern. Conf. on Luminescence Dosimetry, Risø, Denmark, ed. by V. Mejdahl, Risø Rpt. 249 (1971) p. 960

6.38 L. A. DeWerd, T. G. Stoebe: Phys. Med. Biol. **17**, 187 (1972)

6.39 K. F. Petrock, D. E. Jones: Proc. 2nd Intern. Conf. on Luminescence Dosimetry, Gatlinburg, ed. by J. A. Auxier, K. Becker, and E. M. Robinson, Conf. 680920 (1968) p. 652

6.40 L. Bøtter-Jensen: Proc. IAEA Symposium Advances in Physical and Biological Radiation Detectors, Vienna, IAEA-SM-142/20 (1971) p. 113

6.41 A. E. Nash, F. H. Attix, J. H. Schulman: Proc. Intern. Conf. Luminescence Dosimetry, Stanford, ed. by F. H. Attix, Conf. 650637 (1967) p. 244

6.42 M. Srinivasan, L. A. DeWerd: J. Phys. D: **6**, 2142 (1973)

6.43 L. A. DeWerd, R. P. White, R. G. Stang, T. G. Stoebe: J. Appl. Phys. **47**, 4231 (1976)

6.44 L. A. DeWerd: Health Phys. **31**, 525 (1976)

6.45 N. Suntharalingam: "Thermoluminescent Response of Lithium Fluoride to Radiation with Different LET"; Ph. D. Thesis, Univ. of Wisconsin, Madison (1967) (unpublished)

6.46 J. R. Cameron, N. Suntharalingam, C. R. Wilson, S. Watanabe: Proc. 2nd Intern. Conf. on Luminescence Dosimetry, Gatlinburg, ed. by J. A. Auxier, K. Becker, and E. M. Robinson, Conf. 680920 (1968) p. 332

6.47 M. R. Mayhugh, G. D. Fullerton: Health Phys. **28**, 297 (1975)

6.48 A. Dhar, L. A. DeWerd, T. G. Stoebe: Health Phys. **25**, 427 (1973)

6.49 G. E. Chabot, Jr., M. A. Jimenez, K. W. Skrable: Health Phys. **34**, 311 (1978)

6.50 L. F. Kocher, L. L. Nichols, G. W. R. Endres, D. B. Shipler, A. J. Haverfield: Health Phys. **25**, 567 (1973)

6.51 N. Suntharalingam, J. R. Cameron: Health Phys. **12**, 1595 (1966)

6.52 G.P.Romberg, W.Prepejchal, S.A.Spigarelli: Science **197**, 1364 (1977)

6.53 J.F.Wochos, G.D.Fullerton, L.A.DeWerd: Am. J. Roentgenol. **131**, 617 (1978)

6.54 J.B.Lasky, P.R.Moran: Proc. 5th Intern. Conf. on Luminescence Dosimetry, Sao Paulo, Brazil, ed. by A.Scharmann (1977) p. 451

6.55 M.S.Tite: Contemp. Phys. **11**, 523 (1970)

6.56 M.J.Aitken: Phys. Rep. (Sect. C of Phys. Lett. Netherlands) **40C**, 277 (1978)

6.57 W.F.Libby: *Radiocarbon Dating*, 2nd ed. (University of Chicago Press, Chicago 1955)

6.58 C.W.Ferguson: Science **159**, 839 (1968)

6.59 M.J.Aitken, S.J.Fleming: *Topics in Radiation Dosimetry*, Suppl. 1, ed. by F.H.Attix (Academic Press, New York 1972) Chap. 1, p. 2

6.60 F.E.Rogers: J. Chem. Ed. **50**, 388 (1973)

6.61 S.J.Fleming: *Authenticity in Art, the Scientific Detection of Forgery* (Crane-Russak, New York 1976)

6.62 *TIME* p. 40, September 6, 1971

6.63 S.J.Fleming: Archaeometry **13**, 59 (1971)

6.64 D.W.Zimmerman, M.P.Yuchas, and P.Meyers: Archaeometry **16**, 19 (1974)

6.65 G.B.Dalrymple, R.R.Doell: Science **167**, 713 (1970)

6.66 G.Crozaz, V.Haack, M.Hair, H.Hoyt, J.Kardos, M.Maurette, M.Mujajima, M.Seitz, S.Sun, R.Walker, M.Wittels, D.Woolum: Science **167**, 563 (1970)

6.67 J.A.Edgington, I.M.Blair: Science **167**, 715 (1970)

6.68 D.J.McDougall (ed.): *Thermoluminescence of Geological Materials* (Academic Press, London, New York 1968)

6.69 S.J.Fleming: Archaeometry **12**, 133 (1970)

6.70 D.W.Zimmerman: Science **174**, 818 (1971)

6.71 D.W.Zimmerman: Archaeometry **13**, 29 (1971)

6.72 S.J.Fleming, D.Stoneham: Archaeometry **15**, 229 (1973)

6.73 S.J.Fleming: Archaeometry **15**, 13 (1973)

6.74 A.G.Wintle: Nature **245**, 143 (1973)

Bibliography

B. 1 Thermally Stimulated Luminescence and Conductivity

B.1.1 Analysis of TSL and TSC Measurements

Fillard, J.P., Gasiot, J., Manifacier, J.C.: New approach to thermally stimulated transients — experimental evidence for ZnSe-Al crystals. Phys. Rev. B 18, 4497 (1978)

Harasawa, S.: Supralinearity of thermoluminescence. Oyo Butsuri 46, 1019 (1977)

Harasiewicz, K.H.: Determination of deep trapping levels energy and concentration in LED. Elektronika 18, 453 (1977)

Khare, R.P., Nath, R.: On the analysis of complex TSC patterns. Phys. Status Solidi A 44, 627 (1977)

Kotina, I.M., Novikov, S.R., Pirozhkova, T.I.: Determination of the parameters of impurity centres from the thermally stimulated current of the p-n junction. Phys. Status Solidi A 42, 397 (1977)

Mikho, V.V., Dmitrenko, Z.F.: Use of a thermoluminescence method for studying the parameters of active centers. Kinet. Katal. 19, 720 (1978)

Mlitzke, E.: On the original method of Balarin and Zetzsche. Phys. Status Solidi A 47, K181 (1978)

Nakajima, T.: Theoretical consideration on thermoluminescence response. J. Appl. Phys. 48, 4880 (1977)

Rudolf, G., Becherer, J., Glaefeke, H.: Behaviour of the fractional glow technique with first-order detrapping processes, traps distributed in energy or frequency. Phys. Status Solidi A 49, K121 (1978)

Takeuchi, N., Inabe, K., Kido, H., Yamashita, J.: Formulation of the sensitization in lithium fluoride thermoluminescence based on a simple two-step reaction rate model. J. Phys. C 11, L147 (1978)

B.1.2 Applications of Thermoluminescence

Gorbunov, V.I., Moskalev, Yu.A., Moskalev, A.P.: The contrast function of a thermoluminescent image converter. Defektoskopiya 14, 43 (1978) [English transl.: Sov. J. Nondestr. Test. 14, 132 (1978)]

Romberg, G.P., Prepejchal, W., Spigarelli, S.A.: Temperature exposure measured by use of thermoluminescence. Science 197 , 1364 (1977)

Sheinerman, N.A., Levin, A.S.: Use of thermoluminescence analysis for determining crystallization temperatures of secondary calcites and sulfides. Zap. Vses. Mineral. Ova. 107, 348 (1978)

B.1.3 Experimental TSL Techniques

Gartia, R.K.: Optical bleaching—a technique to clean thermoluminescence peaks. Phys. Status Solidi A *42*, K155 (1977)
Manche, E.P.: Differential thermoluminescence (DTL)—a new instrument for measurement of thermoluminescence. Rev. Sci. Instrum. *49*, 715 (1978)

B.1.4 Nonradiative Capture

Abakumov, V.N., Perel', V.I., Yassievich, I.N.: Capture of carriers by attractive centers in semiconductors. Fiz. Tekh. Poluprovodn. *12*, 3 (1978)
Tarasik, M.I., Yavid, V.Yu., Yanchenko, A.M.: Mechanism of electron capture by radiation effects with E_v + 0.37 eV level in gamma-irradiated Ge:Sb. Fiz. Tekh. Poluprovodn. *11*, 1435 (1977) [English transl.: Sov. Phys. Semicond. (USA) *11*, 845 (1977)]
Vardanyan, R.A.: Cross section for hole capture by a charged dislocation in a semiconductor. Zh. Eksp. Teor. Fiz. *73*, 2313 (1977)

B.1.5 Recombination and Trapping Kinetics

Bowman, S.G.E., Chen, R.: Superlinear filling of traps in crystals due to competition during irradiation. J. Lumin. *18-19*, 345 (1979)
Fain, J., Monnin, M.: A model for the explanation of non-linear effects in thermoluminescence yields. J. Electrostat. *3*, 289 (1977)
Fleming, R.J.: Kinetic order of thermoluminescence. J. Polymer Sci.—Polymer Phys. *16*, 1703 (1978)
Fleming, R.J., Pender, L.F.: Comments on thermoluminescence and continuous distributions of traps. Phys. Rev. B *18*, 5900 (1978)
Jenkins, T.R.: On computing the integral of glow curve theory. J. Comp. Phys. *29*, 302 (1978)
Kanunnikov, L.A.: The order of the thermostimulated recombination reaction. Zh. Prikl. Spektrosk. *28*, 877 (1978)
Krongauz, V.G., Blyakhman, E.A.: A model of stationary recombinational luminescence without thermal transitions. Izv. Akad. Nauk SSSR Ser. Fiz. *41*, 1130 (1977)
Maxia, V.: Entropy production in luminescent processes. Lett. Nuovo Cimento *23*, 89 (1979)
Nakajima, T.: Exothermic model for the thermoluminescence response. J. Appl. Phys. *49*, 6189 (1978)

B.1.6 Thermally Stimulated Conductivity (Experimental Results)

Abdullaev, G.B., Agaev, V.G., Mamedov, N.D., Nani, R.Kh.: Investigation of electron traps in CdGaS4 single crystals. Fiz. Tekh. Poluprovodn. *11*, 14 (1977) [English transl.: Sov. Phys. Semicond. (USA) *11*, 7 (1977)]
Anderson, J.C., Norian, K.H.: Quasi-equilibrium thermally stimulated current process. Solid State Electron. *20*, 335 (1977)
Blasi, C.D., Galassini, S., Manfredotti, C., Micocci, G.: Trapping levels in PbI$_2$. Solid State Commun. *25*, 149 (1978)

Blasi, C.D., Galassini, S., Manfredotti, C., Micocci, G.: Photoelectronic
properties of HgI$_2$. Nucl. Instrum. Methods *150*, 103 (1978)
Böhm, M., Erb, O., Scharmann, A.: Thermally stimulated currents in Ca(NbO$_3$)$_2$.
Phys. Status Solidi A *41*, 535 (1977)
Bordovskii, G.A., Izvozchikov, V.A., Avanesyan, V.T., Bordovskii, V.A.:
Pyroelectric and thermally stimulated currents in system lead oxide-
germanium oxide. Ferroelectrics *18*, 109 (1978)
El-Azab, M.I., Champness, C.H.: Thermally stimulated currents in epitaxially
grown selenium monocrystalline films. Appl. Phys. Lett. *31*, 295 (1977)
Fuyuki, S., Hyakutake, N., Hayakawa, S.: Thermally stimulated current of
CdTe doped with Zn. Jpn. J. Appl. Phys. *17*, 851 (1978)
Gelbart, U., Yacoby, Y., Beinglass, I., Holzer, A.: Study of imperfections
in mercury iodide by the thermally stimulated currents method. IEEE Trans.
NS-*24*, 135 (1977)
Grechushnikov, B.N., Predtechenskii, B.S., Starostina, L.S.: Thermally sti-
mulated conductivity in cuprous oxide. Kristallografiya *21*, 1148 (1976)
Hillhouse, R.W.A., Woods, J.: Plasma growth of rutile crystals and their
photoelectronic properties. Phys. Status Solidi A *46*, 163 (1978)
Katsube, T., Adachi, Y., Ikoma, T.: Trap distribution and memory charac-
teristics of MNOS diodes. Electron. Commun. Jpn. *59*, 131 (1976)
Kawamura, S., Royce, B.S.: Thermally stimulated current studies of electron
and hole traps in single crystal aluminum oxide. Phys. Status Solidi A *50*,
669 (1978)
Lagare, B.B.. Lawangar, R.D., Pawar, S.H.: Thermally stimulated conductivi-
ty in (Ba-Sr)SO$_4$ phosphors. Indian J. Phys. *50*, 1039 (1976)
Lawangar, R.D.: Thermally stimulated conductivity in CaS:Pd phosphors.
Indian J. Phys. *52A*, 223 (1978)
Maeta, S., Mukai, T., Sakaguchi, K.: Measurement of thermally stimulated
currents induced by Q-switched ruby laser in anthracene single crystals.
Jpn. J. Appl. Phys. *16*, 2287 (1977)
Maeta, S., Sakaguchi, K.: Thermally stimulated currents from shallow trap-
ping centers in anthracene single crystals. Oyo Butsuri *47*, 417 (1978)
Martin, G.M., Hallais, J., Poiblaud, G.: Study of experimental conditions
in thermally stimulated current measurements. J. Electrostat. *3*, 223
(1977)
Minami, T., Honjo, K., Tanaka, M.: Thermally stimulated conductivity in
AsS$_{3.5}$Te$_{2.0}$ glass. J. Non-Cryst. Solids *23*, 431 (1977)
Prokopalo, O.I., Fesenko, E.G., Malitskaya, M.A.: Photoelectric phenomena
in single crystals of some perovskite oxides. Ferroelectrics *18*, No. 1-3
(1978)
Raevskii, I.P., Malitskaya, M.A., Prokapalo, O.I.: Photoconductivity and
thermally stimulated conductivity of potassium and sodium niobate single
crystals. Sov. Phys. Solid State *19*, 283 (1977)
Voevoda, G.P., Dubrovenko, M.Ya., Litovchenko, P.G.: Investigation of thermally stimulated currents in p-n structures made of high-resistivity p-
type silicon. Fiz. Tekh. Poluprovodn. *11*, 40 (1977) [English transl.: Sov.
Phys. Semicond. (USA) *11*, 21 (1977)]

B.1.7 Simultaneous Measurement of TSL and TSC

Aboltin, D.E., Grabovskis, V.J., Kangro, A.R., Vitol, I.K.: Thermally sti-
mulated and tunnelling luminescence and Frenkel defect recombination in
KCL at 4.2 to 77 K. Phys. Status Solidi A *47*, 667 (1978)
Barriere, A.S.: Characterization of the localized electronic levels in ionic

compounds by thermoluminescence and thermally stimulated conductivity.
J. Electrostat. *3*, 113 (1977)
Birkle, G.V.B., Gavrilov, F.F., Kitaev, G.A.: Determination of trap para-
meters by combined TSL and TSC measurements. Izv. Vuz. Fiz. *6*, 94 (1977)
Bourgoin, J., Massarani, B., Visocekas, R.: Thermally stimulated lumines-
cence and conductivity in boron-doped diamonds. Phys. Rev. B *18*, 768 (1978)
Dolivo, G., Gaumann, T.: Thermoluminescence, thermally stimulated conducti-
vity and optical absorption of 77 K irradiated methylcyclohexane. Radiat.
Phys. Chem. *10*, 207 (1977)
Lebedeva, N.N., Agaronov, B.S., Berezhnoi, A.A., Mamedov, A., Orbukh, V.:
Photoluminescence, thermoluminescence, and thermostimulated conductivity
in strontium titanate single crystals. Fiz. Tverd. Tela *19*, 3684 (1977)
[English transl.: Sov. Phys. Solid State (USA) *19*, 2153 (1977)]
Zavadovskaya, E.K., Fedorov, B.V., Starodubtsev, V.A.: A comparison of the
charge accumulation and light-sum kinetics during the electron irradiation
of phosphate glasses. Fiz. Khim. Stekla *3*, 83 (1977) [English transl.:
Sov. J. Glass Phys. Chem. *3*, 76-78 (1977)]

B.1.8 Thermally Stimulated Luminescence (Experimental Results)

Halides

Aboltin, D.E., Grabovskis, V.J., Kangro, A.R.: Thermally stimulated and
tunneling luminescence and Frenkel defect recombination in KCl and KBr at
4.2 to 77 K. Phys. Status Solidi A *47*, 667 (1978)
Aguilar, M., Lopez, F.J., Jaque, F.: Relationship between thermoluminescence
and x-ray induced luminescence in alkali halides. Solid State Commun. *28*,
699 (1978)
Alcala, R., Alonso, P.J.: Luminescence in x-irradiated CaF$_2$. Cd. J. Lumin.
20, 1 (1979)
Aluker, E.D., Lusis, D.Yu., Mezina, I.P.: Luminescence and stored light sum
for the activated alkali halides excited in the fundamental absorption
region. Izv. Akad. Nauk SSSR Ser. Fiz. *40*, 1958 (1976)
Ang, T.C., Mykura, H.: Colour centres and thermoluminescence in irradiated
potassium bromide. J. Phys. C *10*, 3205 (1977)
Atobe, K.: Role of F-centers in thermoluminescence processes in reactor
irradiated SrF$_2$ crystals. Phys. Status Solidi A *50*, K77 (1978)
Balasubramanyam, K.: Mechanism of thermoluminescence emission in RbCl-Mg.
J. Lumin. *18*, 901 (1979)
Damm, J.Z., Kachniarz, M., Mugenski, E., Opyrchal, H.: Effects of gamma-
ray irradiation on the properties of Pb^{2+}-doped potassium chloride crystals.
Krist. Tech. *12*, 967 (1977)
Elango, M.A., Zhurakovskii, A.P., Kadchenko, V.N.: Manifestation of the
irradiated-induced energy transfer from the bulk of an insulator to its
surface. Fiz. Tverd. Tela *19*, 3693 (1977) [English transl.: Sov. Phys.-
Solid State *19*, 2158 (1977)]
Elango, M.A., Zhurakovskii, A.P., Kadchenko, V.N.: Luminescence and electron
emission by ionic crystals exposed to ultrasoft x-rays (photon energies
in the range 60-240 eV). Izv. Akad. Nauk SSSR Ser. Fiz. *41*, 1314 (1977)
Fillard, J.P., Gasiot, J., Jimenez, J., Sanz, L.F.: Evidence for refilling
of recombination centers during thermal stimulation in AgBr. J. Electrostat. *3*, 133 (1977)
Früh, R., Brunner, P., Bobleter, O.: Thermoluminescence characteristics of
manganese-doped KCl-KBr-mixed crystal systems. Radiochem. Radioanal. Lett.
31, 135 (1977)

Gartia, R.K.: Thermoluminescence of Z_3 centres in x-irradiated Mg-doped LiF crystals. Phys. Status Solidi A *44*, K21 (1977)

Gartia, R.K., Acharya, B.S., Ratnam, V.V.: Effect of re-irradiation on Z_1 centres. Phys. Status Solidi A *48*, 235 (1978)

Gartia, R.K., Ratnam, V.V., Mathur, B.K.: Thermoluminescence studies of trapping centers in KCl. Indian J. Phys. *50*, 1009 (1976)

Gindina, R.I., Ploom, L.A., Maaroos, A.A., Pyllusaar, Yu.V.: Colour centres in high-purity KCl crystals. Zh. Prikl. Spektrosk. *27*, 520 (1977)

Gon, H.B., Rao, K.V., Bose, H.N.: Study of F centres and thermoluminescence of $KCl:BaTiO_3$ composite crystals subjected to dc field and later x-ray irradiation. Indian J. Pure Appl. Phys. *15*, 148 (1977)

Herreros, J.M., Jaque, F.: Luminescence emission in NaCl-Cu x-irradiated at LNT and RT. J. Lumin. *18*, 231 (1979)

Ikeya, M., Schwan, L.O., Miki, T.: Migration and stabilization of hydrogen atoms in KCl. Solid State Commun. *27*, 891 (1978)

Inabe, K., Takeuchi, N.: Effect of impurity aggregation on the thermoluminescent decay of x-irradiated $KCl:Sr^{2+}$. J. Phys. C *10*, 3023 (1977)

Inabe, K., Takeuchi, N.: The role of Cu^+ in the thermoluminescence of x-irradiated NaCl. Jpn. J. App. Phys. *17*, 831 (1978)

Inabe, K., Takeuchi, N.: Effect of impurity aggregation on the shift of thermoluminescence glow peak of x-irradiated $KCl:Sr^{2+}$. Jpn. J. Appl. Phys. *17*, 1549 (1978)

Jain, V.K.: The dependence of thermoluminescence intensity on heating rate and the deep trap in lithium fluoride. Jpn. J. Appl. Phys. *17*, 949 (1978)

Joshi, R.V., Joshi, T.R.: Effect of Tl concentration on the thermoluminescence of deformed NaCl:Tl phosphors. Z. Naturforsch. *32A*, 663 (1977)

Kadchenko, V.N., Elango, M.: Diffusion parameters of hot holes created by ionizing radiation in NaCl(Ag). Phys. Status Solidi A *46*, 315 (1978)

Kos, H.J., Nink, R.: Mechanism of restored thermoluminescence in lithium fluoride. Phys. Status Solidi A *44*, 505 (1977)

Kos, H.J., Mieke, S.: Effect of deformation on the thermoluminescence of titanium doped lithium fluoride. Phys. Status Solidi A *50*, K165 (1978)

Kuzakov, S.M., Martynovich, E.F., Parfianovich, I.A.: The mechanism for x-ray luminescence of rare-earth ions in lanthanum chloride. Izv. Akad. Nauk SSSR Ser. Fiz. *41*, 1380 (1977)

Lapshin, A.I., Kurnikova, V.V.: Some singularities of alkali halide phosphors activated with Eu^{2+} ions by the explosive loading method. Zh. Prikl. Spektrosk. *28*, 95 (1978)

Lopez, F.J., Jaque, F., Fort, A.J., Agullo-Lopez, F.: Thermoluminescence and electron spin resonance after room temperature x-ray irradiation of $NaCl:Mn^{2+}$. J. Phys. Chem. Solids *38*, 1101 (1977)   ·

Lushchik, Ch.B., Gindina, R.I., Maaroos, A.A., Ploom, L.A.: Formation of cation defects in KCl crystals by irradiation. Fiz. Tverd. Tela *19*, 3625 (1977) [English transl.: Sov. Phys. Solid State *19*, 2117 (1977)]

Mariani, D.F., Rivas, J.L.A.: Thermoluminescence in KI, KBr, NaCl and NaF crystals irradiated at room temperature. J. Phys. C *11*, 3499 (1978)

Mehta, S.K., Merklin, J.F., Donnert, H.J.: Thermoluminescence-related Z centers in LiF:Mg,Ti. Phys. Status Solidi A *44*, 679 (1977)

Moharil, S.V., Deshmukh, B.T.: Reflectance spectra and thermoluminescence of NaBr coloured in an electrodeless discharge. Pramana *9*, 411 (1977)

Moharil, S.V., Deshmukh, B.T.: Reflectance spectra and thermoluminescence of alkali halides coloured in an electrodeless discharge. Pramana *9*, 537 (1977)

Moharil, S.V., Deshmukh, B.T.: Deformation coloration of KF. Radiat. Eff. *33*, 101 (1977)

Mukherjee, M.L.: Effect of quenching on the thermal glow curves from x-ray irradiated KCl and KCl:Pb single crystals. J. Mater. Sci. *13*, 336 (1978)

Oczkowski, H.L.: Thermoluminescence of Ag(Cl,I) mixed crystals. J. Lumin. *17*, 113 (1978)

Parfianovich, I.A., Alekseeva, E.P., Sotserdotova, G.V.: A role of impurity-vacancy pairs in the thermoluminescence of LiF-Mg. Izv. Akad. Nauk SSSR Ser. Fiz. *41*, 1350 (1977)

Pashuk, I.P., Pidzyrailo, N.S., Khapko, Z.A.: Luminescence of CsCaCl3 single crystals. Izv. Vuz Fiz. *8*, 121 (1977)

Pung, L.A., Lushchik, A.Ch., Khaldre, Yu.Yu.: ESR and recombination lumi-nescence for KCl and KBr crystals containing Frenkel cation defects. Izv. Akad. Nauk SSSR Ser. Fiz. *40*, 1952 (1976) [English transl.: Bull. Acad. Sci. USSR Phys. Ser. *40*, 159 (1976)]

Rao, D.R., Das, B.N.: Regenerated thermoluminescence (RTL) of KCl. Bull. Am. Phys. Soc. *24*, 254 (1979)

Rascon, A., Rivas, J.L.A.: Thermoluminescence and colour centre thermal stability in KCl:Ca and KCl:Sr irradiated at room temperature. J. Phys. C *11*, 1239 (1978)

Sastry, S.B.S., Balasubramanyam, K.: Role of lead centers in the thermolu-minescence emission of irradiated NaCl doped with lead. J. Lumin. *15*, 267 (1977)

Sastry, S.B.S., Balasubramanyam, K.: Thermoluminescence of irradiated RbCl and RbCl:Sn crystals. Phys. Status Solidi B *90*, 375 (1978)

Sastry, S.B.S., Sapru, S.: Thermoluminescence of gamma-irradiated rubidium chloride doped with europium. Phys. Status Solidi A *48*, K189 (1978)

Sivasankar, V.S., Whippey, P.W.: Thermoluminescence and optical-absorption of x-irradiated NaBr. Bull. Am. Phys. Soc. *24*, 254 (1979)

Stevels, A.L.N.: Thermoluminescence of UV irradiated CsI:Na. Philips J. Res. *33*, 133 (1978)

Takeuchi, N., Adachi, M., Inabe, K.: Thermoluminescence center in x-irra-diated NaCl-Cu single crystals. J. Lumin. *18*, 897 (1979)

Tanimura, K., Okada, T.: Effects of Na^+ impurity on the self-trapped hole in KBr crystals. Tech. Rep. Osaka Univ. *27*, 351 (1977)

Tanimura, K., Okada, T.: Thermoluminescence and reactions of freed inter-stitial and trapped-hole centers in KBr crystals below room temperature. J. Phys. Soc. Jpn. *43*, 1982 (1977)

Taylor, G.C., Lilley, E.: The analysis of thermoluminescent glow peaks in LiF (TDL-100). J. Phys. D *11*, 567 (1978)

T'Kint de Roodenbeke, A., Decamps, E.A.: Study by thermoluminescence of the traps in monocrystalline CdI2. J. Phys. C *11*, L819 (1978)

Tsuboi, T.: The 475 nm emission of KCl:Tl crystals. Physica B and C *93*, 379 (1978)

Zareba, A., Nadolny, A.J., Krukowska-Fulde, B.: Thermoluminescence of Mn centres in CdF2 crystals. Phys. Status Solidi A *44*, K83 (1977)

Zhurakovskii, A.P., Gluskin, E.S., Elango, M.A.: Investigation of the decay of x-ray excitations in NaCl irradiated with synchrotron radiation. Fiz. Tverd. Tela *20*, 1097 (1978) [English transl.: Sov. Phys. Solid State *20*, 633 (1978)]

Metal Oxides

Babinskii, A.V., Trepakov, V.A., Krainik, N.N., Melekh, B.A.: Growth and thermoluminescence in BaZrO3 monocrystals. Pis'ma V Zh. Tekh. Fiz. *3*, 1159 (1977)

Bagdasarov, Kh.S., Pasternak, L., Sevast'yanov, B.K.: Radiation color cen-ters in YAG crystals. Sov. J. Quant. Electron. *7*, 965 (1977)

Bernhardt, H.J.: Studies of the colour of lead molybdate crystals. Phys. Status Solidi A *45*, 353 (1978)

Bessonova, T.S., Stanislavskii, M.P., Sobko, A.I.: Concentration dependence of radiative-optical effects in ruby. Zh. Prikl. Spektrosk. *27*, 238 (1977)

Cooke, D.W., Roberts, H.E., Alexander, Jr., C.: Thermoluminescence and emission spectra of UV-grade Al$_2$O$_3$ from 90 to 500 K. J. Appl. Phys. *49*, 3451 (1978)

Doi, A.: Thermoluminescent and paramagnetic centers in germanium dioxide. Jpn. J. Appl. Phys. *16*, 925 (1977)

Duvarney, R.C., Garrison, A.K.: An EPR-endor study of the (Li) center in crystalline and ceramic BeO. Phys. Status Solidi A *42*, 609 (1977)

Fierens, P., Tirlocq, J.: Application of thermoluminescence method for study of dicalcium silicate hydration rates. Cem. Concr. Res. *8*, 397 (1978)

Gartia, R.K., Ratnam, V.V.: Chemical instability of CaO-Y phosphor-thermoluminescence study. J. Mater. Sci. *14*, 747 (1979)

Gartia, R.K., Ratnam, V.V.: Thermoluminescence studies in Al$_2$O$_3$ using optical bleaching stimulation. Indian J. Pure Appl. Phys. *16*, 53 (1978)

Hofstaetter, A., Oeder, R., Scharmann, A., Schwabe, D.: Paramagnetic resonance and thermoluminescence of the PbWO$_4$/PbMoO$_4$ mixed crystal system. Phys. Status Solidi B *89*, 375 (1978)

Iacconi, P., Lapraz, D., Caruba, R.: Traps and emission centers in thermoluminescent ZrO$_2$. Phys. Status Solidi A *50*, 275 (1978)

Kaplenov, I.G.: Interaction between intrinsic and impurity emission in CaWO$_4$ phosphors. Izv. Akad. Nauk SSSR Ser. Fiz. *41*, 1387 (1977)

Kristianpoller, N., Rehavy, A.: Luminescence centers in Al$_2$O$_3$. J. Lumin. *18*, 239 (1979)

Kyarner, T.N.: Thermal stability of hole centres and hole recombination luminescence of doped MgO, CaO and SrO. Festi NSV. Tead. Akad. Fuus. Inst. Uurim. *47*, 93 (1977)

Lakshmanan, A.R., Shinde, S.S., Bhatt, R.C.: Ultraviolet-induced thermoluminescence and phosphorescence in Mg$_2$SiO$_4$:Tb. Phys. Med. Biol. *23*, 952 (1978)

Lin, S., Vetter, R., Ziemer, P.L.: Thermoluminescence and microwave induced thermoluminescence fading of rare-earth-doped barium titanate ceramics. Radiat. Eff. *38*, 67 (1978)

Luthra, J.M., Sathyamoorthy, A., Gupta, N.M.: A thermoluminescence study of defect centres in MgO. J. Lumin. *15*, 395 (1977)

Mehta, S.K., Sengupta, S.: Annealing characteristics and nature of traps in Al$_2$O$_3$ thermoluminescent phosphor. Phys. Med. Biol. *22*, 863 (1977)

Metha, S.K., Sengupta, S.: Photostimulated thermoluminescence of Al$_2$O$_3$ (Si, Ti) and its application to ultraviolet radiation dosimetry. Phys. Med. Biol. *23*, 471 (1978)

Moharil, S.V., Kamavisdar, V.S., Deshmukh, B.T.: Isothermal stability of colour centres in microcrystalline NaBr. Cryst. Lattice Defects *8*, 15 (1978)

Mori, K.: Transient colour centres caused by UV light irradiation in yttrium aluminium garnet crystals. Phys. Status Solidi A *42*, 375 (1977)

Murthy, K.B.S., Sunta, C.M., Khatri, D.T., Soman, S.D.: Sensitisation of thermally stimulated exoelectron emission and thermoluminescence of BeO discs. J. Phys. D *11*, 561 (1978)

Newton, R.D., Sibley, W.A.: Deformation induced damage in MgO. Phys. Status Solidi A *41*, 569 (1977)

Sathyamoorthy, A., Luthra, J.M.: Mechanism of thermoluminescence in magnesium oxide. J. Mater. Sci. *13*, 2637 (1978)

Sergeev, V.M., Leoniuk, N.I., Belov, N.V.: Thermoluminescent investigation of yttrium-aluminium borate crystals. Dokl. Akad. Nauk SSSR *240*, 1347 (1978)

Sharma, L.K.: Thermoluminescence studies of "gamma" irradiated nickel-doped alumina. Ind. J. Pure Appl. Phys. *15*, 142 (1977)

Shoaib, K.A., Hashmi, F., Bukhari, S., Ali, M.: Thermoluminescence from
   x-ray-irradiated stabilized ZrO_2 single crystals. Phys. Status Solidi A *10*,
   605 (1977)
Sikora, A.V., Voitsenya, T.I., Gritsyna, V.T., Uvarov, B.S.: Mechanism of
   red thermoluminescence of chromium(3+) doped alumina crystals. Ukr. Fiz.
   Zh. *24*, 299 (1979)
Spurny, Z., Hobzova, L.: A band model of energy levels and thermoluminescence
   mechanism in BeO ceramics. Radiochem. Radioanal. Lett. *29*, 287 (1977)
Takeuchi, N., Inabe, K., Yamashita, J., Nanto, H.: Thermoluminescence of
   magnesium oxide single crystals. J. Soc. Mater. Sci. Jpn. *26*, 868 (1977)
Tomita, A., Tsutsumi, K.: Thermoluminescence of BeO ceramics. Jpn. J. Appl.
   Phys. *18*, 397 (1979)
Trukhin, A.N., Etsin, S.S., Shendrik, A.V., Stuchka, P.: Luminescence centers
   and electronic processes in crystalline and glassy SiO_2-Ag. Izv. Akad.
   Nauk SSSR Ser. Fiz. *40*, 2329 (1976) [English transl.: Bull. Acad. Sci.
   USSR, Phys. Ser. *40*, 34 (1976)]
Tsukuda, Y.: Study on blackness of Y_2O_3 sintered body by thermoluminescence
   analysis. Yogyo-Kyokai-Shi *85*, 469 (1977)
Vainer, V.S., Vainger, A.I.: Investigation of the formation and changes in
   point defects in Y_2O_3 single crystals. Fiz. Tverd. Tela *19*, 528 (1977)
White, G.S., Lee, K.H., Crawford, Jr., J.H.: Effects of gamma-irradiation
   upon the optical behavior of spinel. Phys. Status Solidi A *42*, K137 (1977)

Sulfides

Amiryan, A.M., Gurvich, A.M., Katomina, R.V., Petrova, I. Yu.: Recombi-
   nation processes and emission spectrum of terbium in oxysulphides. Zh.
   Prikl. Spektrosk. *27*, 468 (1977)
Chaudhary, R.K., Kishore, L.: Thermoluminescence of CaS:Zr:Gd phosphors
   excited at room temperature. Ind. J. Pure Appl. Phys. *15*, 521 (1977)
Diwan, P.S., Ranade, J.D.: Phosphorescence and thermoluminescence studies
   of CaS(Bi:Mn) phosphors. Ind. J. Pure Appl. Phys. *15*, 523 (1977)
Elmanharawy, M.S.: Thermoluminescence of ZnS:CdS(Ag:Ni:Co) phosphors. Czech.
   J. Phys. Sect. B *B27*, 822 (1977)
Elmanharawy, M.S., Eid, A.H.: Further investigations into the luminescence
   of silver-activated ZnS:CdS phosphors containing nickel and cobalt. Acta
   Phys. Acad. Sci. Hung., 117 (1978)
Gorban, I.S., Gumenyuk, A.F., Grishchenko, G.A., Tychina, I.I.: Trapping
   levels in tetragonal zinc diphosphide crystals. Fiz. Tekh. Poluprovodn.
   12, 410 (1978) [English transl.: Sov. Phys. Semicond. *12*, 238 (1978)]
Hook III, J.W., Drickamer, H.G.: High-pressure studies of thermolumines-
   cence of doped ZnS phosphors. J. Appl. Phys. *49*, 2503 (1978)
Kanari, P.S., Balkrishna, S., Misra, G.C.: Photoluminescence in ZnS (Cu,Pt)
   phosphors. Ind. J. Pure Appl. Phys. *16*, 5 (1978)
Poryvkina, L.V.: Charge transformation $Eu^{2+}Eu^{3+}$ in CaS and CaO. Eesti NSV.
   Tead. Akad. Fuus. Inst. Uurim. *47*, 121 (1977)
Proskura, A.I.: Participation of the recombination barrier in the storage
   of excitation by ZnS-Cu crystal phosphors. Opt. Spektrosk. *42*, 1121 (1977)
   [English transl.: Opt. Spectr. *42*, 645 (1977)]
Rao, R.P., Rao, D.R., Banerjee, H.D.: Studies on the TL and decay of BaS:Cu
   phosphors. Mater. Res. Bull. *13*, 491 (1978)
Shalgaonkar, C.S., Lawangar, R.D., Pawar, S.H.: Thermoluminescence of
   CaS:Bi:Dy phosphors. Mater Res. Bull. *12*, 525 (1977)
Vaidya, S., Diwan, P.S., Ranade, J.D.: Thermoluminescence and fluorescence
   studies of CaS(Pb+Mn) phosphors. Ind. J. Pure Appl. Phys. *16*, 486 (1978)

Voolaid, Kh.I., Ots, A.E.: Nature of the deep capture centers in CaS and ZnS phosphors. Izv. Akad. Nauk SSSR Ser. Fiz. *40*, 2310 (1976) [English transl.: Bull. Acad. Sci. USSR, Phys. Ser. *40*, 17 (1976)]

Phosphates, Sulphates, etc.

Avezov, A.D., Vakhidov, S.A., Gasanov, E.M., Novozhilov, A.I.: Some luminescence properties of synthetic mica fluorphlogopite. Phys. Status Solidi A *44*, K87 (1977)

Bapat, V.N.: Thermoluminescence process in $CaSO_4$:Eu. J. Phys. C *10*, L465 (1977)

Iacconi, P., Caruba, R.: Influence of OH⁻ on thermoluminescent emission of synthetic hydroxyled zircon $Zr(SiO_4)_{1-x}(OH)_{4x}$. C.R. Hebd. Seances Acad. Sci. Ser B *285*, 227 (1977)

Karaseva, L.G., Bondarenko, G.P., Gromov, V.V.: Investigation of optical active centres of the irradiated lithium niobate. Radiat. Phys. Chem. *10*, 241 (1977)

Lapraz, D., Baumer, A., Keller, P., Turco, G.: Thermoluminescent properties of $Ca_{10}(PO_4)_6(OH)_2$:Mn hydroxyapatite. C.R. Hebd. Seances Acad. Sci. Ser. B *285*, 291 (1977)

Merzlyakov, A.T., Krongauz, V.G.: A study of emission and trapping centers in a new x-ray phosphor:europium-activated barium orthophosphate. Izv. Akad. Nauk SSSR Ser. Fiz. *41*, 1384 (1977)

Otero, M.J.: Luminescence of terbium ions in zinc borate crystals. Opt. Pura Apl. *9*, 175 (1976)

Pawar, S.H.: Crystal growth and thermoluminescence of dysprosium-doped calcium sulfate. Ind. J. Pure Appl. Phys. *16*, 1034 (1978)

Sakaguchi, M., Ohta, M., Nakazato, K., Kondoh, T.: Thermoluminescence characteristics of binary sulfate phosphors. J. Electrochem. Soc. *124*, 1272 (1977)

Shinde, S.S., Shastry, S.S.: Relative thermoluminescence response of dysprosium-doped calcium sulfate to alpha and gamma rays. Int. J. Appl. Radiat. Isot. *30*, 75 (1979)

Sivasankar, V.S. Whippey, P.W.; Thermal stability of some radiation damage products in x-irradiated $NaClO_3$. Can. J. Phys. *57*, 128 (1979)

Tomita, A., Tsutsumi, K.: Emission spectra of thermoluminescence in $CaSO_4$. Jpn. J. Appl. Phys. *17*, 453 (1978)

Tuduri, M., Montojo, M.T.: Thermically induced light emission in triglycine sulphate crystals. Ferroelectrics *21*, 369 (1978)

Zimoglyad, I.S., Kazankin, O.N.: The question of thermostimulated luminescence of calcium halophosphate. Zh. Prikl. Spektrosk. *28*, 909 (1978)

Glasses

Doil, A.: Radiation-induced centers in lithium disilicate glass. Jpn. J. Appl. Phys. *17*, 279 (1978)

Kornienko, L.S., Rybaltovskii, A.O., Chernov, P.V.: Electron-paramagnetic resonance and thermoluminescence of gamma-irradiated vitreous silica. Fiz. Khim. Stekla *2*, 396 (1976)

Kornienko, L.S., Rybaltovskii, A.O., Chernov, P.V.: Formation of radiation colour centres in transparent dielectrics by ultraviolet laser-plasma radiation. Fiz. Tverd. Tela *19*, 918 (1977) [English transl.: Sov. Phys. Solid State *19*, 535 (1977)]

Vorob'ev, A.A., Zavadovskaya, E.K., Starodubtsev, V.A.: Irradiation effects and the physical properties of aluminophosphate glasses and the energy of the chemical bonds. Fiz. Khim. Stekla *3*, 54 (1977) [English transl.: Sov. J. Glass Phys. Chem. *3*, 48 (1977)]

Diamond, Carbides

Sobolev, E.V., Eliseev, A.P.: Thermo-stimulated luminescence and phosphorescence of natural diamonds at low temperatures. Zh. Strukt. Khim. *17*. 933
(1976) [English transl.: J. Struct. Chem. *17*, 799 (1976)]
Sobolev, E.V., Eliseev, A.P.: The energy levels of the N9 and ND1 centers
in diamonds. Zh. Strukt. Khim. *17*, 935 (1976) [English transl.: J. Strukt.
Chem. *17*, 802 (1976)]
Vakulenko, O.V., Shutov, B.M.: Thermoluminescence of silicon-carbide excited
by light outside the fundamental absorption region. Sov. Phys. Semiconductors *12*, 603 (1978)

Organic Compounds

Abkowitz, M., Prest, W.M., Luca, D.J., Pfister, G.: Thermocurrent spectra
of gamma-phase-containing PVF2 films. Appl. Phys. Lett. *34*, 19 (1979)
Aramu, F., De Pascale, T., Maxia, V., Spano, G.: On the thermoluminescence
build-up in molecular crystals. J. Lumin. *16*, 97 (1978)
Chernokolev, A.T., Kikushkin, A.K., Solntsev, M.K.: Effect of the redox
properties of the medium on the thermoluminescence of chloroplasts from
higher plants. Biofizika *23*, 554 (1978)
Chernokolev, A.T., Kukushkin, A.K., Solntsev, M.K.: Thermoluminescence study
of processes related to electron-transport in higher plant chloroplasts.
Vestn. Mosk. Univ. Ser. Fiz. Astronomii *20*, 27 (1979)
David, C., Proumen-Demiddel, A., Geuskens, G.: Formation and decay of
radical ions of naphtalene and emission of thermoluminescence in irradiated poly(methylmethacrylate) Radiat. Phys. Chem. *11*, 63 (1978)
Desai, T.S., Tatake, V.G., Sane, P.V.: Origin of low-temperature thermoluminescence in nucleic-acid bases, excitation and emission spectral
studies. Photochem. Photobiol. *26*, 459 (1977)
Desai, T.S., Tatake, V.G., Sane, P.V.: Characterization of low-temperature
thermoluminescence band-ZV in leaf, explanation for its variable nature.
Biochim. Biophys. Acta *462*, 775 (1977)
Fleming, R.J., Pender, L.F.: Electron trap distributions in organic polymers - an additional set of characterisation parameters. J. Electrostat.
3, 139 (1977)
Hashimoto, T., Shimada, H., Sakai, T.: Thermoluminescence from gamma-ray
irradiated normal paraffins. Nature *268*, 225 (1977)
Inoue, Y., Yamashita, T., Kobayashi, Y., Shibata, K.: Thermoluminescence
changes during inactivation and reactivation of oxygen-evolving system
in isolated-choroplasts. FEBS Lett. *82*, 303 (1977)
Kazakov, V.P., Korobeinikova, V.N., Afonichev, D.D.: Radiothermoluminescence and temperature deactivation of UO_2^{2+} luminescence in dimethyl
sulphoxide. Zh. Prikl. Spektrosk. *29*, 633 (1978)
Kieffer, F., Klassen, N.V., Lapersonne-Meyer, C.: Electron tunnelling from
matrix-trapped biphenyl anions to biphenyl cations. J. Lumin. *20*, 17 (1979)
Korobeinikova, V.N., Tolstikov, G.A., Kazakov, V.P., Lerman, B.M.: Thermoluminescence of plastic crystals (1,4-diazobicyclo (2.2.2.) octane and
adamantane). Zh. Prikl. Spektrosk. *29*, 266 (1978)
Kuleschow, I.W., Remisowa, A.A.: Influence of structural factors on melts
and freezing of polyvinylidene fluoride. Plaste Kautsch. *24*, 635 (1977)
Linkens, A., Vanderschueren, J.: Experimental studies of the relationship
between thermoluminescence and molecular relaxation processes in polymers.
J. Electrostat. *3*, 149 (1977)
Nakamura, S., Sawa, G., Ieda, M.: Anomalous luminescence from oxidized poly-
ethylene in a high temperature region. J. Appl. Phys. *48*, 3626 (1977)
Padhye, M.R., Tamhane, P.S.: Thermoluminescence of poly(ethyleneterephtalate)

films. Angew. Makromol. Chem. *67*, 79 (1978)

Radhakrishna, S., Rama Krishna Murthy, M.: Methyl (methacrylate). J. Polym. Sci. Polym. Phys. Ed. *15*, 987 (1977)

Radhakrishna, S., Rama Krishna Murthy, M.: Thermoluminescence of irradiated polystyrene. J. Polym. Sci. Polym. Phys. Ed. *15*, 1261 (1977)

Rafikov, S.R., Korobeinikova, V.N., Lotnik, S.V.: Radiothermoluminescence of poly(methylmethacrylate) sparsely cross-linked by dimethacryloxy-methylanthracene. Vysokomol. Soedin. Ser. A *20*, 766 (1978)

Samoilov, S.M., Aulov, V.A.: Radiothermoluminescence investigation of "beta" and "gamma" relaxation transitions in copolymers of ethylene. Polym. Sci. USSR *18*, 1124 (1976)

Suzuoki, Y., Yasuda, K., Mizutani, T., Ieda, M.: Thermoluminescence in high density polyethylene and its oxidation effects. Jpn. J. Appl. Phys. *16*, 1339 (1977)

Takai, Y., Mori, K., Mizutani, T., Ieda, M.: Thermoluminescence from photo-excited polyethylene terephthalate. J. Polym. Sci. Polym. Phys. Ed. *16*, 1861 (1978)

Zlatkevich, L., Shepelev, M.: Study of latex film structure by radiothermo-luminescence. J. Polym. Sci. Polym. Phys. Ed. *16*, 427 (1978)

Zyball, A.: Thermoluminescence spectrum of polymers. Prog. Colloid and Polym. Sci. *64*, 185 (1978)

B.1.9 Effect of Electric Fields on TSL and TSC

Kiveris, A.Yu., Kudzhmauskas, Sh.P. Pipinis, P.A.: Mechanism of electric-field liberation of electrons from traps in ZnS:Pb crystal phosphors. Fiz. Tverd. Tela *19*, 3485 (1977) [English transl.: Sov. Phys. Solid State *19*, 2039 (1977)]

Nouailhat, A.: Electric field effect on the mechanisms of recombination in KI doped with divalent ions. J. Lumin. *17*, 185 (1978)

Takai, Y., Mori, K., Mizutani, T., Ieda, M.: Field quenching of thermolumi-nescence from photoexcited polyethylene terephthalate (PET). J. Phys. D *11*, 991 (1978)

Vorob'ev, Yu.V., Il'yashenko, A.G., Sheinkman, M.K.: Anomalously strong electric-field-induced release of carriers from trapping levels in high resistivity GaAs:Cr. Fiz. Tekh. Poluprovodn. *11*, 791 (1977) [English transl.: Sov. Phys. Semicond. *11*, 456 (1977)]

B. 2 Application of Thermally Stimulated Luminescence to Archaeology and Geology

B.2.1 Authentication with the Use of Thermoluminescence

Becker, K., Goedicke, C.: A quick method for authentication of ceramic art objects. Nucl. Instrum. Methods (Netherlands) *151*, 313-316 (1978)

Flemming, S.J.: Thermoluminescence authentication of ancient ceramics and bronzes. Bull. Am. Phys. Soc. *24*, 29 (1979)

Shaplin, P.D.: Thermoluminescence and style in the authentication of ceramic sculpture from Oaxaca, Mexico. Archaeometry *20*, 47-54 (1978)

B.2.2 Thermoluminescence for Archaeological Uses

Aitken, M.J.: Archaeological involvements of physics. Phys. Rep. Phys. Lett. Sect. C (Netherlands) *40* C, 277-351 (1978)

Bailiff, I.K., Bowman, S.G.E., Mobbs, S.F., Aitken, M.J.: The phototransfer technique and its use in thermoluminescence dating. J. Electrostat. (Netherlands) *3*, 269-280 (1977)

Higashimura, T., Ichikawa, Y.: Thermoluminescence dating of pottery. J. Atomic Energy Soc. Jpn. *20*, 224-228 (1978)

Melcher, C.L., Zimmerman, D.W.: Thermoluminescent determination of prehistoric heat treatment of chert artifacts. Science *197*, 1359-1362 (1977)

Walker, R.M., Yuhas, M.P., Zimmerman, D.W.: The radioactive inclusion method of thermoluminescence dating of ceramic objects. Proc. Applicazione dei metodi nucleari nel campo delle opere d'arte, Congresso Internationale (1973) pp.483-492 (1976)

Warren, S.E.: Thermoluminescence dating of pottery. Archaeometry *20*, 71-72 (1978)

Wintle, A.G., Aitken, M.J.: Thermoluminescence dating of burnt flint: Application to a lower palaeolithic site, Terra Amata. Archaeometry *19*, 111-130 (1977)

Zimmerman, D.W.: Review of radiation dosimetry and dating by thermally stimulated processes. J. Electrostat. (Netherlands) *3*, 257-268 (1977)

B.2.3 Thermoluminescence for Geological Uses

Dreimanis, A., Hutt, G., Raukas, A., Whippey, P.W.: Dating Methods of Pleistocene deposits and their problems: thermoluminescence dating. Geoscience Canada *5*, 55-60 (1978)

Huntley, D.J., Bailey, D.C.: Obsidian source identification by thermoluminescence. Archaeometry *20*, 159-170 (1978)

Hutt, G., Punning, J., Smirnow, A.: Thermoluminescent dating in its application to geology. Eesti NSV Tead. Akad. Toim. Keem. Geol. *26*, 284-288 (1977)

Huxtable, J., Aitken, M.J., Bonhommet, N.: Thermoluminescence dating of sediment baked by lava flows of the chaine des Puys. Nature *275*, 207-209 (1978)

Levy, P.W., Holmes, R.J., Ypma, P.J., Chen, C.C., Swiderski, H.S.: New thermoluminescence techniques for mineral exploration. Nuclear Techniques and Mineral Resources 1977 pp.523-538 (1977)

Miki, T., Ikeya, M.: Thermoluminescence and ESR dating of akiyoshi stalactite. Jpn. J. Appl. Phys. *17*, 1703-1704 (1978)

Mitin, S.N., Kononov, O.V., Pashin, V.N.: Use of anhydrite thermoluminescence for the paleotemperature reconstruction of oil and gas basins. Dokl. Akad. Nauk SSSR *243*, 1283-1285 (1978)

Romberg, G.P., Prepejchal, W., Spigarelli, S.A.: Temperature exposure measured by the use of thermoluminescence. Science *197*, 1364-1365 (1977)

Shaw, G.H.: Interpretation of low velocity zone in terms of presence of thermally activated point defects. Geophys. Res. Lett. *5*, 629-632 (1978)

Vaz, J.E., Sifontes, R.S.: Radiometric survey using thermoluminescence dosimetry in the cerro impacto (Venezuela) thorium deposit. Mod. Geol. *6*, 147-152 (1978)

Vorob'ev, A.A., Salñikov, V.N.: Radio emission and anomalous changes in electrical conductivity in heated rock and mineral specimens. Sov. Min. Sci. *12*, 461-470 (1976)

B.2.4   Application of Thermoluminescence to Geological Considerations in Space

Bagolia, C., Doshi, N., Lal, D.: Preatmospheric size of the barwell meteorite: Cosmic-ray track, fusion crust and thermoluminescence studies. Nucl. Track Detect. (GB) 2, 29-35 (1978)

Durrani, S.A.: Charged-particle track analysis, thermoluminescence and microcratering studies of lunar samples. Philos. Trans. R. Soc., London A 285, 309-317 (1977)

Forman, M.: IO may have a bright dawn terminator. Nature 275, 519-520 (1978)

McKeever, S.W.S., Sears, D.W.: Thermoluminescence and terrestrial age of the Estacado Meteorite. Nature 275, 629-630 (1978)

Vaz. J.E., Sears, D.W.: Artificially-induced thermoluminescence gradients in stony meteorites. Meteoritics 12, 47-60 (1977)

B.2.5   Thermoluminescence of Geological Minerals

Baumer, A., Lapraz, D.: Relations between starting mixture and hydroxyapatites synthesized by hydrothermal method in the $CaO-P_2O_5-H_2O$, thermoluminescence study. Bull. de Mineralogie 101, 53-56 (1978)

Gomaa, M.A., Eid, A.M.: Thermoluminescence applications of heated sand. Atomkernenergie 29, 290-292 (1977)

Ivaldi, J.P.: New data on thermoluminescence of alpine barren quartzose lodes and their age of crystallization. C.R. Acad. Sci. Ser. D 288, 457-460 (1979)

Kroh, J., Stradowski, C.: Electron trap depth in irradiated alkaline ice. Radiochem. Radioanal. Lett. 36, 233-234 (1978)

Leach, B.F., Fankhauser, B.: Characterization of New-Zealand obsidian sources by use of thermoluminescence. J. R. Soc. N.Z. 8, 331-342 (1978)

May, R.J.: Thermoluminescence dating of Hawaiian alkalic basalts. J. Geophys. Res. 82, 3023-3029 (1977)

Meakins, R.L., Clark, G.J., Dickson, B.L.: Thermoluminescence studies of some natural and synthetic opals. Am. Mineral. 63, 737-743 (1978)

Moss, A.L., McKlveen, J.W.: Thermoluminescent properties of topaz. Health Phys. 33, 649-650 (1977)

Moss, A.L., McKlveen, J.W.: Thermoluminescent properties of topaz. Health Phys. 34, 137-140 (1978)

Nambi, K.S.V., Mitra, S.: Thermoluminescence investigations of natural calcite crystals of differing genesis. Thermochim. ACTA 27, 61-67 (1978)

Nassau, K.: The origins of color in minerals. Am. Mineral. 63, 219-229 (1978)

Prokic, M.: Analysis of the thermoluminescence glow curves of natural barite. J. Phys. Chem. Sol. 38, 617-622 (1977)

Sekulic, M., Milosevic, O.: The thermoluminescence (TL) of barytes with a large magnitude of total emitted light. Mod. Geol. 6, 247-250 (1978)

Sergeev, V.M., Ventslovaite, E.I.: Determination of relative formation temperatures of natural crystals by thermoluminescent method. Dokl. Akad. Nauk SSSR 241, 1182-1185 (1978)

Wintle, A.G.: Thermoluminescence dating study of some quaternary calcite - potential and problems. Can. J. Earth Sci. 15, 1977-1986 (1978)

Wintle, A.G.: Detailed study of a thermoluminescent mineral exhibiting anomalous fading. J. Lumin. 15, 385-393 (1977)

Wintle, A.G.: Thermoluminescence dating of minerals-traps for the unwary. J. Electrostat. (Netherlands) 3, 281-288 (1977)

B.2.6   Thermoluminescence of Quartz

Aloisi, J.C., Charlet, J.M., Wiber, M.: Is thermoluminescence of detrital quartz grains influenced by their exoscopy properties - study about some samples of Lions Gulf, Mediterranean area. Rev. Geogr. Phys. Geol. Dyn. *19*, 251-258 (1977)

Bell, W.T., Zimmerman, D.W.: The effect of HF acid etching on the morphology of quartz inclusions for thermoluminescence dating. Archaeometry *20*, 63-65 (1978)

David, M., Ganguly, A.K.: Change in TL sensitivity of quartz due to stress. Proc. Intern. Symp. Radiation Phys., pp. 231-233 (1977)

David, M. Ganguly, A.K., Sunta, C.M.: Thermoluminescence of quartz, I. Glow curve and spectral characteristics. Indian J. Pure Appl. Phys. *15*, 201-204 (1977)

David, M., Sunta, C.M., Ganguly, A.K.: Thermoluminescence of quartz, II. Sensitization by thermal treatment. Indian J. Pure Appl. Phys. *15*, 277-280 (1977)

David. M., Sunta, C.M., Bapat, V.N., Ganguly, A.K.: Thermoluminescence of quartz, III. Sensitization by pre-gamma exposure. Indian J. Pure Appl. Phys. *16*, 423-427 (1978)

Durrani, S.A., Groom, P.J., Khazal, K.A.R., McKeever, S.W.S.: The dependence of the thermoluminescence sensitivity upon the temperature of irradiation in quartz. J. Phys. D *10*, 1351-1361 (1977)

Durrani, S.A., Khazal, K.A.R., McKeever, S.W.S.: Studies of changes in the thermoluminescence sensitivity in quartz induced by proton and gamma irradiations. Radiat. Eff. *33*, 237-244 (1977)

Fuller, G.E., Levy, P.W.: Thermoluminescence of natural quartz. Bull. Am. Phys. Soc. *23*, 324 (1978)

Hutt, G., Vares, K., Smirnov, A.: Thermoluminescent and dosimetric properties of quartz from quaternary deposits. Eesti NSV. Tead. Akad. Toim., Keem., Geol. *26*, 275-283 (1977)

Hutt, G., Smirnov, A.V.: Dosimetric properties of natural quartz and prospects for using it for thermoluminescent dating of geological substances. Izv. Akad. Nauk. SSSR Ser. Fiz. *41*, 1367-1369 (1977)

Ji-Ling Li, Jing-Xian Pei, Zai-Zhung Wang, Yan-Chou Lu: The thermoluminescence of quartz silts from loess and dating of the loessial layers. Kexue Tongbao *22*, 498-502 (1977)

Mazeran, R.: Thermoluminescence in quartz-containing uranium veins of the Baurbonnaise mountain. C.R. Hebd. Seances Acad. Sci. Ser. D. *285*, 633-636 (1977)

Morsi, M.M., Gomaa, M.A.: Thermoluminescence response of quartz and some silicate-glasses to gamma-rays and UV radiations. Central Glass and Ceramic Res. Inst. Bull. *24*, 43-48 (1977)

Pogorelov, Y.L., Matrosov, I.I., Mashkovtsev, R.I.: Study of radiation defects in quartz induced by alpha-particles by EPR and thermoluminescence methods. Izv. Vyssh. Uchebn. Zaved. Fiz. *22*, iss. 2, 110-112 (1979)

Semenov, K.P., Fotchenkov, A.A.: Effectiveness of the action of various forms of radiation on quartz. Kristallografiya *22*, 571-578 (1977)

Sutton, S.R., Zimmerman, D.W.: Thermoluminescence dating: Radioactivity in quartz. Archaeometry *20*, 67-69 (1978)

B.2.7   Thermoluminescence of Zircon

Jain, V.K.: Some aspects of the thermoluminescence of zircon (sand) and zirconia. Indian J. Pure Appl. Phys. *15*, 601-605 (1977)

Jain, V.K.: Thermoluminescence glow curve and spectrum of zircon (sand). Bull. de Mineralogie *101*, 358-362 (1978)

B. 3 Application of Thermally Stimulated Luminescence to Dosimetry

B.3.1 Thermoluminescence Instrumentation

Bailiff, I.K., Morris, D.A., Aitken, M.J.: A rapid-scanning interference spectrometer: Application to low-level thermoluminescence emission. J. Phys. E *10*, 1156-1160 (1977)

Botter-Jensen, L.: A simple, hot N_2-gas TL reader incorporating a post-irradiation annealing facility. Nucl. Instrum. Methods (Netherlands) *153*, 413-418 (1978)

Carter, A.C.: The use of colour transparency film for estimation of thermoluminescence spectra. Phys. Med. Biol. *22*, 1022-1024 (1977)

Flage, Jr., R.L.: Design improvement for TLD card reader. Health Phys. *36*, 84-85 (1979)

Flemming, R.J.: The dependence of thermoluminescence intensity on heating rate. Jpn. J. Appl. Phys. *16*, 1289 (1977)

Früh, R., Brunner, P., Bobleter, O.: Automatic TLD-readers-comparison of three commercially available instruments. Atomkernenergie *31*, 127-130 (1978)

Gorbunov, V.I., Sviryakin, D.I., Egorenko, Y.A.: Luminescent converters in radiation flaw detection. Defektoskopiya *12*, 48-53 (1976)

Horowitz, Y.S., Freeman, S.: Response of ^{6}LiF and ^{7}LiF thermoluminescent dosimeters to neutrons. Nucl. Instrum. Methods *157* (2) 393-396 (1978)

Hunt, G.F.: Microprocessor controlled, personnel thermoluminescence dosimetry reader. IEEE Trans. NS-*26*, 773 (1979)

Koci, J. Hava, L., Spurny, Z.: Simple vacuum tweezer for thermoluminescent dosimetry. Jaderna Energie *24*, 60 (1978)

Manche, E.P.: Differential thermoluminescence (DTL) - A new instrument for measurement of thermoluminescence with suppression of blackbody radiation. Rev. Sci. Instrum. *49*, 715-717 (1978)

Otsuka, I., Onuma, I., Otsuka, H.: A versatile, highly sensitive thermoluminescence dosimeter reader. Rep. Inst. Phys. Chem. Res. *54*, 49-55 (1978)

Onuma, I., Ikegami, Y., Otsuka, I., Otsuka, H.: Compact gamma-ray irradiator for TLD. Rep. Inst. Phys. Chem. Res. *54*, 56-59 (1978)

Piesch, E., Burgkhardt, B., Sayed, A.M.: Activation and damage effects in TLD 600 after neutron irradiation. Nucl. Instrum. Methods *157*, 179-184 (1978)

Szabo, B., Szabo, P.P., Makra, S., Vagvolgyi, J., Soos, J.: TLD-04B-universal apparatus for measurement of thermoluminescent materials. Report KFKI-77-34 (1977) Hungarian Acad. Sci., Budapest

Szabo, B., Szabo, P.O., Makra, S., Vagvolgyi, J., Soos, J.: The TLD-04B thermoluminescent reader for research and routine dosimetry applications. KFKI-77-33 (1977), Hungarian Acad. Sci., Budapest

Yarza, J.C.: Influence of heating element design on the performance of LiF: PTFE Thermoluminescent dosimeters. Phys. Med. Biol. *23*, 164-170 (1978)

B.3.2 Thermoluminescent Dosimetric Materials

Aypar, A.: Studies on thermoluminescent $CaSO_4$:Dy for dosimetry. Intern. J. Appl. Radiation Isotopes *29*, 369-372 (1978)

Bakulin, Y.P., Kostyukov, N.S., Antonova, N.P., Korchagin, A.S.: Thermoluminescent and dosimetric characteristics of electroceramic samples. Izv. Akad. Nauk SSSR Ser. Fiz. *41*, 1360-1362 (1977)

Balasubrahmanyam, V., Measures, M.P.: Internal background build-up measurement in CaF_2:Mn thermoluminescent dosimeters. Health Phys. *32*, 317-318 (1977)

Burgkhardt, B., Singh, D., Piesch, E.: High-dose characteristics of CaF_2 and $CaSO_4$ thermoluminescent dosimeters. Nucl. Instrum. Methods *141*, 363-368 (1977)

Burgkhardt, B., Piesch, E., Singh, D.: High-dose characteristic of LiF and $Li_2B_4O_7$ thermoluminescent dosimeters. Nucl. Instrum. Methods *148*, 613-617 (1978)

Carter, A.C.: Possible use for industrial diamonds in radiation dosimetry. Ind. Diamond. Rev., 239-241 (1977)

Chaudhri, M.A.: Activation analysis of dosimetric LiF (TLD-100) with cyclotron-produced neutrons. Int. J. Appl. Radiat. Isot. *29*, 683-686 (1978)

DeWerd, L.A.: Handling techniques for thermoluminescent dosimeters. Health Phys. *31*, 525-527 (1976)

Driscoll, C.M.H.: Studies of the effect of LET on the thermoluminescent properties of thin lithium fluoride layers. Phys. Med. Biol. *23*, 777-781 (1978)

Gammage, R.B., Cheka, J.S.: Further characteristics important in the operation of ceramic BeO TLD. Health Phys. *32*, 189-192 (1977)

Gavrilov, F.F., Betenekova, T.A., Terentëv, G.I.: X-ray detectors based on the mixed crystals LiH-LiF, LiH-LiCl, and the crystals LiH-Ce. Izv. Akad. Nauk SSSR Ser. Fiz. *41*, 1358-1359 (1977)

Hashmi, F.H., Shoaib, K.A., Bukhari, S.J.H., Ali, M., Husain, A.: Stabilized ZrO_2 as a thermoluminescent dosimeter for high gamma doses. Health Phys. *32*, 325-328 (1977)

Pin-Chieh Hsu, Chia-Liang Tseng, Pao-Shan Weng: Spurious thermoluminescence of calcium sulfate due to ambient atmosphere during irradiation. Nucl. Instrum. Methods *147*, 529-534 (1977)

Pin-Chieh Hsu, Chia-Liang Tseng, Pao-Shan Weng: Ambient atmosphere effect for lithium fluoride thermoluminescent dosimeters irradiated in a nuclear reactor mixed radiat... Nucl. Instrum. Methods *157*, 147-153 (1978)

Jain, V.K., Kathuria, S.P.: Thermoluminescence response of LiF to gamma-rays. Proc. Intern. Symp. Radiation Phys., 227-230 (1977)

Lakshmanan, A.R., Chandra Bhuwan, Bhatt, R.C.: Gamma radiation induced sensitization in $CaSO_4$:Dy TLD phospor. Nucl. Instrum. Methods *153*, 581-588 (1978)

Lakshmanan, A.R., Shinde, S.S., Bhatt, R.C.: Ultraviolet-induced thermoluminescence and phosphoresence in Mg_2SiO_4: Tb. Phys. Med. Biol. *23*, 952-960 (1978)

Lin, S., Vetter, R.J., Ziemer, P.L.: Thermoluminescence and microwave induced thermoluminescence fading of rare-earth-doped barium-titanate ceramics. Radiat. Eff. *38*, 67-71 (1978)

Mieke, S., Nink, R.: LiF-Ti as a material for thermoluminescence dosimetry (TLD). J. Lumin. *18*, 411-414 (1979)

Mikado, T., Tomimasu, T., Yamazaki, T., Chiwaki, M.: Thermoluminescence response of Mg_2SiO_4:Tb in electron fields. Nucl. Instrum. Methods *157*, 109-116 (1978)

Morgan, T.J., Brateman, L.: The energy and directional response of Harshaw TLD-100 thermoluminescent dosimeters in the diagnostic x-ray energy range. Health Phys. *33*, 340-341 (1977)

Nakajima, T., Murayama, Y., Matsuzawa, T., Koyano, A.: Development of a new highly sensitive LiF thermoluminescence dosimeter and its applications. Nucl. Instrum. Methods *157*, 155-162 (1978)

Nakajima, T., Murayama, Y., Matsuzawa, T.: Preparation and dosimetric properties of a highly sensitive LiF thermoluminescent dosimeter. Health Phys. *36*, 79-82 (1979)

Paun, J., Iozsa, A., Jipa, S.: Dosimetric characteristics of alkaline-earth tetraborates radiothermoluminescent detectors. Radiochem. Radioanal. Lett. *28*, 411-421

Petridou, C.H., Christodoulides, C., Charalambous, S.: Non-radiation induced thermoluminescence in pre-irradiated LiF (TLD-100). Nucl. Instrum. Methods *150*, 247-252 (1978)

Pradham, A.S., Ayyanger, K.: Radiation dosimetry by photostimulated luminescence of CaSO₄:Dy. Int. J. Appl. Isot. *28*, 534-535 (1977)

Pradham, A.S., Kher, R.K., Dere, A.: Photon energy dependence of CaSO₄:Dy embedded teflon TLD discs. Int. J. Appl. Radiat. Isot. *29*, 243-245 (1978)

Prokic, M.: Improvement of the thermoluminescence properties of the non-commercial dosimetry phosphors CaSO₄:Dy and CaSO₄:Tm. Nucl. Instrum. Methods *151*, 603-608 (1978)

Srivastava, J.K., Supe, S.J.: Stability of CaSO₄:Dy thermoluminescent phosphor in high dose region. Nucl. Instrum. Methods *155*, 233-235 (1978)

Mehta, S.K., Sengupta, S.: Annealing characteristics and nature of traps in Al₂O₃ thermoluminescent phosphor. Phys. Med. Biol. *22*, 863-872 (1977)

Mehta, S.K., Sengupta, S.: Photostimulated thermoluminescence of Al₂O₃ (Si,Ti) and its application to ultraviolet radiation dosimetry. Phys. Med. Biol. *23*, 471-480 (1978)

Simons, G.G., Emmons, L.L.: Evaluation of gamma-ray response calculations for ⁷LiF TLDS. Nucl. Instrum. Methods *160*, 79-85 (1979)

Takenaga, M., Yamamoto, O., Yamashita, T.: Lithium borate activated with copper and silver for TLD. J. At. Energy Soc. Jpn. *19*, 543-549 (1977)

Townsend, P.D., Wintersgill, M.C.: Nonlinear response of TL in LiF dosimeter material to mixed dose rates. Health Phys. *35*, 498-500 (1978)

Wintersgill, M.C., Townsend, P.D.: The nonlinear response of thermoluminescence in LiF dosimeter material to sequential irradiations at different dose rates. Radiat. Eff. *38*, 113-118 (1978)

Yamaoka, Y.: Effects of grain size on thermoluminescence and thermally stimulated exoelectron emission of LiF crystals. Health Phys. *35*, 708-711 (1978)

Yamazaki, T., Tomimasu, T., Mikado, T., Chiwaki, M.: On the supralinearity of Mg₂SiO₄:Tb thermoluminescent dosimeters. J. Appl. Phys. *49*, 4929-4932 (1978)

B.3.3 Thermoluminescent Mechanisms

Bartlett, D.T., Sandford, D.J.: Incompatibility of sensitization and re-estimation of lithium fluoride thermoluminescent phosphor. Phys. Medicine Biol. *23*, 332-334 (1978)

Bradbury, M.H., Lilley, E.: Precipitation reactions in thermoluminescent dosimetry crystals (TLD-100) and ionic conduction. J. Phys. D. *10*, 1261-1266 (1977)

Bradbury, M.H., Lilley, E.: Effect of solution treatment temperature and ageing on TLD-100 crystals (ionic conduction measurement). J. Phys. D. *10*, 1267-1274 (1977)

Budd, T., Marshall, M., Peaple, L.H.J., Douglas, J.A.: The low and high-temperature response of lithium fluoride dosimeters to x-rays. Phys. Med. Biol. *24*, 71-80 (1979)

Burgkhardt, B., Herrera, R., Piesch, E.: The effect of post-irradiation annealing on the fading characteristic of different thermoluminescent materials. Part I. Experimental results. Nucl. Instrum. Methods *155*, 293-298 (1978)

Burgkhardt, B., Piesch, E.: Effect of post-irradiation annealing on fading characteristics of different thermoluminescent materials. Part II. Optimal treatment and recommendations. Nucl. Instrum. Methods *155*, 299-304 (1978)

Cooke, D.W.: Thermoluminescence mechanism in LiF (TLD-100) - extension of Mayhugh-Christy model. J. Appl. Phys. *49*, 4206-4215 (1978)

Cooke, D.W., Alexander, C.: Thermoluminescence mechanism in LiF (TLD-100) from 90 to 500 K. Bull. Am. Phys. Soc. *23*, 202 (1978)

Cooke, D.W., Roberts, H.E., Alexander, Jr., C.: Thermoluminescence and emission spectra of UV-grade Al_2O_3 from 90 to 500 K. J. Appl. Phys. *49*, 3451-3457 (1978)

Grude, M.M., Ekmane, A.Y.: Variation of the parameters of thermally stimulated luminescence from LiF detectors. Izv. Akad. Nauk SSSR Ser. Fiz. *41*, 1354-1357 (1977)

Jain, V.K., Kathuria, S.P.: Z_3 center thermoluminescence in LiF TLD phosphor. Phys. Status Solidi A *50*, 329-333 (1978)

Lakshmanan, A., Bhatt, R.C.: Gamma-ray induced sensitization and residual thermoluminescence in common TLD phosphors. Int. J. Appl. Radiat. Isot. *29*, 353-358 (1978)

Lucas, A.C., Kapsar, B.M.: Alteration of the fading of the thermoluminescence of LiF. Health Phys. *35*, 914 (1978)

Nakajima, T.: Theoretical consideration on thermoluminescence response. J. Appl. Phys. *48*, 4880-4885 (1977)

Nakajima, T.: Influence of absorbed dose on spectrum of thermoluminescence of CaF_2:Dy. Jpn. J. Appl. Phys. *16*, 1061-1062 (1977)

Oliveri, E., Fiorella, O., Mangia, M.: High-dose behavior of $CaSO_4$:Dy Thermoluminescent phosphors as deduced by a continuous model for trap depths. Nucl. Instrum. Methods *154*, 203-205 (1978)

Takeuchi, N., Inabe, K., Kido, H., Yamashita, J.: Formulation of the sensitization in LiF thermoluminescence based on a simple two-step reaction rate model. J. Phys. C *11*, L147-L149 (1978)

Wall, J.A.: Annealing of TLDs at low light levels. IEEE Trans. NS-*25*, 1093 (1978)

Yamazaki, T., Tomimasu, T., Mikado, T., Chiwaki, M.: Supralinearity of Mg_2SiO_4:Tb thermoluminescent dosimeters. J. Appl. Phys. *49*, 4929-4932 (1978)

B.3.4   General Thermoluminescence Applications to Dosimetry

Anderson, M.E., Crain, S.L.: A combination TL-film personnel neutron dosimeter. Health Phys. *36*, 76-79 (1979)

Archer, B.R., Glaze, S.A., North, L.B., Bushong, S.C.: Desk-top computer assisted processing of thermoluminescent dosimeters. Health Phys. *33*, 150-154 (1977)

Barinov, A.L., Bochvar, I.A., Gimadova, T.I., Keirim-Markus, I.B.: Curve path with hardness of thermoluminescent glass dosimeters for photons with up to 6 MeV energy. Prib. Tekh. Eksp. *78*, 75-79 (1978)

Bowlt, C.: Thermally stimulated effects in dielectrics and their application to radiation dosimetry. Contemp. Phys. *17*, 461-482 (1976)

Charles, M.W., Khan, Z.U.: Implementation of the ICRP recommendation on skin dose measurement using thermoluminescent dosimeters. Phys. Med. Biol. *23*, 972-975 (1978)

Dennis, J.A., Marshall, T.O., Shaw, K.B., Kendall, G.M.: NRPBs new record-keeping service and thermoluminescent dosimeter. Br. J. Non-Destr. Test. *19*, 234-238 (1977)

Eckwerth, A., Ewen, K., Fischer, P.G.: Determination of the quality of radiation the dose-rate in radiodiagnostics and radiotherapy by means of thermoluminescence dosimeters.Strahlentherapie *155*, 114-116 (1979)

Eisenlohr, H.H., Jayaraman, S.: IAEA-WHO cobalt 60 - teletherapy dosimetry service using mailed LiF dosimeters - a survey of results obtained during 1970-75. Phys. Med. Biol. *22*, 18-28 (1977)

Ewen, K., Fischer, P.G.: Dose measurement in accordance with section 13 of the x-ray regulations with the aid of thermoluminescence dosimetry. Roentgenpraxis *30*, 215-217 (1977)

Furuta, Y., Tanaka, S.: Wide utilization of thermoluminescent dosimeters in the field of atomic energy. J. At. Energy Soc. Jpn. *20*, 559-566 (1978)

Gubatova, D.Y., Balode, G.Y., Frolova, A.V., Astakhova, I.V.: Dosimetry of long-wave x-ray radiation by the thermoluminescent method. Izv. Akad. Nauk SSSR Ser. Fiz. *41*, 1336-1341 (1977)

Horowitz, Y.S.: Response of ^{6}LiF and ^{7}LiF thermoluminescent dosimeters to neutrons. Nucl. Instrum. Methods *157*, 393-396 (1978)

Keirim-Markus, I.B.: Luminescent dosimetry of x-rays. Izv. Akad. Nauk SSSR Ser. Fiz. *41*, 1342-1345 (1977)

Mason, E.W., McKinlay, A.F., Saunders, D.: The re-estimation of absorbed doses of less than 1 rad measured with lithium fluoride thermoluminescent dosimeters. Phys. Med. Biol. *22*, 29-35 (1977)

Morgan, T.J., Brateman, L.: The energy and directional response of Harshaw TLD-100 thermoluminescent dosimeters in the diagnostic x-ray energy range. Health Phys. *33*, 339-341 (1977)

Niewiadomski, T., Jasinska, M., Ryba, E.: Comparative investigations of characteristics of various TL dosimeters. Nukleonika *18*, 535-549 (1973)

Niewiadomski, T.: Comparative investigations of characteristics of various TL dosimeters, Part II: Low-dose measurements.Nukleonika *21*, 1097-1109 (1976)

Oberhofer, M: High-level photon dosimetry with thermoluminescent materials. Atomenergie *31*, 209-216 (1978)

Otsuka, I., Otsuka, H.: Low-dose measurements by·thermoluminescence dosimeter. Rep. Inst. Phys. Chem. Res. *54*, 71-78 (1978)

Panzer, W., Regulla, D.F.: Solid state medical radiation dosimetry. II. Clinical applications. Roentgenpraxis *30*, 228-236 (1977)

Pologrudov, V.V., Karnaukhov, E.N., Parfianovich, I.A.: Light sum storage in dosimetric crystals during the linear stage. Izv. Akad. Nauk SSSR Ser. Fiz. *41*, 1346-1349 (1977)

Ruden, B.I., Bengtsson, L.G.: Accuracy of megavolt radiation dosimetry using thermoluminescent lithium fluoride. ACTA Radiol. Ther. Phys. Biol. *16*, 157-176 (1977)

Sabel, M., Ruff, A., Weishaar, J.: Thermoluminescent dosimetry for radiation exposure of breast during film and xeromammography. Fortschr. Geb. Roentgenstr. Nuklearmed. *128*, 616-622 (1978)

Shaver, I.KH., Krongauz, V.G.: Composite x-ray and gamma-ray detectors. Izv. Akad. Nauk SSSR Ser. Fiz. *41*, 1333-1335 (1977)

Wald, J., DeWerd, L.A., Stoebe, T.G.: Long term recycling characteristics of LiF (TLD-100) dosimeter material. Health Phys. *33*, 303-310 (1977)

Wochos, J.F., Fullerton, G.D., DeWerd, L.A.: Mailed thermoluminescent dosimeter determination of entrance skin exposure and half-value layer in mammography. Am. J. Roentgenol. *131*, 617-619 (1978)

B.3.5 Thermoluminescence Applied to Neutron Dosimetry

Anderson, M.E., Crain, S.L.: A combination TL-film personnel neutron dosimeter. Health Phys. *36*, 76-79 (1979)

Ayyangar, K., Chandra, B., Pradhan, A.S., Lakshmanan, A.R.: Additivity of alpha and gamma response in $Li_2B_4O_7$: Mn TLD phosphor. Health Phys. *35*, 568-569 (1978)

Bendel, W.L.: Displacement and ionization fractions of fast neutron kerma in TLDs and Si. IEEE Trans. NS-*24*, 2516-2520 (1977)

Bhatt, R.C., Lakshmanan, A.R., Chandra, B., Pradhan, A.S.: Fast neutron dosimetry using sulphur activation in $CaSO_4$:Dy TL dosimeters. Nucl. Instrum. Methods *152*, 527-529 (1978)

Douglas, J.A., Marshall, M.: The responses of some TL albedo neutron dosimeters. Health Phys. *35*, 315-324 (1978)

Furuta, Y., Tanaka, S.: Neutron Dosimetry by thermoluminescence dosimeter. Proc. Intern. Symp. Radiation Phys., 209-218 (1977)

Henson, A.M., Thomas, R.H.: Measurement of the efficiency of LiF thermoluminescent dosimeter of heavy ions. Health Phys. *34*, 389-390 (1978)

Horowitz, Y.S.: The thermal neutron sensitivity of LiF (TLD-700; Harshaw): The effect of sample size and batch origin. Phys. Med. Biol. *23*, 340-342 (1978)

Horowitz, Y.S., Dubi, A.: Comment on the use of thermoluminescence dosimeters as thermal neutron detectors. Nucl. Instrum. Methods *146*, 455 (1977)

Horowitz, Y.S., Freeman, S.: Response of LiF-6 and LiF-7 thermoluminescent dosimeters to neutrons incorporating thermoluminescent linear energy-transfer dependency. Nucl. Instrum. Methods *157*, 393-396 (1978)

Horowitz, Y.S., Ben Shahar, B., Dubi, A., Pinto, H.: Thermoluminescence in CaF_2: Dy and CaF_2: Mn induced by monoenergetic parallel beam, 81.0 meV diffracted neutrons. Phys. Med. Biol. *22*, 500-510 (1977)

Hsu, P., Tseng, C., Li, S., Weng, P.: Measurement of thermal neutrons and gamma-rays in a mixed radiation field based on capture gamma or self-irradiation of thermoluminescent dosimeters. Nucl. Instrum. Methods *154*, 561-566 (1978)

Iga, K., Yamashita, T., Takenaga, M., Yasuno, Y., Oonishi, H., Ikedo, M.: Composite TLD based on $CaSO_4$:Tm for γ-rays, x-rays, β-rays and thermal neutrons. Health Phys. *33*, 605-610 (1977)

Kapsar, B.M., Lucas, A.C.: Neutron and gamma-ray detection using thermoluminescence of CaF_2-Tm. Health Phys. *33*, 648 (1977)

Lakshmanarv, A.R., Bhatt, R.C.: Thermal neutron dosimetry with cadmium covered $CaSO_4$:Dy. Int. J. Appl. Radiat. Isot. *28*, 665-666 (1977)

Mohammadi, H., Ziemier, P.L.: Thermoluminescence dosimetry of fast neutrons using silver activated lithium borate phosphors. Nucl. Instrum. Methods *155*, 503-506 (1978)

Otterberg, J.E., Cipolla, S.J.: Neutron dose monitoring using a multiple lithium fluoride TLD badge. Health Phys. *33*, 256-258 (1977)

Pradhan, A.S., Bhatt, R.C., Lakshmanan, A.R., Chandra, B.: A thermoluminescent fast-neutron dosimeter based on pellets of $CaSO_4$:Dy mixed with sulphur. Phys. Med. Biol. *23*, 723-729 (1978)

Puite, K.J., Crebolder, D.L.J.M.: Luminescence dosimetry in photon and fast neutron beams. Phys. Med. Biol. *22*, 1136-1145 (1977)

Rinard, P.M., Simons, G.G.: Calculated neutron sensitivities of CaF_2 and LiF-7 thermoluminescent dosimeters. Nucl. Instrum. Methods *158*, 545-549 (1979)

Rogers, D.W.O., Walsh, M.L., Orr, B.H., Teekman, N.: Albedo-dosimeter responses to monoenergetic neutrons. Health Phys. *33*, 251-254 (1977)

Rossiter, M.J., Lewis, V.E., Wood, J.W.: The response of thermoluminescence dosimeters to fast (14.7 MeV) and thermal neutrons. Phys. Med. Biol. *22*, 731-736 (1977)

Singh, D., Burgkhardt, B., Piesch, E.: A passive neutron spectrometer and dosimeter using LiF: Mg, Ti thermoluminescent detectors. Nucl. Instrum. Methods *142*, 409-415 (1977)

Takenaga, M.: Thermoluminescent response to thermal neutrons of mixture of $CaSO_4$:Tm and non-luminous ^{6}LiF. J. Nucl. Sci. Technol. *14*, 292-299 (1977)

Tymons, B.J., Tnyn, J.W.N.: Personnel neutron dosimetry by means of cellulose nitrate film combined with LiF as both radiator and TLD. Health Phys. *32*, 547-549 (1977)

Wang, T.K., Weng, P.S., Hsu, P.C.: Measurement of reactor thermal-neutrons with dysprosium activated calcium-sulfate thermoluminescent dosimeters. J. Nucl. Sci. Technol. *15*, 72-75 (1978)

Yamashita, T., Iga, K., Takenaga, M., Yasuno, Y., Oonishi, H.: Composite TLD based on $CaSO_4$:Tm for gamma-rays, x-rays, beta-rays and thermal neutrons. Health Phys. *33*, 605-610 (1977)

Zeman, G.H., Snyder, G.I.: Neutron dose precision using LiF TLD. Health Phys. *36*, 75-76 (1979)

B.3.6 Application of Thermoluminescence to Dosimetry of Electrons and β Particles

Ayyangar, K., Chandra, B., Pradhan, A.S., Lakshmanan, A.R.: Additivity of alpha and gamma response in $Li_2B_4O_7$:Mn TLD phosphor. Health Phys. *35*, 568-569 (1978)

Fregene, A.O.: LiF response to high energy x-rays and electrons relative to ^{60}Co photons. Int. J. Appl. Radiat. Isot. *28*, 965-966 (1977)

Fregene, A.O.: The influence of the composition of LiF-TLD materials on their sensitivity to high energy electrons. Phys. Med. Biol. *22*, 372-374 (1977)

Gantchew, M.G.: The influence of the composition of LiF-TLD materials on their sensitivity to high energy electrons. Phys. Med. Biol. *22*, 374-376 (1977)

Jain, V.K.: Graphite mixed $CaSO_4$:Dy for beta dosimetry. Phys. Med. Biol. *23*, 1000-1001 (1978)

Laconi, A., Brancato, G, De Maria, M., Torino, G.: Radiothermoluminescence dosimetry of 10 MeV electrons. Radiaz. Alta Energ. *15*, 201-211 (1976)

Lasky, J.B., Moran, P.R.: Thermoluminescent response of LiF (TLD-100) to 5-30 keV electrons and the effect of annealing in various atmospheres. Phys. Med. Biol. *22*, 852-862 (1977)

Mikado, T., Tomimasu, T., Yamazaki, T., Chiwaki, M.: Response of magnesium silicate thermoluminescence phosphor to high-energy electrons. Bull. Electrotech. Lab. *42*, 29-43 (1978)

O'Brien, K.: Monte Carlo calculations of the energy response of lithium fluoride dosimeters to high energy electrons (< 30 MeV). Phys. Med. Biol. *22*, 836-851 (1977)

Pradhan, A.S., Bhatt, R.C.: Graphite-mixed $CaSO_4$:Dy teflon TLD discs for beta dosimetry. Phys. Med. Biol. *22*, 873-879 (1977)

Shiragai, A.: An approach to an analysis of the energy response of LiF-TLD to high energy electrons. Phys. Med. Biol. *22*, 490-499 (1977)

Wintle, A.G., Aitken, M.J.: Absorbed dose from a beta source as shown by thermoluminescence dosimetry. Int. J. Appl. Radiat. Isot. *28*, 625-627 (1977)

B.3.7 Personnel and Environmental Dosimetry

Anderson, M.E., Crain, S.L.: A combination TL-film personnel neutron dosimeter. Health Phys. *36*, 76-79 (1979)

Chabot, G.E., Jr., Jimenez, M.A., Skrable, K.W.: Personnel dosimetry in the U.S.A. Health Phys. *34*, 311-321 (1978)

Furuta, Y., Tanaka, S.: Wide utilization of thermoluminescence dosimeter in field of atomic energy. J. Atomic Energy Soc. Jpn. *20*, 559-566 (1978)

Grey, L.J., Bowlt, C.: An attempt to use thermally stimulated currents in human nail to estimate dose in cases of accidental exposure to ionizing radiation. Phys. Med. Biol. *23*, 759-760 (1978)

Hsu, P.C., Tseng, C.L., Weng, P.S.: Ambient atmosphere effect for lithium-fluoride thermoluminescent dosimeters irradiated in a nuclear-reactor mixed radiation field. Nucl. Instrum. Methods *157*, 147-153 (1978)

Johnson, T.L., Robinson, R.L., Luersen, R.B.: Fading of LiF - teflon dosimeters used in identification badges. Health Phys. *32*, 31-32 (1977)

Jones, A.R.: Application of an automated thermoluminescent dosimetry system to environmental "Gamma" dosimetry. At. Energy Can. Ltd. AECL. Rep. N5835 (1977)

Makra, S., Szabo, P.P., Benke, I., Biro, T.: Application of individual dosimeters based on film badges and thermoluminescent dosimeters. Report: KFKI-77-36 Hungarian Acad. Sci. Budapest (1977)

Manzanas, M.J., Lanzos, E., Frias, S.: Gauging and implementation of a personal dosimetry system by thermoluminescence. Radiologia *19*, 91-96 (1977)

O'Brien, K.: Response of LiF thermoluminescence dosimeters to the ground-level cosmic-ray background. Int. J. Appl. Radiat. Isot. *29*, 735-739 (1978)

Osvay, M., Biro, T.: Aluminum oxide thermoluminescent dose meter used for dosimetry after accidental gamma radiation. Handling of Radiation Accidents, p.493-500 (1977)

Petridou, C.H., Christodoulides, C., Charalamecus, S.: Non-radiation induced thermoluminescence in pre-irradiated LiF (TLD-100). Nucl. Instrum. Methods *150*, 247-252 (1978)

Plato, P.: Testing and evaluating personnel dosimetry services in 1976. Health Phys. *34*, 219-223 (1978)

Singh, D., Bhatt, R.C., Madhvanath, U.: Simpled dosimetry system for nuclear accidents. Int. J. Appl. Radiat. Isot. *28*, 513-519 (1977)

Tymons, B.J., Tnyn, J.W.N.: Personnel neutron dosimetry by means of cellulose nitrate film combined with LiF as both radiator and TLD. Health Phys. *32*, 547-549 (1977)

Weng, P., Hsu, P.: Concrete shielding for thermoluminescent dosimeter monitoring post to reduce background interference. Nucl. Instrum. Methods *145*, 433-435 (1977)

Will, W.: Evaluation accuracy of thermoluminescent dosimeters for personnel and environmental dosimetry. Kernenergie *21*, 110-112 (1978)

B.3.8   Application of Thermoluminescence to Other Studies

Bassi, P., Busoli, G., Rimondi, O.: A practical dosimeter for UV light. Nucl. Instrum. Methods *143*, 195-197 (1977)

Bojsen, J., Moeller, U., Christensen, P., Lippert, J.: Storage telemetry of radionuclide tracers by implantable thermoluminescent dosimeters. Am. J. Physiol. *233*, E479-E482 (1977)

Buckman, W.G., Payne, M.R.: Photo-stimulated thermoluminescence of lithium fluoride as an ultraviolet radiation dosimeter. Health Phys. *31*, 501-504 (1976)

Cullen, M.J., Malone, J.F., O'Connor, M.K.: Prospective radioiodine dosi-

metry in thyrotoxicosis by a thermoluminescent technique. Ann. Endocrino-
logie _38_, A 54 (1977)

Doi, A., Kanie, T., Naruse, N.: A thermoluminescent radiography. Jpn. J.
Appl. Phys. _16_, 2289-2290 (1977)

Gorbachenko, G.M., Zverev, S.A., Kushin, V.V., Lyapidevskii, V.K.: Application of thermoluminescent detectors in the spectrometry of pulsed x-rays
from hot plasma. Ivz. Akad. Nauk SSSR Ser. Fiz. _41_, 1321-1325 (1977)

Johnson, W.R., Bass, R.B., Littig, G.A., Robinson, F.L.: Spatial distribution of fission produced energy. _Proc.5th Int.Conf. React. Shielding_,
(Sci. Press, Princeton, NJ 1977) p.889

Moghissi, A.A., Carter, M.W.: Comments on a survey of luminescent clocks in
households. Health Phys. _33_, 107-108 (1977)

O'Connor, M.K., Cullen, M.J. Malone, J.F.: The influence of thyroid geometry on the response of LiF and $CaSO_4$ thermoluminescent discs to ^{131}I
and ^{125}I irradiation. Phys. Med. Biol. _23_, 712-722 (1978)

Scarpa, G., Moscati, M., Furetta, C.: Use of thermoluminescent dosimeters
for measuring dose distribution with high spatial resolution in partial
body irradiation of small animals. Br. J. Radiol. _52_, 75-77 (1979)

Takeuchi, N., Inabe, K., Yamashita, J., Nakamura, S.: Thermoluminescence of
MgO single crystals for UV dosimetry. Health Phys. _31_, 519-521 (1976)

B. 4 Thermally Stimulated Depolarization Currents

B.4.1 General

van Turnhout, J.: "Thermally Stimulated Discharge of Electrets", in
Electrets, ed. by G.M. Sessler, Topics in Applied Physics, Vol. 33
(Springer, Berlin, Heidelberg, New York 1979) Chap. 2

B.4.2 TSDC Involving Dipolar Processes

Annenkov, Yu.M., Balkokov, D., Pichugin, V.F.: Effect of M^{3+} impurities on
relaxation processes in alkali halide crystals. Izv. Vyssh. Uchebn. Zaved.
Fiz. [English transl.: Sov. Phys. J. _20_, 155 (1977)]

Gross, B.: Some aspects of the theory of thermally activated processes in
dielectric and viscoelastic materials. J. Electrostat. _3_, 43-51 (1977)

Gupta, N.P., Jain, K., Mehendru, P.C.: Continuous relaxation spectrum of
dipoles in VC : VAc copolymer films. J. Chem. Phys. _69_, 1785-1786 (1978)

Hor, A.M., Jacobs, P.W.M.: Thermal depolarization in crystals of rubidium
bromide doped with Ba^{2+} and Sn^{2+}. Phys. Status Solidi A _44_, 725-729 (1977)

Ikeya, M.: Ionic thermo-current studies of Mn^{2+} -vacancy complexes associated with F^- ion in NaCl. J. Phys. Soc. Jpn. _45_, 1313-1319 (1978)

Kaestner, S.: Contribution to the theory of the determination of the relaxation spectrum of polymers from electric discharge measurements at chrono-
logically increasing temperature. Plaste Kautsch. _24_, 747-750 (1977)

Kirk, K.L., Innes, R.M.: An evaluation of the ITC technique in monitoring
the relaxation behaviour of impurity-vacancy dipoles in monocrystalline
sodium chloride. J. Phys. C _11_, 1105-1121 (1978)

Shindo, K.: Thermally stimulated currents in dielectrics. Shiga Daigaku
Kyoiku Gakubu Kiyo, Shizenkagaku _27_, 1-8 (1977)

Weber, G., Tormala, P.: Dipole and space charge polarization in polymethyl-
methacrylate using thermally stimulated discharge. Colloid Polymer Sci.
256 , 1137-1139 (1978)

B.4.3 Correlation Between Dipolar TSDC and Conventional Dielectric
       Measurements

Bondeau, A., Noyel, G., Huck, J.: Relaxation diélectrique aux très basses
fréquences de l'éthyl-2-hexanol-1. C.R. Acad. Sc. Paris C *286*, 273-276
(1978)

B.4.4 TSDC Involving Space Charge Processes

Anderson, J.C.: Thermally stimulated currents in thin film transistors. Proc.
7th Intern. Vac. Congr. and 3rd Intern. Conf. Solid Surfaces (Vienna 1977)
pp.553-556
Bekeris, Yu.Ya., Kalnynya, R.P., Feltyn', I.A.: Thermally stimulated de-
polarization currents in plasma-grown silica films. Latv. Psr. Zinat.
Akad. Vestis Fiz. Teh. Zinat. Ser. (USSR) *2*, 16-21 (1978)
Gupta, H.M.: On the theory of thermal dielectric relaxation. J. Phys. C *10*,
L429-L431 (1977)
Hor, A.M., Jacobs, P.W.M.: On thermally stimulated space-charge decay in
tin (2+) ion-doped sodium chloride and potassium chloride crystals. J.
Electrochem. Soc. *125*, 430-432 (1978)
Jain, V.K., Gupta, C.L., Jain, R.K., Tyagi, R.C.: Thermally stimulated
currents in pure and copper-doped polyvinyl acetate. Thin Solid Films *48*,
175-186 (1978)
Kessler, A.: Thermally stimulated space charge decay in Sn^{2+} -doped sodium
chloride and potassium-chloride crystals. J. Electrochem. Soc. *125*, 2084
(1978)
Kikineshi, A.A., Mikla, V.I., Mikhal'ko, I.P., Semak, D.G.: Special characteristics of investigations of thermally stimulated depolarization of
chalcogenide glass. Fiz. Tekh. Poluprovodn. [English transl.: Sov. Phys.
Semicond. *11*, 1010 (1977)]
Kojima, K., Maeda, A., Takai, Y., Ieda, M.: Thermally stimulated currents
form polyethylene terephtalate due to injected charges. Jpn. J. Appl. Phys.
17, 1735-1738 (1978)
Mehendru, P.C., Jain, K., Mehendru, P.: Space charge induced relaxation in
polymer thin films. Proc. Nucl. Phys. Sol. State Phys. Symp., Ahmedabad
(1976) *19* C, pp.97-99
Mizutani, H., Hanamoto, K., Saji, M.: Thermally stimulated depolarization
currents in $Si_{12}Ge_{10}As_{30}Te_{48}$ glass. Bull. Nagoya Inst. Technol. Jpn. *28*,
353-356 (1976)
Shindo, K.: Thermally stimulated currents in dielectrics. Shiga Dougaku
Kyoiku Gakubu Kiyo, Shizenkagaku *27*, 1-8 (1977)
Singh, R., Datt, S.C.: Electrical conduction and persistent internal polarization in sisal wax. J. Appl. Phys. *49*, 4191-4194 (1978)
Weber, G., Tormala, P.: Dipole and space-charge polarization in polymethyl-
methacrylate using thermally stimulated discharge. Colloid Polymer Sci.
256, 1137-1139 (1978)
Zielinski, M., Samoc, M.: An investigation of the Poole-Frenkel effect by
the thermally stimulated current technique. J. Phys. D *10*, L105-L107 (1977)

B.4.5 TSDC Involving Interfacial Processes

Tanaka, T., Hayashi, S., Hirabayashi, S., Shibayama, K.: Thermally stimulated depolarization current study of polystyrene composites containing
mica flaskes. J. Appl. Phys. *49*, 2490-2493 (1978)

B.4.6 TSDC Involving Dipolar and Space Charge Processes

Shindo, K.: Thermally stimulated currents in dielectrics. Shiga Daigoku
Kyoiku Gakubu Kiyo, Shizenkagaku *27*, 1-8 (1977)

B.4.7 Correlation Studies of FI-TSC with Other Physical Methods

Berticat, P., Chatain, D., Monpagens, J.C., Lacabanne, C.: Thermally stimulated current and creep in amorphous poly(ethylene therphtalate). J.
Macromol. Sci. Phys. B *15*, 549-565 (1978)
Chatain, D., Lacabanne, C., Monpagens, J.C.: Thermostimulated creep and depolarization thermocurrent in low density polyethylene. Makromol. Chem.
178, 583-593 (1977)
Foldes, E., Pazonyi, T., Hedvig, P.: Effect of processing on the thermomechanical and dielectric relaxation properties of plasticized poly (vinyl
chloride) compounds. J. Macromol. Sci. Phys. B *15*, 527-548 (1978)
Hampe, A.: Comparison of thermally stimulated currents of electron-irradia-
ted and electrical field-polarized polyethylene. Progr. Colloid Polym.
Sci. *62*, 154-160 (1977)
Stupp, S.I., Carr, S.H.: Polarization phenomena in ion-doped polymers. Am.
Chem. Soc. Div. Org. Coat. Plast. Chem. Preprints *37*, no.2 (1977) paper
presented at Am. Chem. Soc. Natl. Meet., 174th, Chicago (1977)
Thurzo, I., Doupovec, J., Vlasak, G.: How to distinguish between uniform
and space charge polarization in high resistivity glasses - A thermally
stimulated depolarization analysis. Acta Phys. Slovaca *27*, 206-218 (1977)

B.4.8 TSDC and Surface Phenomena

Ehrburger, F., Donnet, J.B.: Study of the dielectric properties of water
and methyl alcohol adsorbed on aerosil by the thermally stimulated depolarization method. J. Colloid Interface Sci. *66*, 405-414 (1978)
Ohara, K.: Detecting changes in the condition of the surface layer and the
bulk of rubbed polymer films by thermally stimulated currents. Wear. *48*,
409-411 (1978)

B. 5 Exoemission

B.5.1 Mechanical Deformation Induced EE

Baxter, W.J.: "Exoelectron Emission from Solids", in *Research Techniques in Nondestructive Testing*, Vol. III (Academic Press, New York 1977) pp.395-428

Baxter, W.J., Rouze, S.R.: The effect of oxide thickness on photostimulated exoelectron emission from aluminum. J. Appl. Phys. *49*, 4233-4237 (1978)

Chrustalev, Ju.A., Krotova, N.A.: Emission von Elektronen hoher Energie bei der Zerstörung verschiedener fester Körper. Krist. Tech. *13*, 1077-1081 (1978)

Dickinson, J.T., Bräunlich, P., Larson, L.A., Marceau, A.: Characteristic emission of negatively charged particles during tensile deformation of oxide-covered aluminum alloys. Appl. Surf. Sci. *1*, 515-537 (1978)

Komai, K.: Exoelectron emission from metal surfaces with the progress of fatigue damage. Trans. Jpn. Inst. Met. *19*, 119-124 (1978)

Larson, L.A., Dickinson, J.T., Bräunlich, P., Snyder, D.B.: The emission of neutral particles from anodized aluminum surfaces during tensile deformation. J. Vac. Sci. Techn. *16*, 590 (1979)

Wortmann, J.: Optisch stimulierte Elektronenemission von tribomechanisch aktiviertem Aluminium. Phys. Status Solidi A *46*, 69-76 (1978)

B.5.2 TSEE

Brunsmann, U., Scharmann, A.: Untersuchungen zur Exoelektronenenergieverteilung an NaF-Oberflächen im Ultrahochvakuum. Vak. Tech. *26*, 78-80 (1977)

Scharmann, A., Wiessler, U.: TSEE und Austrittsarbeit von oxidiertem Be-Einkristalloberflächen. Vak. Tech. *27*, 235-237 (1978)

Svitov, V.J., Krylova, J.V.: Thermostimulated exoemission and adsorption centres of BeO-ceramics. Radiat. Eff. *35*, 29-34 (1978)

Tomita, A., Nakamura, H., Kamada, M., Tsutsumi, K.: Thermally stimulated exoelectron emission of evaporated LiF films. Jpn. J. Appl. Phys. *18*, 389-390 (1979)

Yamoaka, Y.: Effects of grain size on thermoluminescence and thermally stimulated exoelectron emission of LiF crystals. Health Phys. *35*, 708-711 (1978)

Subject Index

Acceptor 8, 10
Activation energy 6, 43, 55, 107, 116 ff.
   determination, dipolar process 161 ff.
   determination, space-charge process 188, 192
   distribution 151 ff., 170 ff.
Adiabatic electron-lattice coupling 52
Admittance spectroscopy 115
Aggregation of ionic dipoles 217
Aging 209 ff., 217
Air gap 142, 178, 207
Akali halides 46, 54, 143, 187, 198, 209
Amorphous semiconductor 18
Anharmonicity, lattice 58
Annealing 287 ff.
   pre-irradiation 287
   post-irradiation 288
Anomalous fading 296
Archeological
   artifacts 291, 293, 294
   dating 293 ff.
   fine-grain technique 294
   predose technique 294
   quartz inclusion technique 294, 295
   radioactive inclusion technique 294
   substraction technique 293
Arrhenius equation 4, 105
Attempt-to-escape frequency 19 ff.
Auger effect 51 ff.
Authenticity in art 293

Band
   bending 102
   edge 7 ff.
   gap 7 ff.
Barrier polarization 183, 189, 198, 204, 206
Black body radiation 22
Black body signal 215
Boltzmann
   distribution 15
   factor 49

Capacitance
   bridge 129

   meter 128, 129
   profiling 122
   signal 102
   spectroscopy 96
   transient 8, 46, 97 ff., 103, 110 ff.
Capture 51, 53, 54
   coefficient 11, 12, 60, 61, 68
   cross section 11, 12, 15, 21, 43, 49, 65, 66, 70, 122 ff.
   radius 50, 54
   rate 11, 12, 46, 51, 122, 123
Carrier depletion 101
   capture 120 ff.
Chemical potential 7
Circuit, equivalent 147, 183, 185
Cleaning technique 162
Coefficient
   thermal emission 12
   thermal release 12
Coemission 228, 229
Cole-Cole distribution 154, 155, 157, 168
Color center 217
Complexity of exoemission 230
Conductance 115
Conduction band 7 ff.
Conductivity 188
   thermally stimulated 35, 67 ff.
Cooling rate 150, 157, 167, 198
Coordinate
   diagram 41, 55
   lattice 55
Correlator 111
Coulomb-states 125
Current
   phototransient 46
   transient 46, 105

Dark emission 238, 244
Dating
   archaeological *see* Archaeological dating
   geological *see* Geological application
   lava flow 294
   radio carbon 291, 292
   tree ring 292

Applied Physics

A monthly journal

Board of Editors

S. Amelinckx, Mol; **V. P. Chebotayev,** Novosibirsk;
R. Gomer, Chicago, IL; **P. Hautojärvi,** Espoo;
H. Ibach, Jülich; **V. S. Letokhov,** Moskau;
H. K. V. Lotsch, Heidelberg; **H. J. Queisser,** Stuttgart;
F. P. Schäfer, Göttingen; **K. Shimoda,** Tokyo;
R. Ulrich, Stuttgart; **W. T. Welford,** London;
H. P. J. Wijn, Endhoven

Coverage

application-oriented experimental and theoretical
physics

Solid-State Physics	*Quantum Electronics*
Surface Science	*Laser Spectroscopy*
Solar Energy Physics	*Photophysical Chemistry*
Microwave Acoustics	*Optical Physics*
Electrophysics	*Optical Communications*

Special Features

rapid publication (3–4 months)
no page charges for concise reports
microform edition available

Languages
mostly English

Articles

original reports, and short communications
review and/or tutorial papers

Manuscripts

to Springer-Verlag (Attn. H. Lotsch), P.O. Box 105 280
D-6900 Heidelberg 1, FRG

**Springer-Verlag
Berlin
Heidelberg
New York**

Place North-America orders with:
Springer-Verlag New York Inc., 175 Fifth Avenue,
New York, N.Y. 10010, USA

H. Bilz, W. Kress

Phonon Dispersion Relations in Insulators

1979. 162 figures in 272 separate illustrations.
Approx. 240 pages
(Springer Series in Solid-State Sciences,
Volume 10)
ISBN 3-540-09399-0

Contents:
Summary of Theory of Phonons: Introduction. Phonon Dispersion Relations and Phonon Models. – Phonon Atlas of Dispersion Curves and Densities of States: Rare-Gas Crystals. Alkali Halides (Rock Salt Structure). Metal Oxides (Rock Salt Structure). Transition Metal Compounds (Rock Salt Structure). Other Cubic Crystals (Rock Salt Structure). Caesium Chloride Structure Crystals. Diamond Structure Crystals. Zinc-Blende Structure Crystals. Wurtzite Structure Crystals. Fluorite Structure Crystals. Rutile Structure Crystals. ABO_3 and ABX_3 Structures. Layered Structure Crystals. Other Low-Symmetry Crystals. Molecular Crystals. Mixed Crystals. Organic Crystals. – References. – Subject Index.

O. Madelung

Introduction to Solid-State Theory

Translated from the German by B. C. Taylor

1978. 144 figures. XI, 486 pages
(Springer Series in Solid-State Sciences,
Volume 2)
ISBN 3-540-08516-5

Contents:
Fundamentals. – The One-Electron Approximation. – Elementary Excitations. – Electron-Phonon Interaction: Transport Phenomena. – Electron-Electron Interaction by Exchange of Virtual Phonons: Superconductivity. – Interaction with Photons: Optics. – Phonon-Phonon Interaction: Thermal Properties. – Local Description of Solid-State Properties. – Localized States. – Disorder. – Appendix: The Occupation Number Representation.

V. M. Fridkin

Photoferroelectrics

1979. 63 figures, 3 tables. X, 174 pages
(Springer Series in Solid-State Sciences,
Volume 9)
ISBN 3-540-09418-0

Contents:
Introduction. – The Thermodynamics of Photoferroelectrics. – The Microscopic Theory of Photoferroelectric Phenomena. – Screening of Spontaneous Polarization. – Photoferroelectric Phenomena and Photostimulated Phase Transitions. – The Anomalous Photovoltaic Effect in Ferroelectrics. – The Photorefractive Effect in Ferroelectrics. – Screening Phenomena. – References. – Subject Index.

Electrets

Editor: G. M. Sessler

1979. 205 figures, 27 tables.
Approx. 420 pages
(Topics in Applied Physics, Volume 33)
ISBN 3-540-09570-5

Contents:
G. M. Sessler: Introduction. – *G. M. Sessler:* Physical Principles of Electrets. – *J. van Turnhout:* Thermally Stimulated Discharge of Electrets. – *B. Gross:* Radiation-Induced Charge Storage and Polarization Effects. – *M. G. Broadhurst, G. T. Davis:* Piezo- and Pyroelectric Properties. – *S. Mascarenhas:* Bioelectrets: Electrets in Biomaterials. – *G. M. Sessler, J. E. West:* Applications.

Springer-Verlag
Berlin
Heidelberg
New York